HANDBOOK OF TOPICAL ANTIMICROBIALS

HANDBOOK OF TOPICAL ANTIMICROBIALS

INDUSTRIAL APPLICATIONS IN CONSUMER PRODUCTS AND PHARMACEUTICALS

EDITED BY
DARYL S. PAULSON
BioScience Laboratories, Inc.
Bozeman, Montana, U.S.A.

CRC Press is an imprint of the
Taylor & Francis Group, an **informa** business

CRC Press
Taylor & Francis Group
6000 Broken Sound Parkway NW, Suite 300
Boca Raton, FL 33487-2742

First issued in paperback 2019

ISBN-13: 978-0-8247-0788-0 (hbk)
ISBN-13: 978-0-367-39582-7 (pbk)

Visit the Taylor & Francis Web site at
http://www.taylorandfrancis.com

and the CRC Press Web site at
http://www.crcpress.com

Preface

Over the years, topical antimicrobials have become commonly used to help reduce the risk of infectious disease. From the humble beginnings of Semmelweis, who showed that surgically associated infection could be significantly reduced by simply washing one's hands, to the fast microbial kills and long-lasting effects of current topical antimicrobial products, the industry has progressed. Thus, this text should be viewed as a contribution to an ongoing effort.

The book has several parts. The first consists of three overviews of the topical antimicrobial industry. These include a philosophical view from the editor, the perspectives of a Food and Drug Administration (FDA) officer, and a discussion of healthcare industry regulations. Part II is devoted to describing aspects of the specific antimicrobial compounds used as active drugs in topical antimicrobial products. The third part addresses various medical applications of antimicrobial formulations used as healthcare personnel handwashes, surgical washes/scrubs, and preoperative skin preparations, as well as several novel areas of application. The fourth part presents discussions of concerns faced by topical antimicrobial suppliers in assuring that the role of food handlers in infectious disease transmission is reduced. Finally, Part V addresses specific testing concerns, test methods, and the appropriateness of using certain test methods to assess the value of specific products.

This book has taken several years to produce, and I thank the contributing authors for their dedication, professionalism, and persistence in writing their various chapters. I dedicate it to the following scientists who have been instrumental in forming my views, and who have been guiding lights along the way. Carl Bruch, Ph.D., was my mentor at Skyland Scientific Services. Elaine Larson, R.N., Ph.D., has been an inspiration to me since the beginning of my career in the early

1980s. Mary Bruch and Arthur Peterson, Ph.D., the creative minds behind the "glove juice" procedure used in surgical scrub and healthcare personnel handwash evaluations, were my motivation to enter this field and ultimately link microbiology with biostatistics. Finally, Albert Sheldon, Ph.D., has always been an inspiration to me in his diligence to ensure that business interests do not override good science.

I have been influenced by each of the chapter authors in this book and appreciate their contribution to the field. I am indebted to the staff at BioScience Laboratories, Inc., for keeping things running as I wrote and collected the chapters. I am particularly indebted to John Mitchell, Ph.D., for his help in assembling this book. Additionally, I owe a debt of gratitude to Ms. Tammy Anderson, who provided valuable assistance in assembling and retyping every chapter for formatting and editing. I also thank the staff at Marcel Dekker, Inc., particularly Maria Allegra and Brian Black.

Daryl S. Paulson

Contents

Contributors

Shamin A. Ansari Colgate-Palmolive Company, Piscataway, New Jersey, U.S.A.

Christopher M. Beausoleil BioScience Laboratories, Inc., Bozeman, Montana, U.S.A.

Ward L. Billhimer The Procter & Gamble Company, Cincinnati, Ohio, U.S.A.

Mary Bruch Micro-Reg, Inc., Hamilton, Virginia, U.S.A.

Nancy Bruno Purdue Pharma, Stamford, Connecticut, U.S.A.

Renate Cassillis Purdue Pharma, Stamford, Connecticut, U.S.A.

Michael J. Dolan GoJo Industries, Akron, Ohio, U.S.A.

Eleanor J. Fendler GoJo Industries, Akron, Ohio, U.S.A.

George E. Fischler The Dial Corporation, Scottsdale, Arizona, U.S.A.

David W. Hobson, Ph.D. Chrysalis Biotechnology, Inc., San Antonio, Texas, U.S.A.

David K. Jeng, Sc.D. Allegiance Healthcare Corporation, A Cardinal Health Company, McGraw Park, Illinois, U.S.A.

Rhonda D. Jones Scientific & Regulatory Consultants, Inc., Columbia, Indiana, U.S.A.

Bruce H. Keswick The Procter & Gamble Company, Cincinnati, Ohio, U.S.A.

Barry Michaels Georgia-Pacific Corporation, Palatka, Florida, U.S.A.

Daryl S. Paulson, Ph.D. BioScience Laboratories, Inc., Bozeman, Montana, U.S.A.

Robert F. Reder Purdue Pharma, Stamford, Connecticut, U.S.A.

Steven Ripa Purdue Pharma, Stamford, Connecticut, U.S.A.

Robert I. Roth The Weinberg Group, Inc., Washington, D.C., U.S.A.

Syed A. Sattar University of Ottawa, Ottawa, Canada

Lawton Anthony Seal, Ph.D. Healthpoint, Ltd., San Antonio, Texas, U.S.A.

Marc Shaffer Clinical Research Laboratories, Inc., Piscataway, New Jersey, U.S.A.

Albert T. Sheldon, Jr., Ph.D. Food and Drug Administration, Rockville, Maryland, U.S.A.

Edward B. Walker, Ph.D. Weber State University, Ogden, Utah, U.S.A.

Kathy F. Wiandt The Procter & Gamble Company, Cincinnati, Ohio, U.S.A.

Ronald A. Williams GoJo Industries, Akron, Ohio, U.S.A.

Part I
Overview

Part I presents aspects of antimicrobial product evaluations from three important perspectives. First, a chapter on developing effective antimicrobial products provides a broad view approach in developing topical antimicrobials. These views include marketing, science, and quality assurance. They include objective aspects, including antimicrobial effectiveness, label claims, and quality control; subjective aspects, such as perceived ease of use, feel, fragrance; and shared professional beliefs, such as value of an iodophor, a chlorhexidine gluconate, or triclosan products.

The second chapter is an FDA perspective on topical antiseptic drug products. While it is not an FDA monograph, it does provide valuable insight into the Food and Drug Administration's view of effective test evaluations.

The third chapter is another perspective of categorizing and, hence, evaluating topical antimicrobial products from an industry perspective.

It is strongly suggested that the reader carefully consider these three distinct perspectives, yet recognize that they are not separate. They are related and, thus, are perspectives which those working in the industry should consider.

1

Introduction to Topical Antimicrobials and Their Applications

Daryl S. Paulson
BioScience Laboratories, Inc., Bozeman, Montana

I. DEVELOPING EFFECTIVE TOPICAL ANTIMICROBIALS

The intense levels of competition present in the topical antimicrobial market, as well as the ever-tightening U.S. Food and Drug Administration's (FDA) product performance standards, demand that antimicrobial manufacturers produce products that meet market requirements [1]. Present market requirements are not just antimicrobial effectiveness; they also include low skin irritation potential, ease of use, aesthetically pleasing, and various other ''soft'' attributes. If these have been addressed, the product more than likely has been developed with adequate care. But too often manufacturers ignore important factors and merely get a product to market to compete with those of competitors. In the end, this approach often fails; the product never really is accepted in the market [2]. Because the goal is to introduce products into the market that will be successful, manufacturers are urged to develop products from a multidimensional perspective. At least four factors should be addressed: societal, cultural, personal objective, and personal subjective to address human requirements. Let us look at this multidimensional perspective in greater detail.

A. Societal Requirements

Societal requirements include conforming to regulating agencies standards such as the FDA, the Federal Trade Commission (FTC), and the Environmental Protection Agency (EPA), as well as the rules, laws, and regulations they enforce. Before designing a product, it is critical to understand the current legal regulations governing the product, the product's antimicrobial components and their levels, as well as product stability and toxicological concerns. For example, a New Drug Application (NDA) is required in order to market a regulated drug product. For over-the-counter (OTC) products, the active drug and its level must be both allowable and within allowable ranges. Additionally, the FDA's Tentative Final Monograph for OTC products or the recommendations of the Cosmetic Toiletries and Fragrance Association (CTFA) and the Soap and Detergent Association (SDA) must be addressed.

B. Cultural Requirements

Cultural requirements are very important but are often ignored. Cultural and subcultural requirements are shared values, beliefs, goals, and the world views of a society or subgroup of society [3]. Shared values such as perceived antimicrobial "effectiveness" have great influence on consumers [5]. These values are generally of two types: manifest and latent. Manifest (surface) values are conscious to the consumer. For example, a consumer buys an antimicrobial soap to be "cleaner" than s/he can be using a nonantimicrobial soap. But deeper and more fundamental values are also present. These are referred to as latent values and are generally unconscious to the consumers in that the consumer is not aware of them. In this case, "cleaner" may mean to the consumer such things as being accepted, valued, loved, and worthwhile as a person, spouse, and/or parent.

Most manifest and latent values we share can be magnified by manufacturers' advertising campaigns [7]. For example, if a homemaker perceives that s/he is taking better care of the children by having them use antimicrobial soaps (a manifest value), and if s/he feels more valuable, more loveable, and/or more needed by his/her family, etc. (latent values), s/he will be motivated to purchase the product. Finally, much of what consumers believe to be true is not grounded in objective reality [8]. Most of these beliefs are formed from their interpretation of mass media reports, opinions of experts and explanations of phenomena from various notorieties [6,9].

C. Personal Objective Attributes

These include the physical components of a product—its application, its antimicrobial actions, or its irritation effects on skin and the environment (e.g., staining clothing, gowns, and bedding). It is important that products be designed with the end-user in mind [2]. Hence, products must be easy to use and to open (if in a container) and must be effective for their intended use (e.g., by food-servers, home consumers, medical personnel, and surgical personnel).

Personal Subjective Perceived: need value ease of use	*Personal Objective* irritation potential fragrance feel drying feel perceived quality
Cultural (intersubjective) general desire of the market for the product perceived need requirements	*Societal (interobjective)* laws, rules, mandates, regulations concerning antimicrobial topics

Figure 1 Interaction between values and behaviors.

D. Personal Subjective Attributes

This category includes the personal interpretation of cultural and subcultural ''world views.'' Relative to antimicrobials, these include subjective likes and dislikes of characteristics such as the fragrance and feel of the product, the perceived ''quality'' of the product, and other aesthetic considerations [7]. As with cultural attributes, there are manifest and latent values in this category. Hence, if one likes the ''springtime'' fragrance of a consumer bodywash product (manifest), the latent or deeper value may be that it makes one feel younger and, therefore, more physically attractive and desirable.

These four attribute categories are presented in quadrant form in Figure 1. Each quadrant interacts and is interdependent with the other three quadrants. For example, cultural values influence personal values, and vice versa. Cultural and personal values influence behavior and behavior influences values.

II. A CHRONIC PROBLEM: REDUCTIONISM

A chronic problem of product design, reductionism, occurs when personnel such as engineers, chemists, microbiologists, statisticians, or lawyers focus solely on the objective parameters (the two right quadrants) and therefore misinterpret subjective parameters in objective terms: that is, they reduce the subjective domains to objective ones. Equally problematic are marketing personnel, industrial psychologists, and social scientists, who often focus their efforts in the subjective domains and misinterpret objective domains in subjective terms. Reductionism often is rooted in bias. Those who were trained in the physical sciences tend to have incorporated an objective world view, while those trained in social sciences usually have a subjective one. Both world views are accurate, but only partially so, and both must be recognized and integrated into the product's development. This in no way suggests that each specialist should work in all four quadrant domains—that is unrealistic. Each should focus on his or her own domain (e.g., chemists in the two right quadrants, marketing in the two left quadrants). However, the person or group in charge of the product's development must be conscious of all domains and integrate them into design of the product. With this in mind, let us examine how the quadrant model may be of assistance in the design of topical antimicrobial products.

III. PRODUCT CATEGORIES

There are three general categories of topical antimicrobial products recognized by the FDA:

1. Healthcare personnel handwash formulations
2. Surgical scrub formulations
3. Preoperative skin-prepping formulations

However, the CTFA/SDA groups introduced what they term the Healthcare Continuum Model, expanding these three product categories to include three additional ones:

1. Antimicrobial bodywashes
2. Antimicrobial handsoaps
3. Foodhandler handsoaps

Whether these three additional categories will be formally recognized by the FDA remains to be seen, but because they are anchored in personal and cultural values, they are important. It is fair to say that most Americans want to feel clean and be clean. This, to many, means using an antimicrobial bodysoap for showering and bathing, as well as for handwashing.

Moreover, there is considerable pressure on government agencies to make the food we eat safer. Consumers remember clearly the deadly problems created by ingesting *Escherichia coli* O157/H7 in contaminated hamburger [9]. They want governmental protection to prevent this from recurring [10,11].

IV. HEALTHCARE CONTINUUM MODEL

Now let us focus on the categories addressed by the Healthcare Continuum Model. As stated previously, the FDA acknowledges the healthcare personnel handwash, the surgical scrub, and the preoperative skin preparation evaluations. The ''foodhandler handwash'' category exists, but exactly what federal body will control it is not known. And least in the near future, it is doubtful if the FDA will officially recognize the antimicrobial handsoaps and antimicrobial bodywashes. With this said, let us turn our attention to product development for each product category.

V. PREOPERATIVE SKIN-PREPARATION CATEGORY

A. Personal Objective

Preoperative skin preparation solutions are designed to degerm an intended surgical site rapidly as well as to provide a high level of persistent antimicrobial activity—up to 6 hours—post–skin prepping [12]. In terms of the ''quadrant'' model presented earlier (Fig. 1), these requirements belong to the personal objective region (upper right quadrant).

The preoperative skin-preparation product should be truly convenient and easy to use for the surgical staff consumers. Given two products with equal antimicrobial efficacy, if one is easier to use (e.g., has a shorter skin-prep time, can be more easily seen on the skin due to a coloring agent, and/or is a one-step procedure), it will be preferred over the other product.

An often overlooked aspect is customer service. It is ironic that so many firms spend large sums of money to assure that a product meets its intended purpose, but meeting a customer's needs is not accomplished. Orders are delayed or lost, return phone calls are promised but not made, and so on. In short, the customers are ignored. Tom Peters points out that customers who find it hard to do business with manufacturers and vendors do not remain customers, even if the products provided are ''better'' than those of competitors [13]. Service is critically important.

B. Societal Requirements

The documentation requirements for getting a topical antimicrobial product approved to market in this country are fairly straightforward. For example, the product must be manufactured under current Good Manufacturing Practices (cGMPs), and laboratory testing of the product must be performed under Good Laboratory Practices (GLPs). If the product is considered a drug, a New Drug Application (NDA) or an Abbreviated New Drug Application (ANDA) must be filed. Regulatory agencies such as the FDA also require a battery of specific tests to be conducted. These include such things as clinical trials, time-kill, and minimum inhibition concentration (MIC) studies.

C. Cultural Requirements

The products must be perceived as being of high quality. This perception is based upon the ''world view'' of the culture, including shared values, beliefs, and goals, as well as those of any subcultures, when applicable [3].

Cultural and subcultural world views are not easily perceived by those belonging to that culture or subculture. For example, the traditional member of our western, science-based culture may chuckle at New Age proponents who believe in healing crystals, as well as being part of a larger eco-spiritual system. But members of the scientific community in this country often extend the domain of science to their belief systems and are not aware of it. Consider a statement made by Candice Pert, an eminent neuroscientist who discovered the opiate receptors of neuropeptides responsible for the brain's production of morphine-like substances, endorphins, which create the highs or euphoria associated with physical exercise: ''Scientists by nature, are not creatures who commonly seek out or enjoy the public spotlight. Our training predisposes us to avoid any kind of overt behavior that might encourage two-way communication with the masses. Instead, we are content to pursue our truth in windowless laboratories, accountable only to members of our exclusive club'' [14]. There is no evidence that this view is based on scientific reason. The position has nothing to do with science, but everything to do with the world view of Pert's subculture, the scientific community.

Taking an example more relevant to topical antimicrobial product development, if the product ingredients are viewed as artificial, when natural ingredients are more valued, the introduction and acceptance of the product containing artificial ingredients is likely to be impaired. Also, social institutions within a culture—the family, religious groups, social groups, and political groups—are important, and manufacturers should take care not to alienate them with thoughtless advertising. Shared beliefs and values concerning the firm that makes the product are critical. If a manufacturer is perceived by consumers as a quality, caring firm, it will be easier for it to introduce and sell new products than if it is perceived otherwise.

Much of social reality is constructed—made up—by cultures [7]. The degree to which reality is constructed, however, is bound by objective reality. What we believe to be true or real is often only partially true or real.

Sociologists tell us that there are three stages in socially constructed reality: externalization, objectivation, and internalization [3]. Externalization is the initial stage, when a theory or opinion is accepted as tentatively being true. Objectivation is the next stage, when the theory or opinion is accepted as fact. Internalization is when the accepted facts are incorporated into a person's psyche as absolute truth.

It is often easier to use socially constructed reality as an ally in marketing programs than to educate people concerning truth. For example, with household antimicrobial products, when consumers view an advertisement that states the product ''kills 99.9% of all germs,'' they believe that only a very small fraction of the disease-causing microorganisms survive. The method of calculating the percent $\log_{10}$ reduction uses a percent (linear) of an exponential distribution (nonlinear). Yet in truth, an error has been committed—that of applying a linear ''proportion'' measurement (% reduction in microorganisms) to an exponential distribution. If the linear percent measurement was calculated on a linearized exponential distribution, the common proportional interpretation would be correct, but the percent reduction in microorganisms far less than 99.9% [15]. But it sounds better to consumers who will buy according to their constructed reality.

D. Personal Subjective

Individual members of a group (e.g., surgical staff) need to find value and construct positive beliefs about a specific product if it is to be successful. Therefore, the perspective is important. Individuals need to believe that the product was designed with them in mind. Providing them specific examples, comparisons, and test conclusions is a very effective way of externalizing, objectifying, and, finally, internalizing the advertising. But one must be certain that the advertising claims are supported and are grounded in objective reality [2,6,13]. Socially constructed reality is bounded in objective reality. So when people are told that a product is easy to use, they must find this to be so when they use the product or their beliefs will change to match the reality of their experience.

VI. SURGICAL SCRUB CATEGORY

A. Personal Objective

Surgical scrub formulations are designed to remove both transient and normal (resident) microorganisms from the hand surfaces [16]. Surgical scrubs, to be effective when used with or without a scrub brush, must demonstrate immediate, persistent, and residual antimicrobial properties and must be low in skin irritation effects when used repeatedly over a prolonged period of time. The product's immediate antimicrobial efficacy is a quantitative measurement of both the mechanical removal and immediate inactivation of microorganisms residing on the skin surface [17]. The persistent antimicrobial effectiveness is a quantitative measurement of its ability to prevent microbial recolonization of the skin surfaces, either by microbial inhibition or by lethality. The residual efficacy is a measurement of the product's cumulative antimicrobial properties after it has been used repeatedly over time. That is, as the antimicrobial product is used over time, it is absorbed into the stratum corneum of the skin and, as a result, prevents microbial recolonization of the skin surfaces.

B. Societal Requirements

The documentation and legal requirements for surgical scrub products are similar to those for preoperative skin preparations. In general, products must meet the efficacy requirements of clinical trials utilizing human test subjects, as well as a series of in vitro tests, including time-kill and minimum inhibition studies. The actual requirements are presented in the FDA's Tentative Final Monograph of OTC products.

C. Cultural Requirements

In general, these are equivalent to those presented in the discussion of cultural aspects of preoperative skin preparations.

D. Personal Subjective

These include those issues covered in the preoperative skin-preparation portion of this chapter. However, because surgical staff members actually use the product on themselves, the aesthetic attributes tend to be more important than with preoperative skin preparations. One important aspect of development in surgical scrub formulations is a brushless application, which will certainly change the perspective of the presurgical wash process. It is important that relevant, but unknown product attributes be identified by potential product users. This may include subjective testing to evaluate the sensory attributes of the product—its container, its packaging, and other aesthetic concerns—in order to engineer a more desirable product [4]. Some of the areas of interest are presented in Table 1.

Developing a list of attributes from which panelists may select is challenging. One effective way of doing so is through the use of focus groups [4]. Focus groups consist of a small number of surgical staff members [5–10] who literally

Table 1 Sensory Attributes

Sensory evaluation	Acceptance attributes	Performance	Image (nonantimicrobial)
1. Clear	1. Like appearance	1. Lather well	1. Effective
2. Opaque	2. Like fragrance	2. Feels good on hands	2. Unique
3. Strong smell	3. Like texture	3. Does not irritate hands	3. Good for hands
4. Light smell	4. Like feel after wash	4. Conditions hands	4. Won't ruin gloves
5. No smell	5. Like overall	5. Removes oil from hands	5. Won't irritate hands
6. Oily feel	6. Purchase intent	6. Feels clean	6. High-quality product
7. Dry feel			
8. Soft feel			
9. Lathers well			

sit down together and come up with attributes of surgical scrub product important to them [17].

Once the subjective characteristics deemed important for success of a surgical scrub product have been determined, it is important to actually perform preference testing by enrolling a number of different surgical staff personnel in a study and having them subjectively evaluate several versions of a specific surgical scrub product. Other considerations important to this category were covered in the preoperative skin-preparation section concerning personal subjective attributes.

VII. HEALTHCARE PERSONNEL HANDWASH CATEGORY

A. Personal Objective

Healthcare personnel handwash formulations are intended to quickly remove any transient, pathogenic microorganisms picked up on the hands of a healthcare provider from patient A and prevent their passage to patient B [18]. Hence, the product is intended to break the disease transmission cycle at the level of the contaminated healthcare worker's hands by removing those microorganisms.

The product must demonstrate low skin irritation potential upon repeated and prolonged use—20–30 washes per day for 5 consecutive days. Mildness to the hands, however, is usually attained at the price of reduced antimicrobial effectiveness. An optimal product formulation that provides substantive reductions in contaminative microorganisms, yet is gentle to the hands, must be developed.

Mildness can generally be built into the handwash in three ways [18]:

1. By proportionally reducing the irritating active ingredients such as chlorhexidine gluconate, iodophors, or alcohol. For example, instead of the customary 4% level of chlorhexidine gluconate (CHG) found in surgical scrub formulations, a 1% or 2% formulation may be developed.
2. By adding skin conditioners or emollients. These tend to counteract the irritating effect of antimicrobially active compounds, making the product more gentle and mild to skin surfaces.
3. By using a combination of these two methods (i.e., a reduction in the levels of active ingredient and the addition of emollients and skin conditioners).

B. Societal Requirements

Again, the documentation and regulations required to market a healthcare personnel handwash are covered in the social requirements portion for preoperative-prepping products.

C. Cultural Requirements

The healthcare personnel handwash product must be perceived as a highly effective antimicrobial compound, capable of removing the microorganisms with which healthcare personnel may become contaminated [18]. Important shared beliefs/values for a healthcare personnel handwash formulation include the ability to inactivate certain pathogenic microorganism species perceived as being important indices of the product's antimicrobial effectiveness. Many of these beliefs are not grounded in fact but, nevertheless, are important because they are believed to be important. For example, the anaerobic bacterial species *Clostridium difficile* will not grow in the presence of free atmospheric oxygen. However, this organism is capable of producing a spore, and most healthcare personnel handwash formulations are not sporicidal. Many healthcare workers believe that formulations should be able to inactivate *C. difficile* spores if the products are to be used by healthcare personnel. Although disease transmission potential for *C. difficile* is almost nonexistent, the perceived value of demonstrating product effectiveness against *C. difficile* is great.

Another aspect perceived as valuable in healthcare personnel handwash products is mildness to the hands. The product must not irritate the user's hands or make the user perceive the product as harsh. Most healthcare personnel—particularly physicians—are very conscious of their hand-skin integrity. They do not and will not use products that they perceive as harsh. The important attributes to be determined include those presented in Table 1; hence, a focus group of 5–10 representative healthcare personnel should meet for an extended period to determine important attributes that must be built into the product.

D. Personal Subjective

Once general subjective characteristics important for the healthcare personnel handwash product have been determined, it is necessary to actually perform preference testing [5]. The goal is to engineer a product that healthcare personnel prefer over the competition. This is done by enrolling a number of healthcare personnel as panelists to provide subjective evaluation of several configurations of test product and possibly even those of competitors. Other important considerations regarding personal subjective attributes have been provided in the preoperative preparation section.

VIII. FOODHANDLER HANDWASHES

A. Personal Objective

The potential for foodhandlers to be vectors in the transmission of foodborne disease is very significant [19,20]. Contaminating microorganisms are responsible for outbreaks of infectious diseases passed from foodhandlers to consumers via the food they eat. One of the most common sources of this is foodhandlers contaminated with enteric microorganisms from hand contact with their own feces. A problem in the foodhandling arena is that many who handle food do not wash their hands after defecation. In order to compensate for this, many food establishments have required that foodhandlers wear barrier gloves [19]. A vinyl or latex barrier glove that is intact (has no holes, rips, or punctures) will undisputedly provide protection from microbial transmission. But vinyl food-grade gloves, those most frequently used in the food service industry, commonly have preexisting pinhole punctures that compromise the barrier protection. Additionally, food-grade vinyl gloves are easily ripped, torn, or punctured as personnel perform their normal duties, and, in many cases, such damage remains unknown to the wearer. Exposure to heat has also been reported to alter the integrity of barrier gloves significantly, making them brittle and, thus, prone to breakage. Hence, the actual protection provided by barrier gloves is often much less than assumed. For example, in a study conducted at BioScience Laboratories, volunteer human subjects' hands were inoculated with a strain of *Escherichia coli* [19]. The subjects then donned vinyl foodhandler gloves, each of which had four small needle punctures. Within 5 minutes, the outside of the gloved hands was sampled for microbial contamination. The results of this testing demonstrated that significant numbers of *E. coli* can be transferred from the contaminated hands onto the outer surfaces of the gloves if even small holes exist in the gloves.

Wearing gloves may actually serve to increase the potential for disease transmission. As one wears vinyl or latex gloves for an hour or so, the microorganism populations on the hands within increase dramatically because the gloves prevent aeration of the hands, thereby increasing the levels of moisture, nutrients, and various other factors necessary for the growth of microorganisms. This phenomenon has long been known in the medical field, where mandatory handwashes using antimicrobial soap are required prior to gloving. Logically, as population numbers of both resident and contaminating microorganisms increase, so does the potential for disease transmission. Hence, relying solely on barrier gloves to prevent disease is not prudent.

Handwashing has been used for years to prevent foodborne illness [10,20]. Handwashing effectiveness is dependent on two factors: (1) the physical removal of microorganisms and (2) the immediate inactivation of microorganisms through contact with antimicrobial ingredients in the soap. But many antimicrobial ingredients also have the ability to prevent transient microbial recolonization of hand surfaces after handwashing by either microbial inhibition or lethality. If a nonantimicrobial soap is used, only the mechanical removal of microorganisms is significant.

In general, handwashing is very effective in removing contaminating microorganisms, if the handwash is performed correctly. But assuring that foodhandlers perform effective handwashes or wash their hands at all is difficult. First, the hands of foodhandlers can be exposed to different soil loads. The hands may be very greasy for those who work with pork products and high-fat ground beef. Medium grease loads are usually encountered among foodhandlers working with beef and chicken. Finally, for those working with salads and vegetable products, the hands tend to be dry and chafed and not greasy at all.

If foodhandling personnel work exclusively with any one of the three food categories mentioned above, the problem is more easily addressed by use of a heavy degreaser for high-fat content exposure, a medium degreaser for medium grease loads, and products with no degreaser for no grease loads, but an extra supply of skin emollients to help hold moisture and oils on the skin. If the foodhandler works in all grease-level conditions, a medium degreaser soap will probably suffice, using skin conditioning/moisturizing creams between washes as needed.

The antimicrobial activity of foodhandler products should be very high. This is because foodhandlers tend to perform washes less thoroughly than do healthcare personnel, often leaving dirt and grime under the fingernail beds, which then may serve as a contamination source.

B. Societal Requirements

Currently, there is limited regulation of antimicrobial soap products used in the food industry. The USDA's "E" rating system is obsolete, as the regulatory function has passed from the USDA to the FDA. There is at present no foodhandler efficacy study design using human subjects that is accepted by the regulatory agencies. Many manufacturers adopt a modification of the healthcare personnel handwash using *E. coli* as the contaminating microorganism species, instead of *Serratia marcescens*, the contaminating microorganism most commonly used in those evaluations. The toxicological properties of handwash ingredients are very important, because soap residues may be transferred from the hands of foodworkers to the food they handle.

C. Cultural Requirements

In general, these are similar to the healthcare personnel handwash application; the major difference is the high degree to which medical personnel value clean hands. Foodhandlers often have no real shared values about clean hands. Washing hands is just something they must do while on the job. This is probably because foodhandler positions generally are not filled by professionals, but by individuals at the lower socioeconomic level, young people, and, often, nonmotivated individuals.

To a large degree, culturally shared values of foodworkers will have to be instilled by the food industry itself. This will include training (lower right quadrant aspect), which will stimulate a value in ''doing it right'' (lower left quadrant aspect). In addition, in order to reduce the potential for disease transmission from fecal contamination, the following four steps will be useful in changing the shared values/beliefs of these workers and the industry [10,19].

1. Both gloving and handwashing with an effective antimicrobial product should be required for those performing high-risk tasks such as handling, cooking, or wrapping food. While neither is failsafe, it is probable that, when used in combination, they will provide more protection against disease transmission than either used alone. Observations in testing performed in our laboratory showed that population numbers of contaminative *E. coli* actually increased on the gloved hands when glove changes were performed at 1- or 3-hour intervals. However, concurrent testing showed that a thorough handwash using an effective antimicrobial product prior to gloving prevented significant growth of the contaminative microbes on the hand surfaces over the course of 3 consecutive hours of wearing. The obvious conclusion is that, before gloves are put on, a thorough handwash should be performed using an effective antimicrobial soap. Even so, when feasible, no direct hand/glove contact with food should occur, and sanitized serving tongs or other utensils should be used for its manipulation.
2. Mandatory ongoing sanitation training and education should be pursued for all employees. This is particularly necessary with inexperienced and/or unmotivated workers. Emphatically, without the active participation of employees, achieving adequate sanitation standards will be very difficult.
3. A high degree of personal hygiene should be required of foodservice personnel. Uniforms should be clean and changed often, employees should bathe or shower regularly, and they should not perform high-risk tasks when they are ill. High-risk tasks include hand/glove contact with food.
4. A quality control program supervised by qualified personnel should be initiated at each foodservice facility to monitor handwash/gloving practices. Written standard operating procedures (SOPS) should be drafted and all employees formally trained in handwashing/gloving procedures. Training should be documented in employees' training records and handwashing procedures posted in a conspicuous place near sinks used for handwashing.

D. Personal Subjective

Preference testing will be extremely valuable in this area to determine just what types of product attributes foodhandlers prefer [11,18]. The goal is twofold: to build a product that foodhandlers will want to use and to build a product that is preferred over those of competitors. Because relatively little is known concerning the subjective preferences of foodhandlers, it is important that preference testing be done by enrolling a reasonably large number of foodhandlers to evaluate various products sequentially. For example, begin with fragrance. When one or two preferred fragrances are identified, lathering characteristics can be added to the evaluation. Then, with these attributes identified, the feel of the hands after washing can be evaluated. This type of interactive process can be conducted for all other attributes deemed important.

IX. ANTIMICROBIAL HANDWASH AND BODYWASH

A. Personal Objective

Whether these categories become officially acknowledged by the FDA will not matter, because consumers want antimicrobial hand and body soaps. The antimicrobials generally used in these types of products are parachlorometaxylenol (PCMX), triclosan, and isopropyl alcohol, which have all been used for many years. The antimicrobial hand soaps/body soaps must also be of low skin irritation potential.

Perhaps the biggest problem with antimicrobial hand soaps is that the various claims some manufacturers present are misleading and even false. It is critical that manufacturers take the responsibility for manufacturing products that perform well in both clinical trials and in vitro sensitivity testing.

B. Societal Requirements

Because FDA requirements do not exist for antimicrobial hand soaps, regulation of quality is left to the industry, particularly the CTFA and SDA. These organizations have truly taken responsibility for setting realistic standards to which members voluntarily conform. However, there appear to be some concerns, with various manufacturers being unable to assure and verify their label and advertising claims to the satisfaction of the FTC.

C. Cultural Requirements

There is a tremendous need for consumers to feel clean. These shared values, due mainly to effective advertising campaigns, have convinced people that if they use antimicrobial hand/body soaps they will be cleaner than if they do not. Additionally, if they use antimicrobial bodywashes they will not offend others with body odor.

There is nothing wrong with cultural beliefs that antimicrobial soaps are better, as long as manufacturers truly strive to live up to their stated labeling/advertising claims, i.e., support their claims with data collected from valid, statistically complete studies using human test subjects, as well as from in vitro microorganism sensitivity testing to the product.

D. Personal Subjective

Soap manufacturers have made a very strong campaign for antimicrobial soaps. Manufacturers of these consumer products have frequently taken the time to understand personal subjective and cultural values and have used this understanding to launch highly successful marketing campaigns.

X. CONCLUSION

It is important that, in developing topical antimicrobial products, a multidimensional approach be taken. This will help ensure that the resultant product is designed for the specific needs of the market and that those needs are met. In this way, the product is more likely to have a long, useful, and profitable life.

REFERENCES

1. DS Paulson. A broad-based approach to evaluating topical antimicrobial products. In: JM Ascenzi, ed. Handbook of Disinfectants and Antiseptics. New York: Marcel Dekker, 1996, pp. 17–42.
2. DS Paulson. Research designs for the soaps and cosmetics industries: a basic approach. Soaps/Cosmet/Chemi Spec, 11:21–27, 1995.
3. DM Newman. Sociology. Thousand Oaks, CA: Sage, 1997.
4. HR Moskovwitz. Cosmetic Product Testing: A Modern Psychophysical Approach. Vol. III. New York: Marcel Dekker, 1984.
5. DO Sears, LA Peplau, SE Taylor. Social Psychology. 7th ed. Englewood Cliffs, NJ: McGraw-Hill, 1991.
6. DS Paulson. Successfully marketing topical antimicrobial products. Soaps/Cosmet/Chemi Spec, 2:51–58, 1991.
7. R Searle. The Construction of Social Reality. New York: Free Press, 1995.
8. R Kegan. In Over Our Heads. Cambridge, MA: Harvard University Press, 1994.
9. WC Frazier, DC Westhoff. Food Microbiology. 4th ed. New York: McGraw-Hill, 1988.
10. DS Paulson. Foodborne disease: controlling the problem. Environ Health 5:128–136, 1997.

11. DS Paulson. Designing a handwash efficacy program. Pharmaceut Cosmet Qual, 1: 53–65, 1997.
12. DS Paulson. Efficacy evaluation of a 4% chlorhexidine gluconate solution as a full-body shower wash. Am J Infect Control 21:(4):205–209, 1993.
13. T Peters. Liberation Management. New York: Knopf, 1992.
14. CB Pert. Molecules of Emotion. New York: Schribner, 1997, p. 11.
15. J Neter, W Wasserman. Applied Linear Statistics Models. Homewood, IL: Irwin, 1974.
16. DS Paulson. Comparative evaluation of five surgical scrub formulations. Assoc Operating Room Nurses J 60(2):246–256, 1994.
17. DS Paulson. Handbook of Topical Antimicrobial Testing and Evaluation. New York: Marcel Dekker, 1999.
18. DS Paulson. Designing a handwash for healthcare workers. Soaps/Cosmet/ Chemi Spec 6:12–21, 1986.
19. DS Paulson. To glove or to wash: a current controversy. Food Qual 6:56–68, 1996.
20. DS Paulson. A proposed evaluation method for antimicrobial hand soaps. Soaps/ Cosmet/Chemi Spec, 6:47–55, 1996.

2

Food and Drug Administration Perspective on Topical Antiseptic Drug Product Development

Albert T. Sheldon, Jr.
Food and Drug Administration, Rockville, Maryland

This chapter discusses the evolution of and regulatory requirements for assessment of the efficacy of over-the-counter (OTC) antiseptic drug products. The discussion begins with a summary of the salient concepts proposed in the 1974 Federal Register Notice of rule making and concludes with the 1994 Tentative Final Monograph (TFM) [1,2]. The discussion also addresses the relationship that exists between the OTC and New Drug Application (NDA) antiseptic drug product regulatory process and how these processes influence each other. Finally, the chapter concludes with a discussion of the technical issues facing the FDA and the scientific community regarding protocol standardization, interpretation of the data derived from these studies, and the relationship of these outcomes with those in the clinical setting.

Healthcare antiseptic and topical antimicrobial are distinct terms used throughout this document. Healthcare antiseptic is defined as a product used by healthcare professionals and includes surgical hand scrubs, preoperative skin preps, and healthcare personnel handwashes. The term topical antimicrobial includes healthcare antiseptics and products used as skin wound cleansers, skin wound protectants, and antimicrobial soaps used on minor cuts, scrapes, and abrasions. This review addresses the efficacy requirements for healthcare antiseptics only. It does not address the requirements for the assessment of safety or the content and format requirements for labeling.

I. INTRODUCTION

One of the most important contributions to modern medicine is Semmelweis's mandate that physicians' examining fingers be washed with chlorine to prevent puerperal (childbed) fever, thus resulting in decreased morbidity and mortality in the maternity ward [3]. Semmelweis's observation is one of several studies providing the scientific evidence necessary to justify the adoption of topical antimicrobial drug products in the general practice of medicine [4–6]. These studies are significant because they demonstrate the effect of topical antimicrobial drug products when used as intervention strategies in the reduction of the incidence of clinical disease.

Commensurate with the discovery of topical antimicrobial drug products is the development of in vitro and in vivo test methods designed to assess an antiseptic's potential utility in clinical settings. A thesis on in vitro methods development is that published by Kroing and Paul, who describe the influences of temperature, topical antimicrobial drug product concentration, and time on the rate of kill of organisms under evaluation [7]. They demonstrate by modification of these test parameters the influence on test results, thus preventing comparative assessment of the activity of the various compounds under investigation. These observations argue in favor of the establishment of standardized and controlled experimental designs to provide reproducible results. Included in their proposal is the need for neutralization of the topical antimicrobial drug product and development of media that optimized growth of the surviving microorganism to increase accuracy and reduce variability of the results obtained. The most noted early example of the application of these principles is the work of Rideal and Walker and of Chick and Martin in the development of the phenol coefficient method of testing disinfectants (i.e., topical antimicrobial drug products) [8,9]. Recent examples of the application of these principles can be found in standardized test methods developed by the Food and Drug Administration and the American Society for Testing and Materials (ASTM). These methods are commonly used in assessing the efficacy of topical antiseptic drug products [2,10].

II. FDA REGULATORY PROCESS

The regulatory path followed for antiseptic product development is dependent on the active ingredients under evaluation, the indication sought, and the intent to market to healthcare professionals or consumers. One path is the over-the-counter (OTC) Drug Review. The OTC drug review process is an ongoing review of OTC drug products marketed prior to May 11, 1972. The review process will result in the publication of a monograph describing the conditions of safety and effectiveness for the intended use. A second path is the New Drug Application process. In this process, the active ingredient is either an antiseptic not addressed in the Tentative Final Monograph (TFM) or, if addressed in the TFM, a product that seeks a new indication or is terminal sterilized by ionizing irradiation. The NDA product must also demonstrate safety and efficacy and is primarily marketed for use by healthcare professionals. Irrespective of the paths taken, the product must be labeled in compliance with OTC or new drug regulatory requirements.

A. Development of Efficacy Standards for OTC Topical Antiseptic Drug Products

If the active ingredient is found in a product sold on the OTC market and the product label makes drug claims of safety and effectiveness for specific uses, then the agency must develop the conditions that allow assessment of these claims. This process involves the establishment of an expert panel to evaluate these claims. The Food and Drug Administration is required to publish, in the Federal Register (FR), an advanced notice of proposed rule describing a proposal to establish a monograph for such products. The final monograph (FM) describes the types of claims allowed in labeling and the studies required to prove these claims.

OTC regulations require establishment of an expert panel that reviews the safety and effectiveness of the OTC drug category assigned to them. The panel follows specific standards in its assessment of safety and effectiveness. FDA aids the panel by publication of a notice requesting published and unpublished safety and efficacy data and information pertinent to the designated category under evaluation. The data must include labels, a statement of identity, animal and human safety data, and efficacy data. The panel then issues a report to the Commissioner of the FDA describing the conditions for a drug category that is generally recognized as safe and effective and not misbranded (Category I). The report also includes a statement on all active ingredients, labeling claims or other statements, or other conditions reviewed and excludes from the monograph products that are not safe and/or effective or are misbranded (Category II). Finally, a statement for active ingredients, labeling claims, or conditions for which safety and efficacy cannot be established due to lack of sufficient evidence is recommended (Category III).

The Commissioner reviews the conclusions and recommendations of the panel and publishes an advanced notice of proposed rule making containing the panel's recommendations and proposed monograph that describes products considered Category I. The document also describes the safety and efficacy requirements that must be met to move products not included in the proposed monograph into Category I.

After reviewing all comments in response to the proposed order, a Tentative Final Monograph is published that establishes conditions for OTC drugs considered safe and effective and not misbranded. Interested parties are invited to comment on the TFM and information used to establish the FM. Upon publication of the FM, all products not conforming to the safety and efficacy requirements are removed from the market or evidence must be provided to prove that the product now meets the requirements for safety and efficacy specified in the monograph. This information must be submitted through the new drug approval process, as discussed later in this chapter.

If an active ingredient is not addressed in the OTC FR notice, then two options exist: (1) the manufacturer may submit a petition to amend the monograph or (2) the product must be developed as a new drug as described in Section 505 of the Federal Food, Drug, and Cosmetic Act [11]. Currently, the NDA process allows development of an antiseptic for the indications of surgical hand scrub, healthcare personnel handwash, patient preoperative skin prepping, and skin prepping prior to injection or catheterization.

1. Rule Making, A Proposed Monograph for Antiseptics

In 1972, the Commissioner of the FDA announced a proposal to review, by an independent panel of experts, all OTC drugs for safety, effectiveness, and labeling [12]. The 1972 Notice is a call for data from all topical antimicrobial drug product manufacturers so that existent evidence can be used to assess the claims made for these products and to support the safety and efficacy of these products for their intended use. Subsequently, separate notices for OTC topical antimicrobial drug product active ingredients intended for repeated use were issued and addressed in the OTC Topical Antimicrobial Products Federal Register publication [13]. These notices develop the concept for the establishment of proposed requirements for safety and effectiveness that are to be included in the Code of Federal Regulation under Part 333—Topical Antimicrobial Drug Products for Over-the-Counter Drug Products [14].

FDA established the Advisory Review Panel on OTC Topical Antimicrobial Drug Products (Antimicrobial I Panel) in 1974 to review for safety and efficacy the data for active ingredients found in OTC topical antimicrobial drug products. Based on the recommendations of the Antimicrobial I Panel, FDA published an advanced notice of proposed rule making to establish a monograph for OTC topical antimicrobial drug products [1]. This FR Notice describes a proposed monograph that establishes conditions under which OTC topical antimicrobial drugs are evaluated for safety and effectiveness and are not misbranded. Application of these conditions by the panel allowed classification of antiseptics into Categories II, III, and I.

To aid in the classification of active ingredients, the Antimicrobial I Panel adopted seven specific definitions for antimicrobial product categories (Table 1) [1]. These definitions describe, based on intended use, the desired characteristics for each product category and, in some instances, the type of evidence necessary to meet the condition of the definition of efficacy. Also implied in these definitions is the principle that "the reduction of the normal flora to as low a level as possible will have a positive effect on the prophylaxis of disease" [1]. The supposition that reduction of the flora of the skin or hands results in the reduction of disease, although intuitive to healthcare professionals, has not been supported by adequate and well-controlled clinical studies.

Table 1 Product Categories Established by the Antimicrobial I Panel and Definitions for Categorization of Topical Antimicrobial Products

Product categories	Definitions of product categories
Skin topical antimicrobial drug product	A safe, nonirritating, antimicrobial-containing preparation that prevents overt skin infection. Claims stating or implying an effect against microorganisms must be supported by controlled human studies that demonstrate prevention on infection.
Patient preoperative skin preparation	A safe, fast-acting, broad-spectrum antimicrobial-containing preparation that significantly reduces the number of microorganisms on intact skin.
Surgical hand scrub	A safe, nonirritating, antimicrobial-containing preparation that significantly reduces the number of microorganisms on the intact skin. A surgical hand scrub should be broad-spectrum, fast-acting, and persistent.
Healthcare personnel handwash	A safe, nonirritating preparation designed for frequent use that reduces the number of transient microorganisms on intact skin to an initial baseline level after adequate washing, rinsing and drying. If the preparation contains an antimicrobial agent, it should be broad-spectrum, fast-acting, and, if possible, persistent.
Skin wound cleanser	A safe, nonirritating liquid preparation (or product used with water) that assists in the removal of foreign material from small superficial wounds and does not delay wound healing.
Skin wound protectant	A safe, nonirritating preparation applied to small cleansed wounds that provides a protective (physical and/or chemical) barrier and neither delays healing nor favors the growth of microorganisms.
Antimicrobial soap	A soap containing an active ingredient with in vitro and in vivo activity against skin microorganisms.

The Commissioner also concludes that regulations applicable to antiseptic drug products be promulgated to assure application of the same safety and labeling standards as for cosmetic products. These recommendations are not discussed further since they are not germane to the discussion of efficacy requirements for topical antimicrobial drug products.

Antimicrobial Panel I Efficacy Guideline. The Antimicrobial I Panel also recommends guideline and test methods to assess the efficacy of an active ingredient currently classified in Category III. The tests are used to assess the characteristics of the product as defined by the proposed definitions (Table 1). Thus, tests are included to assess the in vitro spectrum of activity of the active ingredient, the determination of the possible emergence of resistance, and the development of an adequate neutralization system. In vivo tests assess the qualitative and quantitative changes in transient and resident flora, methods for enumeration are recommended, and the utilization of principles in the design of adequate and well-controlled clinical investigations are articulated (Table 2) [1]. These principles must be considered and incorporated into the design of the specified protocols used in evaluating the efficacy of topical antimicrobial drug products.

Determination of the skin flora (Other than on the hands). The development of new sampling techniques provides methods to study, qualitatively and quantitatively, the microbial flora of various parts of the body [15,16]. The panel encouraged utilization of these methods in the evaluation of topical antimicrobial drug products. The methods provide information that allows assessment of the

Table 2 Principles of Clinical Protocols Used in the Assessment of the Effectiveness of Topical Antimicrobial Products and Their Active Ingredients

1. A precise statement of the objectives, research goals, and disease state studied.
2. The conduct of studies in laboratory settings.
3. Use of studies with randomization of panelist to each arm. Analysis to demonstrate comparability of Antimicrobial I Panelist in each arm.
4. To account for the placebo effect noted with topical antimicrobial drug products, the study should:
   a. Include a vehicle arm or next best therapy identical in composition to the drug under investigation
   b. Reduce bias by the introduction of a double blinded study design
   c. Define inclusion and exclusion criteria used in selecting study participants
   d. Provide definition of outcome response variables
   e. Consider study design (e.g., sample size, rationale, blinding, etc.)
   f. Ensure completeness of the study by accounting for dropouts, missing data, and how they are treated for evaluation of the final outcome
   g. Consider statistical study design, analysis, and summary

potential shifts of the microbial flora that were thought to occur during repeated use of these products on the skin. Information obtained with these methods is used to gain an understanding of the normal flora of the skin and the desirable and undesirable effects that topical antimicrobial drug products are thought to have on the skin flora.

Isolation of Gram negatives and other organisms from the skin. The Antimicrobial I Panel is concerned with observation in the published literature that uses of topical antimicrobial products with predominantly gram-positive spectrums of activity produce an environment, on repeated use, that selects for a predominantly gram-negative microbial flora [17,18]. These undesirable ecological shifts of the microbial flora encouraged the Antimicrobial I Panel to recommend studies with subjects that are "carriers of gram-negative microorganisms" and antimicrobial drug products with predominately gram-positive spectrums. The intent of these studies is to assess the potential for population shifts.

Effectiveness testing of a surgical hand scrub (glove juice test). Surgical hand scrubs reduce the resident and eliminate the transient flora of the hands of surgeons and surgical personnel, thus reducing the incidence of postsurgical infections. These products also prevent proliferation of the transient flora remaining on the occluded hand for the duration of a surgical procedure. The definition of the indication surgical scrub, provided in Table 1, suggests that the topical antimicrobial drug product reduces significantly the number of microorganisms from the intact skin, is broad spectrum, fast-acting, and persistent. The recommendation of the Antimicrobial I Panel is adoption of a strategy that allows qualitative and quantitative characterization of these characteristics.

The protocol recommended is known as the surgical hand scrub glove juice method. The glove "juice" protocol provides improvements over existing methods suggested by Price [19] and Cade [20] since their handwash basin method can be manipulated by changing the routine, timing, and recovery of bacteria. In addition, the Price and Cade basin method requires sampling of the wash water and is not as desirable a measure of efficacy as the measuring of bacteria remaining on the hands. The instilling of sampling fluid into a glove-occluded hand allows sampling of the microbial flora of the hand directly.

Based on the observations and suggestions of others [21,22], the Antimicrobial I Panel, consultants, and Bruch [23], developed the surgical hand scrub protocol that provides a more reproducible method of product assessment. The product is used as described on the product label so that test results reflect conditions of normal use. The protocol describes inclusion and exclusion criteria, 2-week wash-out periods to allow normalization of the resident flora prior to baseline assessments, sampling techniques, and time intervals for sampling and statistical analysis. Entry into the study requires baseline counts of 1.5×10^6 to 4×10^6, colony-forming units (CFUs) per hand after the hands have been washed with bland soap.

Enumeration is performed at specified time intervals by the glove juice method. The method requires the donning of surgical gloves, instilling of sampling fluid containing buffer, surfactants, and devoid of neutralizers. The gloved hands are massaged, a sample of the glove fluid taken, serially diluted in neutralizer-containing blanks, and plated. Conversely, the glove is removed, inverted into stripping fluid, and the hands rinsed with sampling solution. A sample is taken and enumerated as previously described.

Efficacy is determined by a test of the assumption that the antimicrobial produces a given log reduction at the 1-minute time interval when compared to the same hand baseline count. The other hand is enumerated at hourly intervals up to 6 hours to demonstrate continued suppression of the microbial flora below the established baseline. The procedure is performed 11 times over a 5-day period. Enumeration is performed at the first, second, and eleventh surgical scrub, corresponding to days 1, 2, and 5 of the study. The surgical hand scrub is an indication found in products intended for use by healthcare professionals in hospital-type settings.

Effectiveness testing of healthcare personnel handwash products. These products are designed for use by healthcare professionals and are formulated to prevent the transmission of resident and transient pathogens (e.g., nosocomial) from the hands of the healthcare workers to patients during maintenance activities. Thus, the products are broad spectrum and fast-acting; they reduce the number of transient microorganisms on intact skin to an initial baseline level and, if possible, are persistent. The product must also be nonirritating since it is used multiple times per day.

The main effect of healthcare personnel handwash products is removal of the transient bioburden acquired on the hands during daily activities. Therefore, the protocol is designed to mimic this hand contamination process. Contamination is accomplished either by dipping of the hands into an overnight culture of a marker organism or by allowing test panelists to handle heavily contaminated material to simulate contamination of the hands during maintenance activities. The latter method is the least desirable due to lack of reproducibility. Two marker organisms—*Serratia marcescens* (pigmented strain) and *Bacillus subtilis* var. *niger*—are suggested as candidates for artificial contamination. Once the hands are contaminated and air-dried, the product is used as described on the label and the hands evaluated by an undefined enumeration technique. The process of contamination and washing is repeated 25 consecutive times, with enumeration performed at every fifth interval.

The Antimicrobial I Panel also recommends the Cade [20] and Quinn [24] handwash procedure to assess the efficacy of topical antimicrobial drug products intended for healthcare personnel handwash use. This method provides a measure of the microorganisms removed from the hands as opposed to those remaining on the hands. Unfortunately, the panel did not provide efficacy requirements (e.g., log reductions per unit of time) as a measure of successful outcome for the healthcare personnel handwash indication.

Effectiveness testing of a patient preoperative skin preparation. Patient preoperative skin preparations are designed for use by healthcare professionals to prep the patient's skin prior to invasive surgery. The product is routinely applied once and, by definition, must be a broad spectrum, fast-acting antimicrobial that significantly reduces the number of resident and transient microorganisms on intact skin of the patient.

The Antimicrobial I Panel did not provide a specific protocol for the assessment of the efficacy of products intended for this use but did suggest the incorporation of certain concepts in assessing product efficacy. The sampling procedures recommended are the cup scrubbing [25] or skin stripping techniques or any other appropriate techniques that allow accurate quantitation of the microbial flora at the test site. Any body site can be used but the genital area must be included. Clearly, the panel is concerned with notable differences of the microbial flora found in different parts of the human body.

Since the product must be fast-acting, the Antimicrobial I Panel recommends enumeration 30 minutes after product use during conduct of the clinical simulation study. In addition, since these products are used as "leave-on products" and are not washed off, as are surgical hand scrubs, the amount of active ingredient carried over during enumeration is of concern. Thus, satisfactory demonstration of the neutralization of any carryover antimicrobial during the enumeration procedure is emphasized. Validation of the neutralization system is required.

The efficacy requirement is a three-log reduction at the test site evaluated. However, it is not clear whether the required reduction is measured per milliliter of sampling fluid or per square centimeter of skin surface area. Therefore, a clear recommendation as a measure of effectiveness is not provided for the patient preoperative skin preparation category.

Effectiveness testing of a skin wound cleanser. Skin wound cleanser is a new category created by the Antimicrobial I Panel and defines products designed for general removal of foreign material from small superficial wounds. The Antimicrobial I Panel did not describe methods to assess efficacy of products used as skin wound cleansers. However, the panel is concerned with the possible delay of skin wound healing caused by these products and suggests animal models and hydrometery studies in humans for this evaluation or until adequate human testing procedures can be devised.

Effectiveness testing of a skin wound protectant. Skin wound protectants are preparations applied to small clean wounds and function as protective physical or chemical barriers. The Antimicrobial I Panel suggests assessment of the physical barrier by challenge of the barrier system with fluorescent particles and measurement of the presence of the particles on the opposite side of the barrier.

In principle, this model is considered appropriate for assessment of the barrier capabilities of a product. The panel did not receive a proposal for evaluation of the antimicrobial properties of products containing antiseptic active ingredients.

The Antimicrobial I Panel suggests evaluation of the barrier's ability to function as a supporting structure that enhances growth promotion of microorganisms present in the wound. An animal wound model with wounds inoculated with specific microorganisms is suggested as a means of assessing the potential of the barrier to promote growth. Enumeration of the wound after product application provides the evidence necessary to determine the effect of the barrier on growth promotion.

Also suggested is a human model to assess the potential of the barrier to encourage the growth of microorganisms. A punch biopsy or skin-stripped induced wound is treated with the product and the wound monitored for growth promotion of microorganisms.

Finally, the panel recommends monitoring of anaerobic species, especially where occlusive dressings are used. A statement of effectiveness as measured by the human or animal model was not provided for a product described as a skin wound protectant.

Antimicrobial soap testing. An antimicrobial soap is a product that contains an active ingredient and has in vitro and in vivo activity against microorganisms residing on the skin. Therefore, the efficacy requirements must examine both the in vitro and in vivo efficacy characteristics of the product. The Antimicrobial I Panel did not provide specific recommendations for the assessment of these products but concluded that the Cade [20] and Quinn [24] handwashing procedures could be applicable to these products. A statement describing the efficacy requirement was not provided.

In vitro Susceptibility testing. Five of the product categories (Table 1) described by the Antimicrobial I Panel require assessment of the in vitro spectrum of activity of the active ingredient and the finished formulation. The panel recommends a method that allows estimation of the minimum inhibitory concentration (MIC) of the test organism, but no protocol is recommended to accomplish this assessment. Standardized conditions are recommended, as is a battery of mesophilic clinical strains representing normal skin flora and skin pathogens. Viruses, fungi, and dermatophytes are also suggested for evaluation of the spectrum of the product. A definition of broad spectrum is not provided.

Also required are studies to assess the possible development of resistance to the active ingredient using the serial passage of the test strain(s) in sublethal concentration of the antimicrobial. Finally, the panel recommends the development of techniques for the adequate neutralization of the antimicrobial under investigation so that accurate assessment of in vitro product efficacy can be performed.

Topical Antimicrobial Drug Product Categories. The final task performed by the Antimicrobial I Panel is classification of the active ingredient into appropriate "product categories," based on current knowledge of safety and efficacy. The panel reviewed all active ingredients using the standards for safety, effectiveness and labeling established by regulation in its assessment. The results are summarized in Table 3 and are provided as two columns for each category. The column on the left of each category reflects the recommendation of the panel, and that on the right the final recommendation of the Commissioner of the FDA as described in a later section of this document. Hexachlorophene, fluorosalan, and phenol at concentrations of greater than 1.5% (aqueous alcoholic) and tribromsalan are classified as Category II for all topical antimicrobial product categories described in Table 1. These antimicrobials are not considered safe and/or effective for the intended use(s). A majority of the active ingredients were placed in Category III for most of the indications, suggesting that insufficient data are available on the safety and/or efficacy of most of these antiseptics.

2. The Tentative Final Monograph

The FDA issued numerous TFMs for active ingredients found in OTC topical antimicrobial drug products described in the 1974 proposed notice of rule making. The TFM published in 1978 describes the tentative conclusions made by the Commissioner regarding the safety and efficacy requirements for OTC topical antimicrobial drug products designed for use by healthcare professionals in hospital-type settings.

OTC Topical Antimicrobial Products TFM. The next step in the FDA regulatory review process is publication of the tentative final monograph that establishes the conditions for safety, effectiveness, and proper labeling of OTC topical antimicrobial drug products [26]. This TFM proposal is based on the Commissioner's review of the final report of the Antimicrobial I Panel and all of the comments and reply comments provided in response to the promulgation of this monograph. The proposal provides a restatement, for the purposes of clarity, of the Antimicrobial I Panel's recommendations and conclusion and provides a detailed description of additional safety and efficacy-testing guidelines that must be used in the assessment of Category III ingredients. Successful implementation of these requirements allows reclassification of the Category III active ingredient to Category I status.

This notice also includes modifications of the in vitro and in vivo efficacy testing requirements, the establishment of test methods to assess skin wound healing, and the physical barrier properties of "first-aid" products. In addition, the definitions are modified to provide greater clarity of the desired characteristics for these products and to reflect that some of these products are intended for use

Table 3 Categorization of Active Ingredients by Antimicrobial I Panel and Subsequently by the Commissioner of the FDA

Active ingredient	Antimicrobial wash		Healthcare personnel handwash		Patient preoperative skin prep		Skin antiseptic		Skin wound cleanser		Skin wound protectant		Surgical hand scrub	
Benzalkonium chloride	NA[a]	II[a]	III	III	III	III	III	III	I	I	III	III	III	III
Benzethonium chloride	NA[a]	II[a]	III	III	III	III	III	III	I	I	III	III	III	III
Cloflucarban	III	III	III	II[b]	II	II	II	II	II[b]	II[b]	II	II	II	II
Fluorosalan	II	X	II	X	II	X	II	X	II	X	II	X	II	X
Hexachlorophene	II	X	II	X	II	X	II	X	II	X	II	X	II	X
Hexylresorcinol	NA[a]	II[a]	III	III	III	III	III	III	I	I	III	III	III	III
Iodine complex	NA[a]	II[a]	III	II	III	III	III	III	III	III	III	III	III	III
Methylbenzethonium chloride	NA[a]	II[a]	III	III	III	III	III	III	I	I	III	III	III	III
Nonyl phenoxypoly ethanol–iodine	NA[a]	II[a]	III	II	III	III	III	III	III	III	III	III	III	III
Para-chloro-*meta*-xylenol	III	III	III	III	III	III	III	III	III	III	III	III	III	III
Phenol:														
>1.5% aqueous/alcoholic	II	II	II	II	II	II	II	II	II	II	II	II	II	II
<1.5% aqueous/alcoholic	III	III	III	III	III	III	III	III	III	III	III	III	III	III
Poloxamer iodine complex	NA[a]	II[a]	III	II	III	III	III	III	III	III	III	III	III	III
Povidone-iodine complex	NA[a]	II[a]	III	III	III	III	III	III	III	III	III	III	II	III
Tincture of Iodine	NA[a]	II[a]	II	II	I	I	II	III	II	III	II	III	II	II
Tribromsalan	II	II	X	II	X	II	X	II	X	II	X	II	X	
Triclocaban	III	III	III	II[b]	II	II	II	II	II[b]	II[b]	II	II	II	II
Triclosan	III	III	II	II[c]	II	II[c]	III	III	III	III	III	III	II	II[c]
Triple dye	NA[a]	II[a]	NA[a]	II[a]	NA[a]	II[a]	III	II[a]	NA[a]	II[a]	NA[a]	II[a]	NA[a]	II[a]
Undecoylium chloride-iodine complex	NA[a]	II[a]	III	III	III	III	III	III	III	III	III	III	III	III

[a] Not applicable due to a physical and/or chemical incompatibility for formulation.
[b] Classified in Category III when formulated in a bar soap used with water.
[c] Restricted for use only in neonatal nursery.
Note Each active ingredient category lists two columns. The column on the left represents the conclusions of the Antimicrobial I Panel (1). The column on the right represents the current TFM categorization recommendation (2).

by healthcare professionals while others are intended for use by the general public. The Commissioner considers the test methods described in the TFM final unless the FDA receives a properly supported request to the contrary. Finally, the notice provides information on the reclassification of the active ingredients in Table 3 into Category II, III, and I. Each product category has two columns, and the column on the right side represents the final decision of the Commissioner. The column on the left represents the recommendation of the Antimicrobial I Panel.

The modifications made to the efficacy guidelines for most test protocols are those that provide greater clarity to the methods (i.e., timing of procedures, methods of enumeration, etc.). The modifications do not change, substantively, the concepts or principles that must be incorporated into the study design to assure the conduct of an adequate and well-controlled study. However, one change is substantial in trial design: a requirement for the inclusion of a study arm that allows efficacy assessment of the product's vehicle and its contribution to the overall effect of the finished product. Sorely lacking from the protocols, especially those that measure product effect by reduction of the microbial flora versus a baseline, are quantitative endpoints that describe an acceptable measure of performance. Two exceptions are the protocols for soaps used as deodorants and the preoperative skin prepping products described later.

Tentative Final Monograph for Healthcare Antiseptic Drug Products; Proposed Rule. On June 17, 1994, the FDA issued a proposed rule making in the form of an amended tentative final monograph and establishes conditions under which OTC topical ''healthcare antiseptics'' are generally recognized as safe and effective and not mislabeled [27]. This document amends the 1978 TFM discussed previously by narrowing the scope of that document to include a proposed TFM for topical healthcare antiseptics and addresses products designed for use primarily by healthcare professionals. The categories include surgical hand scrubs, healthcare personnel handwashes, patient preoperative skin preps, skin preps prior to injection, and ''antiseptic handwashes'' for personal use in the home. The antiseptic handwash is a new product category created by the FDA and includes multiuse products intended for use in the home to help prevent cross-contamination from one person to another, especially when caring for invalids or ill family members. This product category has characteristics not unlike those recommended for healthcare personnel handwash products but for use in the household environment. Thus, these products must meet the same efficacy requirements as healthcare personnel handwash products.

The product categories now classified as ''first aid antiseptics'' are addressed in a separate proposed rule for antimicrobials used as skin wound protectants, skin wound cleansers, and skin antiseptics [28]. The first aid antiseptics are primarily used by consumers to clean superficial wounds and for the treatment of minor cuts, scrapes, and abrasions and are not addressed further in this chapter.

The intended uses of healthcare antiseptic products were carefully evaluated and redefined as necessary to reflect the current thinking of the FDA. These uses are based on the recommendations of the Antimicrobial I Panel, comments submitted to the TFM, and the FDA's own regulatory experiences with NDAs. The TFM monograph for healthcare antiseptic drug products provides one of the most comprehensive and detailed attempts at describing protocols and efficacy requirements of products intended for use by healthcare professionals. The preclinical efficacy requirements include an assessment of the in vitro spectrum of activity and time-kill kinetic studies. The clinical assessment requires conduct of the clinical simulation studies either for the healthcare personnel handwash, surgical hand scrub, or patient preoperative skin prepping indications. The studies performed are dictated by the indications desired on the product label.

Clinical Simulation Studies. These protocols allow assessment of each of the characteristics of the proposed indications described in Table 4 [27] and include the basic principles articulated by the Antimicrobial I Panel (Table 2). The design is also based on reviewer experiences gained from the development of topical antibiotic and antimicrobial drug products through the NDA process. Examples of changes or recommendations include the incorporation of a positive control to validate the performance of the study, a test product vehicle arm, and proposed statistical methods of analysis to evaluate the data. Studies must be

Table 4 Product Categories and Definitions Established by the Commissioner of the Food and Drug Administration for Healthcare Antiseptic Drug Products to Aid in Their Categorization as Topical Antimicrobial Products

Product categories	Definitions of product categories
Antiseptic handwash or healthcare personnel handwash	An antiseptic containing preparation designed for frequent use; it reduces the number of transient microorganisms on intact skin to an initial baseline level after adequate washing, rinsing, and drying; it should be broad-spectrum, fast-acting, and, if possible, persistent.
Patient preoperative skin preparation	A fast-acting, broad-spectrum, and persistent antiseptic-containing preparation that significantly reduces the number of microorganisms on intact skin.
Surgical hand scrub	An antiseptic-containing preparation that significantly reduces the number of microorganisms on the intact skin; it is broad-spectrum, fast-acting, and persistent.

double-blinded wherever possible and evaluator-blinded at a minimum. In addition, every effort is made to assure that variability is reduced and procedure manipulations minimized to prevent the introduction of bias into the study. Also included in the proposed protocols are the clinical efficacy requirements, measured as log reductions from preestablished baseline values that must be met to demonstrate efficacy of the products for each indication. Finally, validation of the neutralization system is performed using the methods and materials used in the clinical simulation study.

3. Preclinical In Vitro Studies

The in vitro spectrum of activity study allows us to assess the potential utility of an antiseptic by providing information that cannot be obtained from the required clinical simulation studies. A standardized method is used to assess the MIC of the test product, vehicle control, and positive control versus a battery of clinically relevant organisms [29]. The battery of organisms selected for this assessment is based on the intended use of the product in professional healthcare settings. Since healthcare antiseptic drug products are one aspect of the intervention strategy developed to minimize the spread of nosocomial pathogens, it stands to reason that the battery of pathogens evaluated must reflect, where possible, this nosocomial population. The National Nosocomial Infectious Disease Surveillance program monitors the prevalence rates of nosocomial infections in the United States [30]. Thus the genera and species identified through this surveillance program are the nosocomial species that must be reflected in the battery of isolates used in the in vitro and time-kill assessment.

Since the components and/or composition of the formulation are known to affect the in vitro efficacy of some antiseptic drug products, it is necessary to evaluate the effect of excepients on the active ingredient [31]. Thus, the in vitro study is performed with both the active ingredient and the finished product in order to assess the effects of the formulation on the spectrum of activity.

The in vitro study in and of itself provides limited information and is but one aspect of the evidence characterizing the potential effectiveness of the antiseptic. It cannot be used exclusively to predict in vivo efficacy but does serve to confirm the expected susceptibility of particular isolates [32,33]. The in vitro spectrum of activity information developed for the product and submitted to the agency is allowed in the product package insert but must be followed by the statement: "The following in vitro data are available but their clinical significance is unknown."

Also required is a study to assess whether the antiseptic is "fast-acting," as described by the definition for an indication (Table 4). The time-kill kinetic study is designed to measure the rate of kill by the antiseptic under controlled conditions. The time-kill kinetic study is performed with a select group of gram-positive and gram-negative microbial species originally used in the in vitro spectrum of activity study. Valuable insights can be gleamed from this data by evaluation of the concentrations required to produce bacteriostatic and/or bactericidal responses. The relationship of the time-kill concentration used in the study and the concentrations used in the clinical simulation studies can help define the utility of the antimicrobial when used in the clinical setting. At the time of printing of the FR Notice, standardized time-kill kinetic protocols were not available and sponsors were encouraged to submit protocols for evaluation prior to initiation of these studies. The FDA does encourage protocol designs that incorporate variables and concepts described for antibiotic drug products [34]. The American Society for Testing and Materials, recognizing a need for a standardized method, is developing a time-kill method to aid in the evaluation of antiseptic drug products [35].

Both the MIC and time-kill kinetic in vitro studies allow us to gain insights into the potential utility of the antiseptic, insights that cannot be gained through the clinical simulation studies alone. The in vitro studies are used in concert with the clinical simulation studies to draw conclusions regarding the potential efficacy of the product under investigation. Alone, each test is limited, but together they help characterize the activity of the antiseptic as defined by definition and provide information required for approval. The limitations of these protocols and assessment of product efficacy are described below in Section III of this chapter.

The FR notice also requires studies to assess the potential for emergence of resistance to the antiseptic under investigation. Investigation of this potential was first suggested by the Antimicrobial I Panel, and in retrospect the recommendation may have greater implications than previously envisioned. The issue of resistance is described in detail later in this chapter. The proposed method described requires serial passage of indicator organisms through increasing concentrations of the antiseptic. A standardized protocol to assess the emergence of resistance is not available at the present time.

4. Assessment of the Efficacy of Topical Antimicrobial New Drug Products

An active ingredient not described by the TFM (Table 3) should be assessed for safety and efficacy under an investigational new drug (IND) application and all pivotal studies submitted to the agency for review as an NDA. An example is the bisguanide class of antimicrobial represented by chlorhexidine gluconate. In addition, any OTC active ingredient that is terminally sterilized by ionizing radiation must be submitted to the FDA as an NDA for assessment of safety and efficacy for the desired indication(s).

The development and standardization of the protocols and the establishment of efficacy requirements described in the TFM are based on experiences gained from the NDA review process using the bisguanide, chlorhexidine gluconate, and terminally sterilized category II iodophors. These NDAs contain preclinical and clinical data in support of the effectiveness for each of the indication(s) sought for the product. The studies are designed and performed to provide evidence to support the characteristics desirable for each indication.

During the early stages of the antiseptic NDA review process, standardized protocols did not exist for the assessment of the efficacy of topical antimicrobial drug products. Each NDA was reviewed independently of others to determine efficacy. However, a regulatory agency requires standardized and reproducible methods by which to assess the efficacy of test products. Therefore, as the NDA review process evolved, the clinical protocols used throughout the NDA review process also evolved into the protocols now recommended in the TFM. The NDA protocols are designed to incorporate the concepts articulated in the monographs that supersede them.

Additionally, each study performed and submitted to the agency includes a positive control: a product that is approved for the indication under evaluation. This control is used to assess the reproducibility of clinical simulation studies. During this time period, two antiseptic products were approved for marketing through the NDA process. One of these is Hibiclens (4% chlorhexidine gluconate, CHG). Since the FDA considered this product safe and effective for the approved indications, the product was used as a control in all future evaluations of other antiseptic products. Sufficient experience was gained with this 4% CHG product using standardized protocols to establish efficacy requirements for future products. In addition, the test product under evaluation is not required to perform better than or as well as the 4% CHG control product but is required to meet the efficacy requirements established for the indication [27]. The FDA selected this approach as opposed to head-to-head comparisons because the number of panelists required to demonstrate differences between two antiseptics would be prohibitively large.

Once the FDA concludes that sufficient evidence is provided in an NDA to justify approval of a product as a healthcare antiseptic, it is logical to conclude that this product can be used as a control in a clinical simulation study designed to assess the efficacy of other test products. In addition, by utilizing standardized versions of the protocols used to approve the control product and by using the approved products as one of the control arms of the study, we can assess the performance characteristics of the test product in a reproducible manner. Using this approach, the agency verifies the performance of the control product and gains knowledge about the reproducibility of the protocol. The FDA's Division of Anti-infective Drug Products evaluates hundreds of studies in INDs and makes recommendations to the Division of Over-the-Counter Drug Products regarding trial designs and efficacy requirements that must be used to assess the efficacy characteristics of new test products. These requirements are described in the TFM.

Irrespective of the indications sought, the applicant must submit the results from preclinical in vitro spectrum of activity and time-kill kinetic studies. The in vitro spectrum of activity study is performed with a standardized protocol using 50 strains each of the genera and species described in the TFM [27,29]. The test product, the test product active ingredient, the test product vehicle, and the control product are used in the evaluation. Sponsors of investigational new antiseptic drugs suggest that this requirement needs reevaluation to determine whether all these data are necessary. Based on data submitted to NDAs, the FDA evaluated this proposal and concluded that the in vitro spectrum must be performed with the test product and 50 strains of the genera and species listed in the TFM. Data are also required for the test product active ingredient, the test product vehicle, and the positive control, but only 10 strains of each genus and species need to be used. The 10 strains must be a subset of the 50 strains originally tested with the test product. The applicant is allowed to use this in vitro spectrum of activity data in the product label, including however, the statement, "The following in vitro data are available but their clinical significance is unknown."

Time-kill kinetic studies are required and are performed using a protocol submitted by the sponsor and approved by the FDA. The principles described in the TFM for the time-kill kinetic study must be incorporated into the design of the protocol. For antiseptic products developed through the NDA process, the FDA also suggests the use of the methods described by the National Committee for Clinical Laboratory Standards [36].

Clinical simulation studies are also required. The sponsor must submit the results of two adequate and well-controlled clinical simulation studies per indication before approval of the NDA is granted. The studies must use the protocols and meet the efficacy requirements for the indication sought as described in the TFM [27]. If the surgical hand scrub and the healthcare personnel handwash indications are sought, then the FDA requires two adequate and well-controlled surgical hand scrub studies and one healthcare personnel handwash study. If the preoperative skin prepping indication is sought, then two adequate and well-controlled studies are required. In any case, different investigators in different geographic regions of the country must perform the clinical simulation studies. Modification of the protocol is strongly discouraged due to the potential effect on outcome.

The requirements for NDA products are identical to those described in the TFM for preclinical and clinical assessment of efficacy. However, the NDA process is more flexible in that as nosocomial patterns change and new efficacy methods emerge, they are more easily incorporated into the regulatory process. The OTC regulatory process does not have this flexibility.

III. EFFICACY TESTING, LIMITATIONS, AND CONTROVERSY

Effectiveness characteristics considered desirable in healthcare antiseptic drug products are described in Table 4, and the requirements for assessment of efficacy depend on the indication sought. The methods used to examine these product characteristics are controversial since many methods are available to assess efficacy and each may produce different outcomes with the same product. Also lacking is interpretation of the data obtained from these test methods. The relationship of the outcome measure of efficacy defined by the test method and its relevance to the clinical setting is not well established. Efforts to close this gap are not forthcoming. Consequently, the FDA must use methods at its disposal to assess product efficacy. Thus, protocols whose purpose is to explore product characteristics and to assess whether the product meets the minimum efficacy requirements for an indication are recommended. Due to limitations associated with each of these test protocols, they cannot be used independently of each other but must be used in concert to gain an overall perspective of product efficacy. The in vitro assays are used to define the use of the antiseptic in a specific instance while the in vivo assays are used to define its use in a particular clinical setting. Ideally, both approaches must produce similar results, but this may not always be the case. Despite these limitations, preclinical and clinical simulation studies attempt to provide predictive evidence of clinical efficacy.

A. In Vitro and Time Kill Kinetic Studies

A desired characteristic of healthcare antiseptic drug products is a broad microbial spectrum that includes nosocomial pathogens encountered in product use settings. Most in vitro spectrum-of-activity data originally submitted to the FDA in response to the proposed rule making are obtained using nonstandardized methods. Thus, an accurate characterization of the in vitro spectrum of activity of an antiseptic active ingredient, or comparisons of the activity of active ingredients in a given class, or the effect of expedients found in the finished product on activity cannot be performed. The use of nonstandardized methods also prevents assessment of the changing patterns of susceptibility of nosocomial pathogens to antiseptic active ingredients.

Thus, the FDA proposes the use of standardized methods to assess the spectrum of activity of antiseptic active ingredients and the finished drug product [29]. The justification for using a standardized method is its reproducibility. The method allows comparison of data derived from different laboratories and the evaluation of a large number of strains so that a current understanding of the spectrum of activity can be defined for the antiseptic under investigation. Data generated from standardized methods are also used to monitor changing susceptibility patterns over time, an aspect that is increasingly becoming more important as reports of resistance to some antiseptics are emerging. The in vitro susceptibility-testing program must be proactive and designed to monitor changes in susceptibility patterns as well as the incorporation of new nosocomial pathogens as they emerge.

In some instances the proposed standardized method cannot be used with a particular active ingredient or formulation due to technical reasons or neutralization of the antiseptic. Therefore, a detailed discussion of the technical problem(s) and data supporting the need to modify the standard method must be submitted to the agency for evaluation and concurrence. An alternate and reproducible susceptibility test method must be provided for evaluation prior to use in assessing the spectrum of an antiseptic.

In vitro susceptibility test methods have several limitations, and these are evident when scientists attempt to use data derived from them to predict the potential efficacy of an antiseptic in clinical settings. The MIC method measures the effect of decreasing concentrations of antiseptic over a defined period of time and measure inhibition of growth. The concentration of drug required to produce the effect is defined as the MIC and is normally several hundred to thousands of times less than the concentration found in the finished dosage form. Extrapolation of this MIC information to predict clinical outcome clearly underestimates the potential utility of the antiseptic.

The MIC study can be used to characterize the bacteriostatic/bactericidal nature of the antiseptic by determining growth in wells above the established MIC. If sampling reveals that growth in wells is no more than one- to twofold higher than the MIC established for the drug, the antiseptic is considered bactericidal. If growth is threefold or more than the MIC, the antiseptic is considered bacteriostatic.

The regulatory agency uses MIC data to gain an understanding of the relationship of the susceptibility patterns of the resident flora evaluated in the preclinical in vitro studies and clinical simulation studies to other nosocomial pathogens evaluated in the spectrum of activity studies. If the MICs of the organisms representing the nosocomial flora are similar to those represented by the resident flora and the time-kill kinetic studies suggest a similar rate of kill, then we assume that the product is likely to work in a nosocomial setting where these pathogens are encountered.

Another characteristic desirable in antiseptics used by healthcare professionals is that they act as microbiologically lethal agents. The spectrum of activity study does not allow evaluation of this characteristic since incubation exceeds 18 hours. However, the time-kill kinetic study does allow assessment of the rate of kill of an antiseptic provided certain variables are controlled [34].

Several issues arise with the conduct and interpretation of time-kill kinetic studies. The first is lack of standardization of the method as described in the

TFM, but efforts are underway to address this issue [35]. Once we have an acceptable method, the next issue is selection of a concentration for testing that is clinically meaningful. The concentrations selected must be extrapolatable to concentrations found on the body site after the product is used as labeled. The TFM suggests a 1:10 dilution ratio for surgical hand scrub and healthcare personnel handwash products designed for use with water. However, such products may also be used neat. In this circumstance, we would recommend testing of the product in diluted form. In addition, new ''leave-on'' products are now seeking entry into the market that contain an antiseptic ingredient and may be formulated with an alcoholic vehicle. Other products are formulated as alcoholic gels and contain no other active ingredient. Should time-kill kinetic studies be performed with undiluted product for each of these scenarios? The agency does seek to have time-kill studies performed with concentrations that are likely to reside on the treated site, but how should that concentration be determined in each case? Should it be based on the alcohol and thus tested neat? Or, for combination products, should the test be performed with the second antiseptic ingredient found in the formulation? How should the concentration to be tested be selected? Must that concentration be based on the whole hand (surgical hand scrub, healthcare personnel handwash) or per centimeter square of skin surface area (preoperative skin prep)? Careful consideration must be given to the intended uses of the antiseptic, concentrations expected at the use site, and the anticipated duration of exposure.

The FDA uses the in vitro preclinical time-kill data to make several assessments of product characteristics. The first is to evaluate the relationship between the MIC of an organism and the rate of kill when assessed at a specific concentration tested. For example, the agency reviewed in vitro MIC data for an antiseptic, and the data suggest that species evaluated were susceptible to the action of the antimicrobial at concentrations substantially lower than the antiseptic product concentration. However, when a select subgroup of isolates with known MICs were tested, the time-kill kinetic studies revealed that the rate of kill, conducted with much higher concentration of drug than the MIC of the organism, was less than expected. The antiseptic was bacteriostatic when measured at the time interval normally used for a healthcare personnel handwash. Thus, the time-kill kinetic study is capable of discerning a characteristic of the antiseptic not discernible by the spectrum-of-activity study.

We then attempt to extrapolate the preclinical results to the clinical simulation study. We assess the rate of kill as measured by the time-kill kinetic method and that measured through the clinical simulation model. This extrapolation is reasonable since the clinical simulation study is a time-kill kinetic study performed on the skin.

The FDA must rely on the in vitro spectrum of activity studies and time-kill kinetic studies to assess the potential utility of the antiseptic versus the genera and species encountered in the hospital setting. The limitations of the clinical simulation studies described next justify the use of these techniques to help characterize the antiseptic. In addition, the agency encountered instances where the in vitro studies are the only source of information that demonstrates the contribution of individual active ingredients found in a formulation. There are instances in the assessment of products with two antiseptics where the sponsors cannot formulate separate test products, each containing the active ingredient alone for use as independent arms of clinical simulation studies. The only methods available to assess the independent contribution of both active ingredients to the product are the in vitro preclinical studies. In these situations, we must rely on the preclinical studies to assess the contribution of each component to the overall efficacy of the product.

B. Clinical Simulation Studies

The clinical simulation protocols recommended in the TFM are designed as standardized methods that mimic clinical and product use conditions. It is critical that studies be performed exactly as described in the TFM protocols. Modification of any of the test parameters influences the quantitative results and subsequently the efficacy outcomes of the study. It is for this reason that the FDA scrutinizes timing of critical events when pivotal studies are under review. For example, it is critical that a timed glove juice sample taken during a surgical hand scrub or a healthcare personnel handwash be placed immediately into a dilution blank containing effective neutralizers. Depending on the antiseptic class under evaluation and the concentration of the finished dosage form, failure to do so results in false interpretation of efficacy because the antiseptic may continue to kill and reduce the bacterial populations during this test interval. In addition, our scientific investigation team visits the study site to ensure the use of good clinical practices (GLPs) and to validate the conduct of the study.

The efficacy requirements established in the TFM also require comment. These protocols are designed to monitor bacterial populations, and if a desired reduction is achieved post-product use, we conclude the product is effective as an antiseptic. However, the relationship of these outcomes and a corresponding reduction in the incidence of nosocomial infections in the healthcare setting where the products are used remains ill defined. The predictive values of these surrogate test methods have not been validated. Healthcare professionals agree that reduction of the resident flora and elimination of the colonizing or transient flora is the desired endpoint. However, determination of that relevant endpoint remains controversial. Yet despite these limitations, professional organizations continue to develop test methods without considering their clinically predictive value.

It is essential to the well-being of patients and the healthcare community that products purporting to reduce the nosocomial infection rates do so. It is just as important that the scientific community help develop test methods to assess antiseptic with justifiable endpoints. Should we expect the same efficacy outcomes for products used in the home setting and those used in the clinical setting? Should efficacy requirements monitor organisms remaining on the hands as opposed to reductions from established baselines? These are questions that must be considered in the establishment of efficacy outcomes.

We must also realize that the clinical simulation models also test efficacy against resident bacterial populations that are not representative of the species encountered in a clinical setting. The subjects used in these studies are not healthcare personnel, and their microbial flora may not represent the types or numbers of organisms found on the hands of healthcare personnel. In addition, the hands of healthcare personnel are more likely to encounter antibiotic-resistant organisms that become transient or colonizers on their hands. Thus, the predictive value of these test methods, given this difference, must also be considered in the evaluation of test products.

The antiseptic drug–manufacturing industry is innovative and designs product formulations to enhance ease of application and efficacy. Some of these products form physical barriers impregnated with active ingredients as they dry. Unfortunately, products of this type were not available when the TFM guidelines were proposed, and they do pose special problems during the clinical simulation trials. For example, film-forming products tested for preoperative skin prepping require the development of diluent capable of penetrating though the film so that quantitation of the microbial flora can be performed. This issue exemplifies the need for flexibility in trial design and must allow for the introduction of new and novel products. The TFM should provide language to address these concerns.

C. Neutralization

Validation of the neutralization system is one of the most critical, if not the most critical, aspects of the preclinical and clinical studies. Recall that assessment of the efficacy of a product in the time-kill kinetic study and the clinical simulation studies is based on the measurement of a bacterial population from an established baseline. Basically, these test methods are designed to capture a snap shot of the time-kill profile so that when a sample is taken, immediate neutralization must take place. Theoretically, this provides an accurate enumeration and subsequent calculation of reduction from a predetermined baseline.

Validation of neutralization must be performed for each of the test protocols under evaluation. One neutralization study may not allow proper characterization of the neutralizer potential unless evidence is provided that shows that the neutralizer concentration is capable of working under all tests conditions used in the evaluation of the antiseptic. Validation of the neutralization system used for the in vitro time-kill study is the easiest because the time-kill study is performed with a known concentration of antiseptic per unit of test material. When a sample is taken and enumerated, the concentration of antiseptic carried over per unit of volume is known. Thus, validation of the neutralization system is performed with the concentration of material found in the enumeration volume used in the time-kill study.

Validation of the neutralization system used in the clinical simulation studies is problematic because the concentration of antiseptic selected in the evaluation must represent the worst-case scenario for each study. How does one approximate the concentration encountered in the sampling solution of healthcare or surgical hand scrub sample? Some sponsors perform clinical simulation pilot studies and use these samples to validate the neutralizer. This approach has merit but assumes that the sample taken from an individual is representative of the worst-case scenario. That is, this approach may underestimate subjects with greater carry-over of antiseptic in a sample and result in assumptions of greater efficacy of the product. We must also consider that each indication may use a different volume of product and duration of exposure. Further, the concentration of antiseptic for leave-on products can be substantially greater than for products traditionally used with water. Thus, the concentration used to validate the neutralizer requires thought. A possible approach is to define, using qualitative and quantitative chemical testing methods, the concentrations that are actually encountered in these preclinical and clinical study samples and then to use these concentrations in the neutralization validation process.

The organism used as a marker in the validation process also requires comment. For the healthcare personnel handwash protocol, the marker organism, *Serratia marcescens* (ATCC #14756), would seem the logical choice. However, this organism is resistant to antiseptic action and may exceed the neutralization potential by increased survival during the study. What organism(s) must we use for the surgical hand scrub or preoperative skin prepping? Does validation of the neutralization system with *S. marcescens* justify use of this neutralization system in the surgical hand scrub or preoperative skin prepping study? The answer depends on numerous factors such as the duration of exposure to the product, the concentration used per exposure, and the concentration we expect to recover per unit of skin or volume of sampling fluid. We must also select an organism that is not overly sensitive to the antiseptic under investigation nor overly resistant to the antiseptic, since both scenarios create difficulties in the interpretation of data.

The temporal sequence of steps described by the protocol used to perform the neutralization validation requires comment. The neutralization validation protocol and data submitted to the agency demonstrate that some studies are performed by the addition of the active ingredient into the neutralizer and then, after a specified time period, the marker organism is added. This procedure may introduce bias in favor of the product under evaluation if the time span between addition of the antiseptic and the marker organism is delayed. The effect is dependent on the antiseptic used, the concentration carried over in the enumeration sample, and the indicator organism used. When clinical simulation studies are performed as recommended in the TFM, the sample containing both the antiseptic and the bacterial population under investigation is obtained simultaneously and added to the neutralizer. The assumption is that quenching of the antiseptic is immediate and allows survival of bacteria carried over in the sample to survive and be enumerated. Thus, the agency requires the addition of the marker organism into the neutralizer before addition of the antiseptic to assure that delays do not unduly influence the results obtained from the neutralization validation experiment. This sequence of events eliminates any bias associated with the introduction of the antiseptic into the neutralizer first.

D. Resistance to Antiseptics

Exhaustive reviews on the genetic and biochemical basis of resistance to antiseptics suggest that they may be conveniently divided into intrinsic and acquired mechanisms of resistance [37–39]. Intrinsic resistance to antiseptics is an innate characteristic of the bacterial genome and is exemplified by permeability barriers such as the cell walls of gram-negative bacteria [31]. Acquired resistance to antiseptics occurs by mutation of target sites or acquisition of genetic material by conjugation, transformation, and transduction that increase the nonsusceptibility of the microorganism to the action of the antiseptic [31].

Recent studies suggest that some antiseptics have defined target sites [40,41] and, by mutation of these target sites, nonsusceptible microorganisms to an antiseptic can be isolated [42,43]. These data suggest that if antibiotics also have a mode of action similar to that of the antiseptic, an organism with reduced susceptibility to the antiseptic may also confer resistance to an antibiotic. A recently characterized mode of action of triclosan in *Escherichia coli* shows that it binds to enoyl-acyl protein reductase, an enzyme involved in fatty acid synthesis [40]. A similar mode of action is described for *Mycobacterium smegmatis*, where strains that have missense mutations in enoyl reductase also resulted in decreased susceptibility to triclosan and resistance to the antituberculosis drug isoniazid [44]. Conversely, a resistant strain originally selected on isoniazid was found to be nonsusceptible to triclosan. These studies demonstrate that both the antiseptic and the antibiotic act on the same target site and that the emergence of resistance to one compound may confer cross-resistance to the other.

Other studies reveal a mechanism of resistance to antiseptics mediated by a multidrug efflux pump [45–47]. This mechanism mediates nonsusceptibility to the antiseptic and resistance to multiple antibiotics by mutation of the repressor/operator region controlling efflux pump gene expression (marA) or by mutation of the efflux pump structural gene resulting in reduced affinity to the antiseptic [37,38].

Thus, laboratory studies have shown that the potential for cross-resistance between antiseptics and some antibiotics exists, prompting professional organizations to question the lack of proven benefit in infection control by antimicrobial-impregnated household products and the potential for the emergence of antiseptic-mediated resistance to useful antibiotics [48,49].

Clearly two distinct issues arise from these observations. The first is whether the development of nonsusceptibility to antiseptics by nosocomial pathogens, skin flora, and other microorganisms results in decreased clinical efficacy of the topical antiseptics used in healthcare settings. The second issue is whether the emergence of resistance to the antiseptics will result in cross-resistance to clinically useful antibiotics. The concern in the latter case is that the wide availability and use of antiseptics and the selection pressure they provide will counterselect for antibiotic resistant pathogens.

The FDA addressed these issues in January 1997 when it convened a panel of experts representing the Division of Anti-Infective Drug Products Advisory Committee and the Division of Over-the-Counter Drug Products Nonprescription Drugs Advisory Committee [50]. The joint panel was charged with assessing the impact of the use of topical antimicrobial wash products on the occurrence of antibiotic and antiseptic resistance with respect to a new proposed classification system for topical antimicrobial wash products [51]. The proposal, the Healthcare Continuum Model, is a classification system that incorporates the three previously described healthcare professional use product categories published in the 1994 TFM. The model also proposes for inclusion the categories of foodhandler handwash, antimicrobial handwash, and antimicrobial bodywash. The concern of the FDA with this proposal is that the three new categories (foodhandler handwash, antimicrobial handwash, and antimicrobial bodywash) will increase the amount of antiseptics in the environment. This increased exposure may result in the selection of microorganisms resistant to the antiseptic and subsequently cross-resistant to antibiotics. Since the Healthcare Continuum Model does not address the issue of resistance, the joint committee meeting was convened to address this issue.

The joint committees concluded that "at this time" the evidence suggests that use of antiseptics as proposed will not have an adverse outcome on the continued efficacy of the antiseptic and that antiseptics have not currently contributed to antibiotic resistance. The joint committee did recommend that the FDA consider the feasibility of targeted surveillance studies to monitor change in the susceptibility of microorganisms from the proposed use settings. This proposal enforces the need for in vitro susceptibility testing to monitor the baseline susceptibility of nosocomial pathogens to the action of antiseptics and to use this information to monitor change in nonsusceptibility of the pathogens.

IV. CONCLUSIONS

The ideal setting for assessment of the efficacy of antiseptics is the clinical settings where these products are used. The desired outcome would be a reduction of the nosocomial infection rates below that historically seen within the study setting. However, many scientists suggest that these studies are not feasible, and we are left with surrogate methods that are assumed to be predictive of an acceptable outcome.

Surrogates are useful in the assessment of antimicrobial products, including antiseptics, provided they are standardized, reproducible, and have known predictive value. The methods proposed in the TFM, although standardized and reproducible, have not been validated to assess their clinical predictive value. In fact, none of the standardized surrogate methods used in antiseptic drug product development have been validated (TFM and ASTM reference). Thus, we are left to extrapolate from the premise that reduction of the microbial flora is a desired outcome, and if it is achieved, the product will be successful in the clinical setting.

I describe many of the limitations of this extrapolation of thought. Considerable debate continues to exist in the scientific community as to the validity of this extrapolation. In addition, the panelists, used as surrogates in the clinical simulation studies, are not representative, microbiologically, of the individuals in clinical settings further compounding the significance and interpretation of these studies.

Therefore, we are left with assumptions that are based on deductive reasoning, and we logically conclude that these protocols allow characterizations of desired product characteristics. Since the scientific community and the FDA generally agree on the characteristics for antiseptic healthcare products, protocols have been developed that we believe are capable of measuring the desired characteristics effectively. What is not agreed upon is the interpretation of the data derived from these studies. The problem is fundamental and is based on the lack of published information that allows correlation of the outcomes produced by the test product in clinical settings and outcomes produced by these products in surrogate models.

Existing data and experiences of experts must be considered when establishing the efficacy outcomes so that data derived from these methods provide insight into the potential utility of these products in clinical settings. We must continue to pursue this characterization of efficacy "value" so that we may have assurances that products evaluated and approved through this regulatory process actually impact on the nosocomial rates of infection in clinical settings or in other settings where these products are likely to be used.

REFERENCES

1. Food and Drug Administration. OTC topical antimicrobial products and drug cosmetics: establishment of monograph and use of certain halogenated salicylanilides as active ingredients. Fed Reg 39(179):33103–33141, 1974.
2. Food and Drug Administration. Tentative final monograph for healthcare antiseptic drug products: proposed rule. Fed Reg 59(116):31402–31452, 1994.
3. IP Semmelweis. The etiology, concept, prevention of childbed fever. Am J Obstet Gynecol 172(1 pt 1):236–237, 1995.
4. OW Holmes. The contagiousness of puerperal fever. N Engl Quart J Med Surg 1503–1530, 1843.
5. GFH Kuchenmeister. On disinfecting action in general, carbolic acid and its therapeutic application in particular. Dtsch Klin 12:123, 1860.
6. J Lister. On the topical antimicrobial drug product principle in the practice of surgery. Lancet 2:353–356, 1867.
7. B Kroing, TL Paul. The chemical foundation of the study of disinfection and the action of poisons. Z Hyg Infekt 25:1–112, 1897.
8. S Rideal, JTA Walker. The standardization of disinfectants. JR Sanit Inst 24:424–441, 1903.
9. H Chick, CJ Martin. The principles involved in the standardization of disinfectants and the influences of organic matter upon germicidal value. J Hyg 8:654–697, 1908.
10. Annual Book of Standards, Section II, Vol. 11.05, E35.15 ASTM, West Conshohocken, PA, 1998.
11. Federal Food, Drug and Cosmetic Act, as amended, 1993.
12. Food and Drug Administration. OTC drugs: proposal establishing rule-making procedures for classification. Fed Reg Notice 37:85–89, 1972.
13. Food and Drug Administration. Request for data and information for OTC antibacterial ingredients. Fed Reg Notice, 37:235, 1972; 37:6775–6776, 1972.
14. 21 Code of Federal Regulations. Part 333, Topical Antimicrobial Drug Products for Over the Counter Human Use, subpart E-Healthcare Topical Antimicrobial Drug Products.
15. DM Updegraff. A cultural method of qualitatively studying the microorganisms in the skin. J Investigation, 43:129–137, 1964.
16. P Williamson. Quantitative estimation of cutaneous bacteria. In: HL Maibach, G Hildick-Smith, eds. Skin Bacteria and Their Role in Infection. New York: McGraw-Hill, 1965.
17. D Taplin, N Zaias Grebell. Environmental influences of the microbiology of the skin. Arch Environ Health 11:546–550, 1965.
18. RA Amonetta, and WE Rosenberg. Infections of the toe webs by gram-negative bacteria. Arch Dermatol 107:71–73, 1973.
19. PB Price. The bacteriology of normal skin flora: a new quantitative test applied to

a study of the bacterial flora and the disinfection action of mechanical cleansing. J Infect Dis 63:301–318, 1938.
20. AR Cade. A method for testing the degerming efficacy of hexachlorophene soaps. J Soc Cosmetic Chem 2:281–290, 1951.
21. CW Walter, R Knudsen. The bacteriologic study of surgical gloves from 250 operations. Surg Gynecol Obstet 129:949–952, 1969.
22. RN Michaud, MB McGrath, WA Gross. Improved experimental model for measuring skin degerming activity on the human hand. Antimicrob Ag Chemother 21:8–15, 1972.
23. M Bruch. Methods for testing antiseptics: antimicrobials used topically in humans and procedures for hand scrubs. In: SS Block, ed. Disinfection, Sterilization, and Preservation. Malvern, PA: Lea & Febiger, 1991, pp. 1028–1046.
24. H Quinn, JG Voss, HS Whitehouse. A method for the in vivo evaluation of skin sanitizing soaps. Appl Microbiol 2:202–204, 1954.
25. P Williamson, AM Kligman. A new method for the quantitative investigation of the cutaneous bacteria. J Invest Dermatol 45:498–503, 1965.
26. OTC topical antimicrobial products: over-the-counter drugs generally recognized as safe, effective and not misbranded. Federal Register Notice 43(4):1210–1249, 1978.
27. Tentative final monograph for healthcare antiseptics drug products. Federal Register Notice Proposed Rule 59(116):31402–31452, 1994.
28. Topical antimicrobial drug products for over-the-counter human use; tentative final monograph for first aid antiseptic drug products. Federal Register Notice 56(140): 33644–33680.
29. National Committee for Clinical Laboratory Standards. Methods for dilution susceptibility tests for bacteria that grow aerobically. Approved standard, fifth edition. NCCLS Document M7-A2 2000.
30. National Nosocomial Infections Surveillance (NNIS) System Report, Data Summary from January 1990–May 1999, Issued June 1999. Am J Infect Control 27:520–532, 1999.
31. EL Larson. American Practitioners of Infection Control guidelines for infection control practices: APIC guideline for handwashing and hand antiseptics in healthcare settings. Am J Infect Control 23:251–269, 1995.
32. JH Platt, PA Bucknall. MIC tests are not suitable for assessing antiseptic handwashes. J Hosp Infect 11:396–397, 1988.
33. F Baquero, C Patron, R Canton, MM Ferrer. Laboratory and in-vivo testing of skin antiseptics: a prediction for in-vivo efficacy? J Hosp Infect 18 (suppl B):5–11, 1991.
34. FD Schoenknecht, LD Sabath, C Thornsberry. Susceptibility tests: special tests. In: EH Lennette, ed. Manual of Clinical Microbiology. Washington, DC: American Society for Microbiology, 1992, pp. 1000–1008.
35. Annual Book of Standards. Section II, Vol. 11.05 Time-kill method proposed standard ASTM West Conshohocken, PA.
36. National Committee for Clinical Laboratory Standards. Methods for determining bactericidal activity of antimicrobial agents. Approved Guideline. NCCLS Document M26-A, 1999.
37. AD Russell. Mechanisms of bacterial resistance to antibiotics and biocides. Prog Med Chem 35:134–197, 1998.

38. AD Russell, MJ Day. Antibiotic and biocide resistance in bacteria. Microbios 85: 45–65, 1996.
39. AD Russell, I Chopra. Understanding Antibacterial Action and Resistance. 2nd ed. Chichester: Ellis Horwood, 1996.
40. LM McMurry, M Oethinger, SB Levy. Triclosan targets lipid synthesis. Nature 394: 531–532, 1998.
41. RJ Heath, YT You, MA Shapiro, E Olson, CO Rock. Broad spectrum antimicrobial biocides target the FabI component of fatty acid synthesis. J Biol Chem 273:30316–30320, 1998.
42. M Satstzu, K Shimizu, N Kono. Triclosan-resistant *Staphylococcus aureus* (letter). Lancet 341:20, 1993.
43. U Tattawasart, JY Maillard, JR Furr, AD Russell. Development of chlorhexidine and antibiotic resistance in *Pseudomonas stutzeri*. Proceedings of the 98th General Meeting of the American Society for Microbiology. 1999, p. 533.
44. LM McMurry, PF McDermott, SB Levy. Genetic evidence that InhA of *Mycobacterium smegmatis* is a target for triclosan. Antimicrob Ag Chemother 43:711–713, 1999.
45. MC Moken, LM McMurry, SB Levy. Selection of multiple-antibiotic resistant (Mar) mutants of *Escherichia coli* by using the disinfectant pine oil: roles of the mar and acrAB locus. Antimicrob Ag Chemother 41:2270–2272, 1997.
46. LM McMurry, M Oethinger, SB Levy. Overexpression on marA, soxS, or acrAB produces resistance to triclosan. FEMS Micro Lett 166:305–309, 1998.
47. R Chuanchuen, H Schweizer. Multidrug efflux pumps and triclosan resistance in *Pseudomonas aeruginosa*. Proceedings of the 100th General Meeting of the American Society for Microbiology, 2000, p. A-31.
48. The Use of Antimicrobial Household Products: APIC 1997. Guidelines Committee Position Statement. APIC News (16) 6:13, 1997.
49. Use of Antimicrobials in Consumer Products (Resolution 506, A-99). Report of the Council on Scientific Affairs. American Medical Association House of Delegates.
50. Food and Drug Administration. Joint meeting of the OTC and DAIDP Advisory Committees. Fed Reg, 62(3):754, 1997.
51. Soap and Detergent Association, Cosmetics, Toiletry, and Fragrance Association. Letter to W. Gilbertson and attachments. FDA Public Docket, 75N-183H, 1995.

3

Healthcare Continuum: A Model for the Classification and Regulation of Topical Antimicrobial Products

George E. Fischler
The Dial Corporation, Scottsdale, Arizona

Marc Shaffer
Clinical Research Laboratories, Inc., Piscataway, New Jersey

Rhonda D. Jones
Scientific & Regulatory Consultants, Inc., Columbia, Indiana

I. INTRODUCTION

It is commonly recognized that bacteria are ubiquitous in our environment. These bacteria are associated with a variety of disease conditions; therefore, it is important to hold the numbers of bacteria in check. As early as 1860, Semmelweis [1] demonstrated that, by reducing the level of bacteria on the skin, through handwashing, a corresponding reduction in the incidence of infection could be observed. Semmelweis recognized the advantage of using an antimicrobial product over water or plain soap and water. Through the years, the use of antimicrobial products in consumer and healthcare settings has grown. Antimicrobial wash products play an important role in improving personal hygiene and can help in the prevention of disease.

In the last 30 years the types and forms of antimicrobial products have evolved and now include bar soaps, liquid soaps, lotions, hand dips, hand sanitizers, foams, and rub products. In the United States the importance and need for antimicrobial products was recognized by an expert panel convened by the Food and Drug Administration (FDA) in 1972, and in two Tentative Final Monographs

(TFM) for Healthcare Antiseptic Drug products subsequently issued in 1978 and 1994 by the FDA [2]. In an attempt to provide a tool to bring clarity to the area of antimicrobial wash products, a coalition of industry companies proposed the Healthcare Continuum Model (HCCM).

The Healthcare Continuum Model provides a flexible framework for characterizing use patterns and situational risks in order to establish efficacy requirements and appropriate labeling. The HCCM offers an initial six categories based on professional healthcare, foodworker, and consumer use patterns and risks. The six categories are preoperative skin preparation, surgical scrub; healthcare personnel handwash, foodhandler (worker) handwash, antimicrobial handwash, and antimicrobial bodywash.

The HCCM suggests an underlying philosophy of defining a product category or indication based on a thorough understanding of use patterns and attendant risks. Health hazards and characteristics of exposure are parameters of the framework. The result in the marketplace would be products that have been formulated for specific uses and thereby assure appropriate levels and types of active ingredients for each scenario. All antimicrobial products are not used for the same purpose, nor should they be. One of the results of implementation of the HCCM would be to avoid the inappropriate use of products, which could result in potentially negative consequences. Professional products, which are not typically formulated for general use, may irritate, stain, or contain complex use directions or warnings. This could result in a decrease in handwashing compliance. As a result of the potential for adverse effects, the process of establishing a category includes balancing the potential for disease acquisition or transmission with the product effectiveness and type of active ingredient.

Characteristics for each identified use pattern should be consistent with the specific risks associated with the use setting. These characteristics may include antimicrobial effectiveness versus transient or resident bacteria, persistent effect, speed of action, and spectrum of activity.

II. BACKGROUND

In a Centers for Disease Control and Prevention (CDC) study spanning the years 1980–1990, 62% of all nosocomial infections were attributed to bacterial pathogens [3]. An undefined number are due to bacteria transferred by and from the hands and skin. Products such as preoperative skin preparations and surgical scrubs have been used in hospital settings for the reduction of nosocomial infections. Products designed for these uses rapidly and dramatically reduce the levels of resident bacteria on the hands or skin immediately prior to invasive surgical procedures. By definition, they should exhibit a persistent effect. Persistence is prolonged or extended antimicrobial activity which acts to prevent or inhibit the growth or regrowth of organisms that remain on the skin after product use and/or the establishment of transient organisms that may contact the skin after product use. Healthcare personnel handwash products have been used by professional healthcare providers to reduce the incidence of nosocomial infections as well as to protect themselves or other patients from the transmission of microorganisms and consequently disease from infected patients. These products are designed to be used by the healthcare professional as many as 50–100 times per day. They should be fast-acting, effective against transient bacteria, and, if possible, mild to the skin.

Healthcare professionals and their patients represent a small but significant segment of the population that require specific antimicrobial wash products. Bacteria can cause infections outside of the clinical setting. An increasing segment of the population are seeking and using antimicrobial products designed as foodhandler handwash products. A study spanning the years 1983–1987 reports that poor personal hygiene by foodhandlers or food service workers is the second leading cause of foodborne illness [4]. Research supports the findings that the U.S. population lacks basic food safety information/skills and engages in food-handling and food-preparation practices that studies have linked to a significant number of foodborne illness outbreaks [5]. Foodhandlers require specific antimicrobial wash products not only to address their own personal hygiene needs, but also to protect those people who consume the food that they have prepared. Products designed for use by foodhandlers are used in restaurants, cafeterias, hotels, schools, hospitals, federally inspected meat and poultry plants, prisons, and airline food-preparation facilities. Today, more people in our society are assuming responsibility for food handling and preparation in the home and elsewhere [5]. Vulnerable sectors of the population more severely affected by foodborne illness are also increasing in size, including immunocompromised persons (e.g., persons with diabetes, cancer, chronic intestinal disease, organ transplants, or human immunodeficiency virus); and persons 65 years and older, a growing proportion of the population who are at increased risk due to normal decline in immune response. Foodhandler handwash preparations need to be fast-acting and effective against transient microorganisms, including foodborne organisms, and, if possible, exhibit persistence.

Every day, consumers are changing diapers, caring for sick, elderly, or invalid family members, preparing family meals, having contact with pets, gardening and performing yard work, having contact with other people, healthy or otherwise, attending daycare, attending school or work, and traveling and enjoying recreational activities. Topical antimicrobial products are used by consumers to provide a variety of end-user benefits. Antimicrobial handwash products help control the transient bacteria consumers acquire from contact with the environment. Consumers are constantly exposed to a variety of bacteria that have the potential to cause infection [6–12]. The transfer of transient bacteria via hands is recognized as a major factor in the spread of disease. Antimicrobial handwash products should have a broad spectrum of activity to reduce the number of these bacteria on the hands, thus reducing the potential for transmission of disease causing organisms. Additionally, antimicrobial activity should exhibit persistence, if possible.

Antimicrobial bodywash products are used for whole body cleansing, to reduce odor, and to help control bacteria, which may help prevent minor skin infections. Products for bodywashing are used for self-health and act as an aid in controlling the risk of pyogenic infection. Antimicrobial bodywash products can control the number of resident and transient microorganisms on the skin [13,14]. Antimicrobial bodywashes need to have an activity spectrum that will target the gram-positive resident flora. They may, in some instances, also target the gram-negative transient flora. A key characteristic of products in this category is persistence to reduce and maintain the microbial flora of the skin below baseline levels between washings.

III. THE HEALTHCARE CONTINUUM MODEL

Underlying the HCCM is a proposed framework for a system of broadly classifying topical antimicrobial use patterns, establishing transmission risks and etiology agents, and determining the pertinent characteristics of a topical formulation dictated by the situational risks. The following section describes the six initial use patterns proposed in the HCCM. The use pattern and attendant risks typical of microbial flora are described. Based upon the HCCM philosophy, formulary attributes and test methodology were proposed. Although the initial categories proposed within the HCCM are well established use patterns, development of new use patterns and new technology will drive the need to apply the flexible, underlying principles of the model to establish new categories in the future.

IV. GENERAL POPULATION PRODUCTS: ANTIMICROBIAL HANDWASH AND ANTIMICROBIAL BODYWASH

In everyday life, consumers encounter situations in which they are exposed to a variety of bacteria that have the potential to cause infection. It is well recognized that good personal hygiene can reduce the risk of infection. Antimicrobial washes can play an important role in improving personal hygiene. Antimicrobial washes are used for whole body cleansing, to reduce odor-causing bacteria, and to help control bacteria that can cause skin infections.

The routine use of personal antimicrobial wash products is beneficial to all. The potential benefits to consumers of antimicrobial washes, in addition to cleaning, are (1) to help reduce the incidence of pyogenic infections [15,16], (2) to help remove transient organisms which are potentially pathogenic [17], and (3) to reduce odor-causing bacteria [15]. Washing with antimicrobial washes or with nonmedicated washes will remove some bacteria from the skin due to the surfactancy of the base and the mechanical action of the washing procedure. However, antimicrobial washes deposit an active ingredient on the skin that can help control the number of organisms that survive and help prevent the colonization of potential pathogens, such as *Staphylococcus aureus*. Washing with nonmedicated products does not provide this persistent antimicrobial activity.

Antimicrobial washes are available in many forms (bars, liquids, gels, etc.) and usually contain a single antimicrobial ingredient. Products applied to the hands during handwashing reduce transient organisms on the hands and can reduce the possibility of disease transmission. Ehrenkranz and Alfonso point out that "a pervasive misconception in infection control circles is that simple handwash reliably prevents hand transmission of transiently acquired bacteria" [14]. Antimicrobial bodywashes can play a role in the prevention of pyogenic infections and help control odor-causing bacteria. Whole body use of antimicrobial products can control the number of organisms on the skin and has been demonstrated, in laboratory studies, to reduce the number of potential pathogens on the skin [18].

A. Antimicrobial Bodywash

Antimicrobial bodywashes are used to control the numbers of bacteria, where appropriate [15]. Because the organisms of potential concern are primarily Gram-positive, the activity of antimicrobial agents used in antimicrobial bodywashes should be suitable for that use and so may be of limited spectrum. Most importantly, this will enable a proportion of the normal flora to remain.

B. Public Health Importance

Skin infections due to gram-positive organisms are recognized as a common and significant public health problem [19,20]. These skin diseases are most commonly caused by staphylococci and streptococci. They include pustules, folliculitis, impetigo, furuncles, and infection of cuts and scrapes. In addition to the staphylococci and streptococci, other gram-positive bacteria such as *Corynebacterium minutissimum* can cause contagious skin infections like erythrasma.

According to data obtained from the National Disease and Therapeutic Index (NDTI), on average from 1992 to 1994 there were approximately 2 million diagnosis visits per year to dermatologists, pediatricians, and general or family practitioners, and others for impetigo, pyoderma, and carbuncles/furuncles. It has been estimated that up to 8% of visits to dermatology clinics are a result of some form of pyoderma [21,22]. Skin-related problems constitute about 6.8% of the visits to general pediatric clinics [23]. Of these, about half are cutaneous infections, some of which can lead to frequent recurrence and intrafamilial spread. A recent survey estimated that 5.5 million office visits per year are due to skin infections. Children overwhelmingly constitute the population at greatest risk for bacterial skin infection, with those under 9 years of age having the greatest incidence [24]. The number of children worldwide with skin infections is estimated to be in the millions and constitutes a significant load on medical services [24].

Atopic dermatitis affects approximately 10–20% of the population. Clinical studies have shown that the skin flora of atopics is quantitatively and qualitatively different from the skin of the normal population [25]. It has been reported that these patients have increased numbers of skin flora and a higher frequency of colonization with *S. aureus*, not only on their skin, but also in their nares [26]. Due to the chronic presence of this organism, it is not surprising that patients with atopic dermatitis experience increased numbers of skin infections. In experimental staphylococcal infections, there is a direct relationship between the numbers of staphylococci applied to normal skin and the likelihood that infection will occur [17]. Many normal daily activities that result in minor cuts and scrapes have the potential to become contaminated with these transferred *S. aureus* and other organisms from environmental sources.

A survey of skin problems in the elderly identified skin infections in noninstitutionalized patients ages 68 and older [27]. It was strongly suggested that skin problems, including infections, were common. Also, bathing and shampooing were often substantially limited, adversely effecting personal hygiene.

Zimakoff and colleagues pointed out the importance of reinfection on the transmission of *Staphylococcus* within families and concluded that a residual effect is desirable to help prevent reacquisition of staphylococci that are shed into the household environment on sheets, towels, etc. [28]. Recently in the United States, special occurrences affecting many households were shown to increase the incidence of skin infections. Quinn et al. reported on medical care of families affected by Hurricane Andrew in 1992 [29]. For 2 weeks following the storm there were noted increases in pediatric dermatological infections, including impetigo, wound infections, and cellulitis.

The breach or destruction of the skin defenses, as a result of abrasions, cuts, burns, or the action of toxic chemicals, inevitably leads to increased colonization of the area by microorganisms. Thus, the likelihood of infection from the reduced defense capacity is increased. However, less severe and often barely detectable changes in the balance of the surface inhabitants of the skin can also increase the susceptibility to infection. These changes can be induced by alterations in host metabolism and other factors that modify the surface environment, even though they do not sever the epidermis [30].

Skin infection with streptococci covers a range from simple colonization to primary and secondary infections. The skin provides an important portal of entry for systemic infection by these organisms. Minor trauma to the skin not prompting specific first aid attention, such as trivial cuts, abrasions, and scratches, may allow streptococci to initiate infection [17]. Staphylococci and streptococci are often found in mixed cultures from pyoderma lesions. There is still controversy over which is the more important pathogen. Most streptococcal skin infections occur among children and their contacts under conditions of overcrowding and poor hygiene, particularly in warm and humid parts of the world. Even in developed parts of the world, streptococci are commonly isolated from many clinical specimens. This is likely due to many factors, including the common asymptomatic carriage of organisms, the minor nature of most infections that are left untreated, and the opportunities for bacterial transmission that exist as people work, live, and play together.

Antimicrobial wash products are used for whole body cleansing, to reduce odor, and to help prevent minor skin infections. The efficacy of the ingredients used in antimicrobial washes against the types of organisms outlined has been well documented [31]. The regular use of these products results in the deposition of a bacteriostatic residue, which can significantly reduce the carriage of these organisms on the skin and play a role in the prevention of disease.

C. Antimicrobial Handwashes

There is consensus in the medical and scientific communities that transfer of transient bacteria via hands is a major factor in the spread of disease [32]. Hands can be viewed as unique in three respects. First, hands, more than any other part of the body, are in constant contact with the environment and, as such, reflect exposure to transient contaminants from many sources. Second, various parts of the hand, such as the nail folds and interdigital spaces, provide specific microenvironments, which can support organisms with varying growth requirements. Third, the flora of the skin of the hands are highly subject to modification because of the exposure to a number of varied household activities.

The importance of the role of handwashing for infection control has been thoroughly reviewed for settings outside the home [32,33]. Although there are no definitive studies in the literature within home settings involving currently marketed products, there are numerous studies suggesting a role for antimicrobial washes in personal hygiene [31]. Regarding the absence of clearly definitive trials, Larson has written [34]: ''I'm not convinced that even the definitive study for which we have been lobbying and waiting would, in fact, influence practice. What we know now from natural experiments, epidemiological studies and experimental models is that clean hands are associated with reduced risk of contact-spread infection in a variety of settings.'' [34]. It can be argued, based on the spread of disease among family members, that the added benefit of residual activity and of increased awareness of the need for handwashing reflected in the purchase of antimicrobial consumer products has a role for their use in the home.

Everyone picks up germs from contact with the environment, and antimicrobial soaps help to control these germs. Broad-spectrum activity is preferable because of the wide variety of potential sources of infection. However, rapid kill is not necessary because all soaps remove a portion of the bioburden on the hands through the detergency mechanism. Nonantimicrobial soaps do not provide the long-lasting persistent effect of antimicrobial soaps [13,35]. Typical consumer handwashing is incomplete. Rarely is it as thorough as in the hospital setting. Because of the lower level of consumer handwashing efficiency, persistence is desirable. Bartzokas et al. [36] reported that the efficacy of an antimicrobial handwash preparation ''was significantly augmented'' with repeated handwashing.

The flora on the hands and influencing factors have been studied. The transient bacteria lie free on the skin or are loosely attached with dirt [37]. The resident flora comprise a stable population in both size and composition. Washing readily removes some transients, but the resident flora are removed more slowly. Obviously, the flora that may be present on the hands as transients are greatly influenced by the activity related to a source of contamination and the environmental conditions. For the consumer, these include, among others, food preparation, contact with pets, gardening and yard work, contact with other people, changing diapers, assisting ill persons, daycare, school, work, travel, and recreation.

Surveys of the bacteria found in the home environment suggested four major sites of household contamination: dry areas (e.g., floors, linens, furniture, clothing), wet areas (e.g., baths, kitchens, sinks, toilets, drains), food, and people [6,7,38,39,40]. In many homes, animals (e.g., pets, farm animals) and outside work (gardening, yard work) should be included. Scott et al. [6] pointed out that, although it is accepted that the risk of infection in the general community is lower than that associated with hospitalized patients, increases in the number of outbreaks of household food-poisoning cases had been observed. In that survey, high bacteria counts were found mainly in wet areas associated with sinks, baths, and diaper pails [6]. High bacteria counts also were frequently recovered from washcloths, dishcloths, and cleaning towels. The survey included isolation for *Escherichia coli*, pseudomonads, *S. aureus*, and streptococci. Marples and Towers further established a model to study how contact transfer of *Staphylococcus* can occur from objects [41]. Borneff et al. examined households for organisms causing infectious enteritis and found 267 of 4683 samples contained staphylococci [42]. These studies support the need for handwashing and the desirability of antimicrobial soaps.

The shift in recent years from home-based child care to group daycare further extends the household environment because of the likelihood of transmission back to the family and subsequent intrafamilial spread [43–47]. Surveys of the daycare center environment have found contamination of surfaces, toys, food areas, diaper-changing areas, and the hands of children and adults [48–52].

A number of other activities encountered at work lead to exposures to potentially harmful microorganisms. These include handshaking, exposure to ill colleagues in meetings, contact with the public, sharing objects such as public toilets, telephones, exercise equipment, and money, as well as other obvious situations such as those encountered by animal handlers or sanitation workers [17], The Mayo Clinic points out that it is critical to wash hands after using the bathroom, handling food, handling money, coughing, sneezing, etc. [53].

Black et al. first demonstrated the effectiveness of handwashing to prevent diarrhea in daycare centers [9]. Following the initiation of a handwashing program in several daycare centers, the incidence of diarrhea among children in the study was significantly and consistently lower (approximately half) than the incidence in the two control centers over the 35-week study period.

It is recognized that it is often difficult to attribute independent specific effectiveness to an intervention-and-control program because they are inherently multifaceted. For example, Butz et al. evaluated the effectiveness of an intervention program in daycare homes that included handwashing education, the use of vinyl gloves, disposable diaper changing pads, and an alcohol-based hand rinse [54]. Symptoms of enteric illness were lower in the intervention homes, but it was not possible to separate out the effects of each component of the intervention.

It can be demonstrated that antimicrobial washes reduce the numbers of organisms on the skin to a greater extent than nonmedicated wash. In addition, model systems have demonstrated the control of potentially pathogenic organisms on the skin [55]. For handwashing to be effective, it is important that any product also be acceptable for regular and frequent use by consumers: ''As the value of frequent handwashing is well established, the choice of soap brand should be made with a view to encouraging frequent handwashing, while maintaining healthy skin'' [56]. Antimicrobial ingredients, deposited on the skin, can also be of benefit when washing is perfunctory or inadequate, leaving behind organisms that can cause infection or be transferred to other skin sites. Appropriate degerming for consumers also does not present the risk of removing resident flora to the point of creating a flora shift. Aly and Maibach showed that prolonged use of antimicrobial soap on skin did not lead to overgrowth of undesirable flora [57]. Using this information collected on the antimicrobial handwash use pattern, the model details suggested product attributes and test methodology.

The above discussion has clearly pointed out that improved hygiene can help prevent skin infections and interrupt the transmission of infectious disease as transferred by the hands. Therefore, the regular use of an antimicrobial soap for personal cleansing has a recognized role in the prevention of disease. Handwashing is repeatedly cited as the most important infection-control measure. It is no less important in the home than in these villages and institutional studies cited above.

D. Foodhandler Handwash

The importance of proper foodhandling during food preparation significantly contributes to the prevention of foodborne illness. As new and more virulent forms of bacterial pathogens appear, and as commercial preparation of food becomes more prevalent, our national strategies for prevention and control of foodborne disease will be increasingly tested.

Recent, well-publicized outbreaks of foodborne illness have reinforced the need to review foodhandling practices and how they are regulated. Two regulatory initiatives of note were the 1993 Food Code published by FDA [58] and the U.S. Department of Agriculture (USDA) regulation proposed in the Federal Register [59]. Both recommended the development of a Hazard Analysis and Critical Control Point (HACCP) system for foodhandling operations. HACCP principles call for the study of any system of food manufacture or preparation to determine the critical control points to protect food quality. HACCP has gained widespread acceptance. These evaluations of food preparation and serving practices identify personal hygiene on the part of foodhandlers as a critical point. The use of hand antimicrobials is cited as a measure available in a concerted effort to break the chain of transmission of disease.

Formerly USDA provided oversight for the use of hand cleaning and sanitizing products for meat and poultry processing plants [60]. However, products were only authorized by USDA if they were labeled for use in an inspected facility [59]. Many antimicrobial products intended for use by foodworkers in restaurants, hotels, schools, hospitals, and grocery stores were not considered for or required to have USDA authorization. As USDA discontinued the program in 1998, the HCCM offers a framework for regulation of this diverse group of product forms and use scenarios.

The cause of foodborne illness is widely recognized to be infection by pathogenic microorganisms [61]. Although preventable, no segment of the population is immune to the acute gastroenteritis caused by these pathogens. The exact symptoms of food poisoning vary but may include vomiting, diarrhea, abdominal pain, and fever and progress to the more severe blood clotting abnormalities, arthritis, kidney failure, autoimmune disorders, and death [62].

As the U.S. population ages and the proportion of immunosuppressed individuals continues to rise, the risk of foodborne illness will become an even greater threat [5]. This reinforces the need to prevent contamination of the nation's food supply. The bacterial organisms typically responsible for food poisoning are spread fecal-(food)-orally. The contamination may occur through an employee's hands transferring the contaminating organisms from raw food to noncontaminated food or prepared/cooked food, which is then consumed [63,64]. Bryan and Doyle reported findings from investigations of two poultry-processing facilities: "[s]almonellae of the same serovars that were on incoming carcasses were found on 30% of hands, 38% of rubber gloves, and 31% of wire gloves of workers" [33]. Microbial contamination may rapidly grow out of control when combined with poor refrigeration, inappropriate storage, or improperly sanitized equipment.

Bacterial food poisoning may be caused in two ways: either by the direct presence of bacteria in consumed food or by the production of a toxin that remains in the food. The organisms primarily responsible for toxins are *S. aureus*, *Bacillus cereus*, *Clostridium botulinum*, *Clostridium perfringens*, and *Vibrio cholera*. Those primarily responsible for direct infection include *Salmonella* spp., *Shigella* spp., *Campylobacter jejuni*, *E. coli*, *Listeria monocytogenes*, *Vibrio* spp., and *Yersinia enterocolitica*. These organisms may be transiently acquired from the soil, air, water, raw foods, hard surfaces, animals, or other contaminated food. They may also represent normal skin inhabitants. With the organisms that produce toxins, it is even more important to prevent their initial contact with food, since the bacteria may be killed by cooking while the toxin remains to affect the unsuspecting consumer [33,65].

The most well-known and frequently diagnosed foodborne pathogen in the United States is Salmonella. Some reports indicate that 25–64% of broiler chickens in the United States are contaminated with this organism [67]. Typically responsible for a short-lived gastroenteritis, which may be life-threatening in high-risk populations (elderly, children, immunosuppressed individuals), the infection may turn deadly if it enters the blood stream [62,68].

The CDC has released data derived from a composite review of several national surveillance systems [5]. These figures indicate that 7–33 million cases of foodborne illnesses occur each year, resulting in 7,000–10,000 deaths. The figures indicate that 17% of these deaths involve meat/poultry products contaminated by pathogenic microorganisms. These deaths are preventable with the appropriate precautions, including proper handwashing with an efficacious product. These numbers are suspected to be inaccurately low due to the voluntary nature of the programs and the innumerable cases that go undiagnosed and unreported.

Some estimates report that 70% of foodborne illnesses occur due to restaurant incidents and 20% occur due to home food preparation incidents. The CDC reports that 20–30% of food-poisoning incidents are the direct result of consumer mishandling of food [69]. With the dramatic increase in the number of restaurants and their popularity in our society, this rate is expected to continue to rise.

Our risk of exposure due to cultural and societal changes is increasing. This, coupled with the aging of the population and the increase in the percentage of immunocompromised citizens, contributes to the risk of foodborne illness. Bean et al. reviewed cases of foodborne disease outbreaks reported to the CDC [4]. These are cases from food service rather than food production. The article states: ''[f]or each year from 1983 to 1987, the most commonly reported food preparation practice was improper storage or holding temperature, followed by poor personal hygiene of the foodhandler.'' They report that, of 127 total reports of bacterial incidents in 1983, 13.4% concluded poor personal hygiene was a contributing factor. The reports also noted from 1983 to 1987 that approximately 25% of cases were in the home, 32% in restaurants, 5% in schools, and 3% in churches [4].

Restaino and Wind provide a survey addressing antimicrobial effectiveness of handwashing for food establishments [70]. They conclude that washing reduces the number of viable organisms remaining (that would be available to contaminate handled food) and that use of an E2 (USDA antimicrobial wash category) level product produces a measurably greater reduction as compared to washing with plain soap. From this understanding of the hand-to-food transmission risks, the model offers specific test methodology to elucidate the minimum performance characteristics.

E. Healthcare Personnel Handwash

Transmission is defined as the conveyance of disease from one person to another. The transfer of transient bacteria via hands is recognized as a factor in the spread of disease [71,72]. Nosocomial infections represent major sources of morbidity and mortality for hospitalized patients and constitute a serious public health problem. These hospital-acquired infections add significantly to the impact of the underlying diseases alone. The most complete study to date of the incidence of nosocomial infection rates in the United States was the Study of the Efficacy of Nosocomial Infection Control (SENIC) Project conducted by the CDC [73]. This study covered patients hospitalized during 1970 and 1975–76. SENIC Project estimates of the incidence of nosocomial infection rates at 6449 acute care hospitals in the United States for the period 1975–76 were 5.7 infections per 100 admissions, 7.18 per 1000 patient-days, and over 2 million total for a 12-month period. The same study estimated that the total rate of infection in 1983, including nursing homes, could approach 4 million cases per year and that related deaths placed nosocomial infections among the top 10 causes of death in the United States. The epidemiology of nosocomial infections has been affected by the introduction of the prospective payment system, which changed the economic basis of U.S. healthcare delivery [74]. Patients admitted to hospitals today tend to be more seriously ill or require sophisticated and, sometimes, high-risk procedures only suitable for inpatients. They are also usually discharged earlier, and care is often continued at home or in nursing facilities.

The sites of nosocomial infection are diverse, and microorganisms transmitted on the hands of healthcare workers may cause infection at these sites. The extensive contact that patients have with healthcare workers and the high concentration of organisms often present in wound drainage, catheter drainage bags, urine, and feces makes them efficient disseminators of these flora, often to the hands of the healthcare workers. Studies have shown that the major reservoir of nosocomial infection in the hospital is the infected or colonized patient, and the primary mode of transmission of organisms between patients is on the hands of medical personnel [75]. Carriage of microorganisms on the hands has been implicated in numerous nosocomial outbreaks [71]. Most transmittable infections are transmitted by the hands of healthcare workers [76].

The importance of handwashing by medical personnel in the prevention of nosocomial infections was recognized over 100 years ago [2]. The literature concerning a causal link between handwashing and infection has been extensively reviewed by Larson and coworkers [32,33,77]. Today, handwashing to remove transient organisms acquired from patients or the environment and to prevent cross-infection is generally regarded as one of the most fundamental infection-control practices. Use of antimicrobial handwash products within this context is similarly widely established. Several agencies and organizations have published guidelines and standards regarding the use of topical antimicrobial products for skin hygiene. These include the Association for Professionals in Infection Control and Epidemiology (APIC) [78], the CDC [79], and the Association of Operating Room Nurses (AORN) [80]. In general, these organizations propose a similar approach to selection and use of antiseptic handwash products based upon infection control considerations.

Antimicrobial handwash products for use in the healthcare setting have been routinely and widely available for decades. The history of these products can be followed by examination of the evolution of active ingredients, regulatory practice, and clinical standards. The modern use of antimicrobial healthcare personnel handwash products has been said to have begun with the use of hexachlorophene following World War II. The decline in the use of hexachlorophene led to the proposal and use of a number of alternate active ingredients for healthcare personnel hand antisepsis. These included *para*-chloro-*meta*-xylenol (chloroxylenol, or PCMX), triclosan, iodine and iodophors, alcohol, triclocarban, tribromsalan, cloflucarban, and others. Chlorhexidine gluconate (CHG) was widely used in Europe and Canada and was introduced in the United States in the mid-1970s.

Healthcare personnel handwashes on the market today are predominantly liquid detergent formulations with sufficient active antimicrobial ingredient levels to achieve targeted organism reductions in both in vitro and in vivo tests as specified in the 1978 TFM. The most commonly used active ingredients are PCMX (0.24–3.75%) and triclosan (up to 1%) [58]. Formulations are typically optimized systems rather than plain soap to which an antimicrobial agent has been added.

The trend in formulation of these products is toward low irritating systems using mild surfactants, emollients, and moisturizers. Contrary to popular belief, antiseptics are not necessarily more irritating to the skin than plain soaps [81,82].

The distinguishing characteristics of a healthcare personnel handwash/antiseptic focus on its intended use as a fast-acting, broad-spectrum antimicrobial antiseptic designed for rapid removal and/or kill of transient skin microorganisms encountered in a healthcare setting. These products are designed for very frequent use, up to 50–100 times per workday. A persistent antimicrobial effect is a desirable characteristic but is not necessary since these products are used frequently throughout the day.

F. Surgical Products: Preoperative Skin Preparations and Surgical Scrubs

The definitive role of antimicrobial surgical scrubs and preoperative preparations in reducing nosocomial infections has been consistently debated over the past 50 years. The nature of the relationship is a complex interaction that comprises the condition of the patient, the transient and resident flora of the patient, the operating room (OR) team, the sterility of the devices involved, and other factors [71].

The hazardous microorganisms are primarily derived from the patient's own resident flora. In addition, patients in medical environments are frequently exposed to a wider variety of pathogens and potentially more antibiotic-resistant organisms than the general population.

During invasive procedures, the primary barrier function of skin is significantly compromised. Examples of invasive procedures include surgery, catheterization, and injection. There are a variety of situational risks associated with invasive medical procedures [77]. For example, incisions into the skin during surgery and placement of IV and other catheters provide the greatest risk of infection, while a lesser risk is realized during injection. The degree of effectiveness of the products should be related to the severity of the risk of infection within the use situation.

The 1994 TFM defined a patient preoperative preparation to be "fast-acting, broad spectrum, and persistent antiseptic-containing preparation that significantly reduces the number of microorganisms on intact skin." The purpose of surgical scrubs and preoperative skin preparations are similar [85]. Lowbury emphasized that "skin disinfection" eliminates the transient microorganisms and reduces or kills the resident flora at the operative site [83]. A preoperative product remains on the skin to offer antimicrobial protection during the procedure and may be used to postoperatively cleanse the wound.

Current surgical scrub formulations typically contain the following active ingredients: chlorhexidine gluconate, chloroxylenol, hexachlorophene, triclosan, iodophors, and alcohol. Active ingredients found in preoperative preparations are commonly formulated with chlorhexidine gluconate, iodine, alcohol, and iodophors [78,84]. Although not covered under the current monograph, many new products are now available which may prove to aid in the prevention of wound infections when used in combination with the preoperative preparation.

In 1994, Larson put aside the controversy of definitively proving the infection prevention benefit of topical antimicrobial soaps and refocused attention in a practical direction. She stated, "Although a definitive, double-blind clinical trial of the effects of handwashing with an antiseptic product on nosocomial infection rates may not be feasible, it appears that, at least in certain high-risk situations, such antimicrobial products are beneficial. Two major dilemmas facing Infection Control Practitioners in healthcare settings today, however, are when to use antiseptic agents and which agents to use" [78].

The 1994 TFM defined a surgical scrub as "an antiseptic containing preparation that significantly reduces the number of microorganisms on intact skin; it is broad spectrum, fast acting, and persistent" [58]. AORN states that the purpose of the scrub is to "[r]emove debris and transient microorganisms from the nails, hands, and forearms; reduce the resident microbial count to a minimum; and inhibit rapid rebound growth of microorganisms" [85].

The nature of the use pattern of these products results in this category holding a high risk of infection for the patient; however, the category risk is of limited scope due to the small percentage of the population who require surgery. The Association of Operating Room Nurses has published practices that address when to use surgical scrubs and preoperative preparations [85].

V. EFFECTIVENESS TEST METHODS

The purpose of surrogate endpoint test methods within this public health framework is to demonstrate that a product is efficacious in reducing risks of infection or acquisition of disease within a given situation. As such, the methods must address the key performance criteria for the product under conditions that simulate use situation(s). The key performance parameter for topical antimicrobial products is effectiveness against a spectrum of bacteria representative of those encountered in the targeted situations. Depending on the situation and task, speed of action and residual activity may also be key parameters.

In an OTC monograph, the characterization of an active ingredient is carried out prior to finalization of the monograph. Minimum inhibitory concentration (MIC) testing and other methods are used to determine the spectrum of antimicrobial activity. As a consequence, testing of an active ingredient to determine its spectrum of activity in a formulation should not be necessary

Several general principles can serve as guidance for determining appropriate test methods.

Standardized, defined, and peer-reviewed test methodologies ensure reliability, reproducibility and comparability of test results.
Appropriate methods should duplicate or simulate actual use conditions, present a minimum of hazard to investigator and subject, be reasonably economical, and be flexible enough to handle a variety of product forms.
Antimicrobial test methods should also utilize a reliable supply of standardized microorganisms.
In situations where product form or ingredients interfere with a method, use of an equivalent method should be allowed provided it meets the general guidelines embodied in the original test method.

ASTM (American Society for Testing and Materials) methods are proposed for testing because they embody the above principles. Use of ASTM procedures ensures that periodic peer review of the methods will maintain their validity, currency, and reproducibility.

To be consistent with an ingredient-based monograph approach, a regimen incorporating both in vitro and in vivo tests is used to demonstrate the speed of action and spectrum of activity of a final formulation and its ability to meet the effectiveness criteria to support product claims and indications. The monograph criteria should correspond to a reduction in risk in a given situation.

An active ingredient listed in an OTC monograph as effective and safe (category I) has had the breadth of its efficacy attributes established during the formal drug review process. Therefore, when such ingredients are used in product formulations, only limited testing is needed to confirm the level of effectiveness of the ingredient at the use concentration and to assess the impacts of other ingredients in the formulation on its effectiveness. Supplemental methods may be used to demonstrate attributes to support other truthful and not misleading statements, not necessarily indications or label claims.

To demonstrate the speed of action and spectrum of activity of a final formulation, an in vitro time-kill methodology is used. This is a suspension test method that demonstrates that there is a reasonable expectation of antimicrobial activity within a time frame that is relevant to the use situation for that product.

Specific in vivo tests are performed on final formulations in order to support specific indications. The surrogate endpoint effectiveness criteria proposed for these tests should be correlated to clinical performance, where possible, or statistical risk models. The criteria should reflect the severity of the risk associated with the performance of the task.

A. Reduction of Transmission Primarily to Oneself (Bodywash)

In vivo testing to support indications for products used to reduce the incidence of minor skin infection must demonstrate either: activity against resident organisms to reduce their numbers to a specified level or the maintenance of bacterial levels remaining after washing to below initial levels. If a decrease in transient skin microorganisms from the gut that are left on the hands after washing is claimed, activity against transient organisms down to effective levels must be demonstrated. The surrogate criteria are based on data from either clinical studies, or microbial risk modeling.

In most cases, criteria for these products will overlap with the criteria for products that interrupt transmission between individuals, and the methodology and efficacy criteria should be the same. Appropriate methods include ASTM E1174 (the Healthcare Personnel Handwash test), the Cade method, and the Cup Scrub. Persistent activity could be demonstrated using the same methods.

B. Reduction of Transmission Between Individuals/ Fomites (Handwash)

In vivo testing must demonstrate that the drug product reduces the number of transient organisms. The effect should be immediate and greater than or equal to surrogate reduction criteria in order to support indications that the use of a product reduces transient organisms on the hand. Efficacy criteria should be reflective of the risks encountered in various settings. They are based on clinical or risk modeling data from studies that look at these specific scenarios. The method of primary utility is ASTM E1174, the Healthcare Personnel Handwash Method. This method allows for evaluation of various product forms and uses, including waterless products. This method is flexible enough to allow the products to be tested under use conditions specified on the label.

In cases where activity against transient and resident flora on the hands needs to be demonstrated, ASTM E1874, the Cup Scrub method, would be appropriate in addition to E1174.

C. Reduction of Transmission During Presurgical/ Surgical, Preinjection Procedures (Surgical Scrub, Preop Prep)

In vivo testing must demonstrate the immediate and significant reduction of the resident and transient flora, reflective of the highest potential for acquisition of disease, and the increased risk associated with breaching the skin barrier. In some cases residual activity over a period of hours should also be demonstrated. ASTM E1173, the Preoperative Preparation method, can be used to demonstrate the in vivo activity of products labeled for preoperative skin preparations and for injection site preparation as well. This method samples primarily resident flora using the Cup Scrub technique. It is flexible to allow evaluation of various product forms under conditions of use. Substantiation of residual activity against resident bacteria can also be demonstrated with this method. In cases where immediate and persistent activity against transient and resident flora on the hands needs to be demonstrated, ASTM E1115, the Surgical Scrub method would be appropriate.

VI. SUMMARY

The six categories proposed in the Healthcare Continuum Model illustrate the underlying principles of defining performance attributes following a thorough understanding of the use pattern, user population, and microbiology. Any model as complex and detailed as the HCCM answers many old questions while stimulating new questions: Are other use patterns in practice that should be included? Should certain categories be combined or refined? Are there additional details concerning the use patterns that need to be addressed? Should additional attributes and test methodology be considered? The regulation of topical antimicrobial products remains a work in progress. The HCCM offers a flexible system to quickly define additional categories to meet the challenges of emerging disease and technology. We have pointed out only a few of the obvious questions that remain and restated the need for a flexible, cohesive regulatory framework.

REFERENCES

1. PJ Miller. Semmelweis Infect Control 3(5):405–409, 1982.
2. Fed Reg 59(149):39888–39896, 1994.
3. WR Jarvis. Nosocomial outbreaks: The Centers for Disease Control's Hospital Infections Program experience, 1980–1990. Am J Med 91(suppl. 3B): 101–106, 1991.
4. NH Bean, PM Griffin, JS Goulding, CB Ivey. Foodborne disease outbreaks, 5-year summary 1983–1987. J. Food Prot 53:711–728, 1990.
5. Fed Reg 60(23):6780–6783, 1995.
6. E Scott, SF Bloomfield, CG Barlow. An investigation of microbial contamination in the home. J Hyg (Cambridge) 89:279–293, 1982.
7. M Roach. How to win at germ warfare. Health July/Aug: 77–80, 1994.
8. C Gerba, C Wallis, J Melnick. Microbiological hazards of household toilets: droplet production and the fate of residual organism. Appl Microbiol 30(2):229–237, 1975.
9. RE Black, AC Dykes, KE Andersen. Handwashing to prevent diarrhea in day care centers. Am J Epidemiol 113:445–451, 1981.

10. TA Cogan, SF Bloomfield, TJ Humphrey. The effectiveness of hygiene procedures for the prevention of cross-contamination from chicken carcases in the domestic kitchen. Appl Microbiol 29:354–358, 1999.
11. LL Gibson, JB Rose, CN Haas. Use of quantitative microbial risk assessment for evaluation of the benefits of laundry sanitation. Am J Infect Control 27(6):S34–S39, 1999.
12. E Scott. Hygiene issues in the home. Am J Infect Control 27[6]:S22–S25, 1999.
13. NJ Ehrenkranz. Bland soap handwash or hand antisepsis? The pressing need for clarity. Infect Control Hosp Epidemiol 13:299–301, 1992.
14. NJ Ehrenkranz, BC Alfonso. Failure of bland handwash to prevent hand transfer of patient bacteria to urethral catheters. Infect Control Hosp Epidemiol 12:654–662, 1992.
15. FN Marzulli, M Bruch. Antibacterial soaps: benefits versus risks. In: HI Maibach, R Aly, eds. Skin Microbiology: Relevance to Clinical Infection. New York: Springer-Verlag, 1981, 125–134
16. D Taplin. Chlorhexidene and clinical infections. In: H Maibach, R Aly, eds. Skin Microbiology: Relevance to Clinical Infection. New York: Springer-Verlag, 1981, 113–124.
17. WC Noble, ed. The Skin Microflora and Microbial Skin Disease. Cambridge: Cambridge University Press, 1993, 73–101, 135–172.
18. R Aly, HI Maibach. In vivo methods for testing topical antimicrobial agents. J Soc Cosmet Chem 32:317–323, 1981.
19. RJ Hay. Skin disease. Bri Med Bull 49(2):440–453, 1993.
20. RR Roth, WD James. Microbiology of the skin: resident flora, ecology, infection. J Am Acad Dermatol 20(3):367–390, 1989.
21. R Aly, HI Maibach. Comparative antimicrobial efficacy of a 2 minute surgical scrub with chlorhexidine gluconate, povidone iodine, and chloroxylenol sponge-brushes. Am J Infect Control 16(4):173–177, 1988.
22. RS Stern, C Nelson. The diminishing role of the dermatologist in the office-based care of cutaneous diseases. J Amer Acad Dermatol 29(5):773–777, 1993.
23. D Ben-Amitai, S Ashkenazi. Common bacterial skin infections in childhood. Pediatr Ann 22:225–227, 231–233, 1993.
24. DL Taplin, L Lansdell, AM Allen, R Rodriguez, A Cortes. Prevalence of streptococcal pyoderma in relation to climate and hygiene. Lancet 1:501–503, 1973.
25. M Lacour, C Hauser. The role of microorganisms in atopic dermatitis. Clin Rev Allergy 11:491–522, 1993.
26. HA Sampson. Atopic dermatitis. Ann Allergy 69:469–479, 1992.
27. S Beauregard, GA Gilchrest. A survey of skin problems and skin care regimens in the elderly. Arch Dermatol. 12:1638–1643, 1987.
28. J Zimakoff, V Thamdrup, W Petersen, J Scheibel. Recurrent staphylococcal furunculosis in families. Scand J Infect Dis. 20:403–405, 1988.
29. B Quinn, R Baker, J Pratt. Hurricane Andrew and a pediatric emergency department. Ann Emerg Med 23(4):737–741, 1994.
30. S Selwyn. Microbial interactions and antibiosis. In HI Maibach, R Aly, eds. Skin Microbiology. New York: Springer-Verlag, 1981, pp. 63–74.
31. BH Keswick, CA Berge, RG Bartolo, DD Watson. Antimicrobial soaps: their role

in personal hygiene. In: Ali et al., eds. Cutaneous Infection and Therapy. New York: Marcel Dekker, 1996, pp. 49–82.
32. E Larson. A causal link between handwashing and risk of infection? Examination of the evidence. Infect Control Hosp Epidemiol 9(1):28–36, 1988.
33. FL Bryan, MP Doyle. Health risks and consequences of *Salmonella* and *Campylobacter jejuni* in raw poultry. J Food Prot 58:326–344, 1995.
34. E Larson, ML Rotter. Handwashing: Are experimental models a substitute for clinical trials? Two viewpoints. Infect Control Hosp Epidemiol 11(2):63–66, 1990.
35. GAJ Ayliffe. The effect of antimicrobial agents on the flora of skin. J Hosp Infect. 1:111–124, 1980.
36. CA Bartzokas, JE Corkill, T Makin. Evaluation of the skin disinfecting activity and cumulative effect of chlorhexidine and Triclosan handwash preparations on hands artificially contaminated with *Serratia marcescens*. Infect Control 8(4):163–167, 1987.
37. GM Rayan, DJ Flournoy. Microbiologic flora of human fingernails. J Hand Surgery 12A(4):605–607, 1987.
38. JE Finch, J Prince, M Hawksworth. A bacteriological survey of the domestic environment. J Appl Bacteriol 45:357–364, 1978.
39. MF Mendes, DJ Lynch. A bacteriological survey of washrooms and toilets. J Hyg (Cambridge) 76:183–190, 1976.
40. SF Bloomfield. A review: the use of disinfectants in the home. J Appl Bacteriol 45: 1–38, 1978.
41. RR Marples, AG Towers. A laboratory model for the investigation of contact transfer of microorganisms. J Hyg (Cambridge) 82:237–248, 1979.
42. J Borneff, JR Wittig, M Borneff, G Hartmetz. Untersuchungen über das Vorkommen von Enteritis-Erregen in Haushalt—eine Pilotstudie. Zbl Bakteriol Microbiol Hyg I Abt 180(2–3):319–334, 1985.
43. RA Goodman, MT Osterholm, DM Granoff, LK Pickering. Infectious diseases and child care. Pediatrics 74:134–139, 1984.
44. LK Pickering, AV Bartlett, WE Woodward. Acute infectious diarrhea among children in day care: epidemiology and control. Rev Infect Dis 8(4):539–547, 1986.
45. AL Morrow, IT Townsend, LK Pickering. Risk of enteric infection associated with child day care. Pediatr Ann 20(8):427–433, 1991.
46. TL Chorba, RA Meriwether, BR Jenkins, RA Gunn, JN MacCormack. Control of a non-foodborne outbreak of salmonellosis: day care isolation. Am J Public Health 77(8):979–981, 1987.
47. M Fornasini, RR Reves, BE Murray, LK Pickering. Trimethoprim-resistant *Escherichia coli* in households of children attending day care centers. J Infect Dis 166(2): 326–330, 1992.
48. E Ekanem, HL Dupont, LK Pickering, BJ Selwyn, CM Hawkins. Transmission dynamics of enteric bacteria in day care centers. Am J Epidemiol 118(4):562–572, 1983.
49. R Van, AL Morrow, RR Reves, LK Pickering. Environmental contamination in child day-care centers. Am J Epidemiol 133(5):460–470, 1991.
50. MT Osterholm, RR Reves, JR Murph, LK Pickering. Infectious diseases and child day care. Pediatr Infect Dis J 11(8): S31–S41, 1992.

51. B Holaday, R Pantell, C Lewis, CL Gillis. Patterns of fecal coliform contamination in day-care centers. Public Health Nurs 7(4):224–228, 1990.
52. DJ Laborde, KA Weigle, DJ Weber, JB Kotch. Effect of fecal contamination on diarrheal illness rates in day-care centers. Am J Epidemiol 138(4):243–255. 1993.
53. Mayo Clinic Health Lett June:4, 1993.
54. AM Butz, E Larson, P Fosarelli, R Yolken. Occurrence of infectious symptoms in children in day care homes. Am J Infect Control 6:347–353, 1990.
55. DD Scala, GE Fischler, BM Morrison, Jr, R Aly, HI Maibach. Evaluation of antimicrobial bar soaps containing triclocarban. Annual Meeting of American Academy of Dermatology, 1994.
56. E Reiss-Levy, E McAlister, M Richards. Handwashing and antiseptics (lett). Med J Aust 140(4):245, 1984.
57. R Aly, HI Maibach. Effect of antimicrobial soap containing chlorhexidine on the microbial flora of skin. Appl Environ Microbiol 31(6):931–935, 1976.
58. FDA Public Health Service. The Food Code, 1993.
59. USDA Food Safety and Inspection Service. List of Proprietary Substances. Publication No. 1419, January 1993.
60. U.S. Department of Agriculture. Guidelines for obtaining authorization of compounds to be used in meat and poultry plants. In: USDA Food Safety and Inspection Service Agriculture Handbook No. 562, Oct. 1974.
61. EC Todd. Factors that contributed to foodborne disease in Canada, 1973–1977. J Food Prot 46(8):737–747. 1983.
62. JL Oblinger, ed. Bacteria associated with foodborne diseases. Food Technol April: 181–200, 1988.
63. RA Shooter, et al. Isolation of *Escherichia coli*, *Pseudomonas aeruginosa*, and *Klebsiella* from food in hospitals, canteens, and schools. Lancet August 21:390–392, 1971.
64. LA Lee, SM Ostroff, HB McGee, DR Johnson, FP Downes, DN Cameron, NH Bean, PM Griffin. An outbreak of shigellosis at an outdoor music festival. Am J Epidemiol 133(6):608–615, 1991.
65. JA Lopes. Food- and water-infective microorganisms. In SS Block, ed. Disinfection, Sterilization, and Preservation. 4th ed. Philadelphia: Lea & Febiger, 1991, 773–787.
66. FL Bryan. Risks of practices, procedures and processes that lead to outbreaks of foodborne diseases. J Food Prot 51(8):663–673, 1988.
67. JC deWit, G Broekhuizen, EH Kampelmacher. Cross-contamination during the preparation of frozen chickens in the kitchen. J Hyg (Cambridge) 83:27–32, 1979.
68. JV Pether, RJ Gilbert. The survival of *salmonella* on finger-tips and transfer of the organisms to foods. J Hyg (Cambridge) 69:673–681, 1971.
69. Code of Federal Regulations, Volume 21, Part 178.1010, 1994.
70. L Restaino, CE Wind. Antimicrobial effectiveness of hand washing for food establishments. Dairy Food Environ Sanit 10(3):136–141, 1990.
71. DG Maki. Control of colonization and transmission of pathogenic bacteria in hospital. Ann Intern Med 89(2):777–780, 1978.
72. J Bryan, J Cohran, E Larson. Handwashing: A ritual revisited. In: WA Rutala, ed. Chemical Germicides in Healthcare International Symposium, 1994. Assoc. For Pro-

fessionals in Infection Control and Epidemiology. Washington, DC: Polyscience Publishers, 1995, pp. 163–178.
73. RW Haley, DH Culver, JW White, WM Morgan, and TG Emori. The nationwide nosocomial infection rate. Am J Epidemiol 121(2):159–167, 1985.
74. EB Keeler, KL Kahn, D Draper, JM Sherwood, LV Rubenstein, EJ Reinish, J Kosecoff, RH Brook. Changes in sickness at admission following the introduction of the prospective payment system. J Am Med Assoc 264(15):1962–1968, 1990.
75. DG Maki. The Use of Antiseptics for Hand Washing by Medical Personnel. J. Chemother 1(suppl 1) 3–11, 1989.
76. TM Bauer, E Ofner, HM Just, H Just, FD Daschner. An epidemiological study assessing the relative importance of airborne and direct contact transmission of microorganisms in a medical intensive care unit. J Hosp Infect. 15:301–309, 1990.
77. E Kretzner, E Larson. Do antiseptic agents reduce surgical wound infections? Preoperative bathing, operative site preparation, surgical hand scrubs. In: WA Rutala, ed. Chemical Germicides in Healthcare International Symposium, 1994. Assoc. For Professionals in Infection Control and Epidemiology. Washington, DC: Polyscience Publishers, 1995, pp. 149–161.
78. EL Larson. Draft APIC guideline for hand washing and hand antisepsis in health care settings. Am J Infect Control 22(5):25A–47A, 1994.
79. FS Garner, MS Favero. CDC guideline for handwashing and hospital environmental control, 1985. Infect Control 7:1–20, 1985.
80. Association of Operating Room Nurses (AORN). Recommended practices: surgical hand scrubs. AORN J 52:830–836, 1990.
81. EL Larson, JJ Leyden, KJ McGinley, GL Grove, GH. Talbot. Physiologic and microbiologic changes in skin related to frequent hand washing. Infect. Control 7(2):59–63, 1986.
82. PD Meers, GA Yeo. Shedding of bacteria and skin squames after hand washing. J Hyg (Cambridge) 81:99–105, 1978.
83. EL Lowbury. Removal of bacteria from the operation site. In Skin Bacteria and Their Role in Infections. New York: McGraw-Hill, 1965, pp. 263–275.
84. H Laufman. Current use of skin and wound cleansers and antiseptics. Am. J. Surgery 157:359–365, 1989.
85. Association of Operating Room Nurses (AORN). 1993 Standards & Recommended Practices. Denver: AORN, Inc., 1993, pp. 129–133.
86. E Larson. Home hygiene: a reemerging issue for the new millennium. Am J Infect Control 27(6):S1–S3. 1999.

Part II
Topical Antimicrobials

In Part II, we discuss various aspects of the topical antimicrobial products currently in common use in the medical, food service, and consumer (personal hygiene) markets. The antimicrobial products of primary interest include iodine complexes (aqueous iodophors and tinctures), aqueous formulations and tinctures of chlorhexidine gluconate, triclosan, and parachlorometaxylenol, alcohol formulations, and quaternary ammonium products. Let us review some general aspects of these topical antimicrobials.

I. IODINE COMPLEXES

Iodine in its pure form is relatively insoluble in water without a solubilizing agent, but it dissolves well in various alcohols to provide an iodine tincture. Tinctures of iodine are used primarily as antiseptics.

By far the most common form of iodine for use as a topical antimicrobial is the iodophor. Iodophors, complexes of elemental iodine (tri-iodine) linked to a carrier, have several advantages: (1) greater solubility in aqueous solution than elemental iodine, (2) sustained release reservoir for the iodine, and (3) reduced equilibrium concentrations of free elemental iodine.

The most commonly used iodophor is povidone iodine, a compound of 1-vinyl-2-pyrrolidinone polymer with available iodine ranging between 9 and 12%.

Iodophors and tinctures of iodine provide excellent immediate antimicrobial action against a broad range of viruses, both gram-positive and gram-negative bacteria, fungi, and various protozoa. In fact, almost all important human disease microorganisms, including enteric bacteria, enteric viruses, protozoan trophozoites and cysts, mycobacteria, spores of *Bacillus* spp., and *Clostridium* spp. and many fungal species are susceptible to free iodine. It should be noted, however, that exposure times and concentrations of available iodine required vary from microorganism to microorganism.

In topical application to skin surfaces (e.g., hands and body surfaces in the inguinal, abdominal, anterior cubital, subclavian, and femoral regions), iodophors and tinctures of iodine providing at least 1% available iodine demonstrate effective immediate and persistent antimicrobial properties.

II. CHLORHEXIDINE GLUCONATE

Chlorhexidine gluconate (CHG) was first synthesized in 1950 by ICI Pharmaceutical in England. CHG has high levels of antimicrobial activity against a wide variety of microorganism species but relatively low levels of toxicity to mammalian cells. Additionally, CHG has a strong affinity for skin and mucous membranes. As a result, CHG has been used as a topical antimicrobial for wounds, skin prepping, and mucous membranes (especially in dentistry), where it provides, by virtue of its proclivity for binding to the tissues, time-extended antimicrobial properties. CHG also has value as a product preservative, including ophthalmic solutions, and as a disinfectant of medical instruments and hard surfaces.

The antimicrobial activity of CHG is pH dependent, with an optimal use range of 5.5–7.0, a nice match with the body's usual range of pH. The relationship between antimicrobial effectiveness of CHG and pH varies with the microorganism, however. For example, its antimicrobial activity against *Staphylococcus aureus* and *Escherichia coli* increases with an increase in pH, but the reverse is true for *Pseudomonas aeruginosa.*

Currently, there is much interest in alcohol tinctures of CHG. These alcohol/CHG products may prove to be highly effective for use as preoperative skin preps, surgical scrubs and healthcare personnel handwash formulations. Additionally, tincture of CHG may be useful both as a preinjection and an arterial/venous catheterization prep. Preparations of alcohol/CHG combine the excellent immediate antimicrobial properties of alcohol with the persistence properties of CHG to provide a clinical performance superior to either alcohol or CHG alone.

III. PARACHLOROMETAXYLENOL

Parachlorometaxylenol (PCMX) is one of the oldest antimicrobial compounds in use, dating back to 1913. It has not been widely used as a surgical or presurgical skin preparation because of its relatively low antimicrobial efficacy.

The initial evaluations from the FDA listed PCMX as a Category III product, meaning that there were not enough data to recognize it as both safe and effective as a topical antimicrobial. Because of this, it has not been extensively used in products for medical applications, such as surgical hand scrubs, preoperative skin preparations, or healthcare personnel handwash formulations.

A number of studies conducted since 1980 have demonstrated PCMX to be relatively safe for human use. After this determination, the number of companies interested in developing PCMX for use in a topical antimicrobial product has increased, and recent studies have shown PCMX to be an effective antimicrobial compound.

Current over-the-counter PCMX formulations demonstrate varying degrees of antimicrobial efficacy, depending upon the specific formulation. It is, at this time, generally agreed that PCMX products provide fair to good antimicrobial activity. Currently, PCMX products are formulated mainly for healthcare personnel handwash applications.

IV. ALCOHOLS

There is considerable debate concerning the antimicrobial effectiveness of alcohol used as a skin antiseptic. The antimicrobial efficacy of alcohol is highly dependent upon the concentration used, as well as the moisture level of the microbial environment treated. The short-chain, monovalent alcohols—ethanol and isopropanol—are probably the most effective for skin disinfection, because they are highly miscible with water, have low skin toxicity and allergenic potential, are fast-acting and are microbicidal, as opposed to microbiostatic.

The microbicidal activity of the alcohols relative to microorganisms is largely a function of their ability to coagulate cellular proteins. Protein coagulation takes place on the cell wall and the cytoplasmic membrane, as well as among the various plasma proteins. The literature suggests, however, that microbicidal effects of alcohols are, too, a result of their leaching effects on lipids.

Alcohols generally are inactive against bacterial spores. And, although there is much debate in the literature concerning the efficacy of alcohols against viruses, there appears to be general agreement that enveloped, lipophilic viruses are more susceptible to inactivation by alcohols than are "non-enveloped" viruses. Lastly, the fungicidal properties of alcohols vary among fungal species, but, in general, alcohols demonstrate a relatively high degree of mycidal/-static activity.

Although alcohols, as topical skin disinfectants, provide excellent immediate antimicrobial activity, they show little persistent activity. Once dried on the skin surfaces, antimicrobial effects of an alcohol have ended. Hence, their value as surgical hand scrub formulations and preoperative skin preparations where persistent antimicrobial properties are important, has been repeatedly challenged. On the other hand, alcohols have been shown to provide adequate antimicrobial results as healthcare personnel handwashes or preinjection skin preparations in removing or killing transient microorganisms.

V. TRICLOSAN

Triclosan, like PCMX, provides varying degrees of antimicrobial efficacy, depending upon the specific formulation and the species of challenge microorganisms. Triclosan has been formulated for a wide range of applications and is currently used in healthcare personnel handwash formulations, in the food industry in handwash products for food handlers, and extensively in consumer product lines, including hand soaps, shower gels, and body cleansers. Triclosan, like PCMX, provides fair to immediate antimicrobial action.

VI. QUATERNARY AMMONIUM COMPOUNDS

Quaternary ammonium chlorides (QACs) occupy a unique niche in the world of antimicrobial compounds. Rather than being a single, well-defined substance, as is the case for many antimicrobially active ingredients, QACs are composed of a diverse, eclectic collection of substances that share a common molecular structure containing a positively charged nitrogen atom covalently bonded to four carbon atoms. This carbon/nitrogen structure is responsible for the name of these antimicrobial compounds and also plays a dominant role in determining their chemical behavior.

The first reports of quaternary ammonium compounds with biocidal activity appeared in 1916. Since that time, QACs have grown in popularity and been utilized extensively as active ingredients in many types of products, including household cleaners, institutional disinfectants, skin and hair care formulations, sanitizers, sterilizing solutions for medical instruments, preservatives in eye drops and nasal sprays, mouthwashes, and even in paper processing and wood preservatives. As a group, QACs are effective across a broad spectrum of microorganisms, including bacteria, certain molds and yeasts, and viruses. However, the specific activity of QACs is as diverse as their range of chemical structures. QAC antimicrobial effectiveness is highly formulation dependent, because a variety of compounds may affect QAC activity. Some components reduce the QAC efficiency, while others may synergize their activity or expand the spectrum of affected microorganisms. This fact has led to some confusion and apparent contradiction in the published literature as to the actual effectiveness of QACs in their role as antimicrobials.

In addition to their antimicrobial activity, QACs also behave as surfactants, assisting with foam development and cleansing action. They are also attracted to the skin and hair, where small amounts remain bound after rinsing. This contributes to a soft, powdery feel to the skin, unique hair-conditioning effects, and long-lasting, persistent activity against microorganisms. These various attributes and multifunctional roles of QACs appeal to formulators and are responsible for their incorporation into many consumer products.

4

Clinical Applications of Povidone-Iodine as a Topical Antimicrobial

Steven Ripa, Nancy Bruno, Robert F. Reder, and Renate Casillis
Purdue Pharma, Stamford, Connecticut

Robert I. Roth
The Weinberg Group, Inc., Washington, D.C.

I. ANTISEPSIS AND THE DEVELOPMENT OF IODINE ANTIMICROBIALS

Antisepsis has its roots in an ancient effort to prevent food spoilage. Smoking of meats, a process that inhibited bacterial-mediated suppuration, was effective because creosote and other phenols inhibited bacterial and fungal growth. For centuries preceding the development of the germ theory of disease and recognition of the pathogenicity of bacteria and other microbial organisms, chemical treatment of wounds had been employed empirically to control suppuration. The early uses of acidic solutions such as wine and vinegar and halogenated compounds containing chlorine (bleaches) and iodine were empiric additions to wound dressings for preventing tissue decay. Many germicidal solutions were in clinical use centuries before microorganisms were discovered and subsequently recognized to negatively impact surgical outcomes and wound healing. When Lister recognized the biological roles of microbes in causing destruction of human tissue in wounds, it was a brief step to using antiseptics, beginning with phenol, for the treatment of hands, surgical instruments, and patient skin. Because the demonstration that sanitization of inert materials used for invasive procedures and aseptic surgical technique greatly improved clinical outcomes, clinicians have come to recognize that a critical component of wound care is infection control.

Skin and mucosal surfaces are normally coated with commensal organisms. Preventing the transition from microbial *colonization*, a normal state of skin, respiratory, and gastrointestinal surfaces, to microbial *infection* is important for successful healing of wounds and avoiding systemic illness. It is generally accepted that optimal healing occurs when bacterial counts are no greater than 10^5 per gram of tissue [1]. Under normal healthy circumstances this limit is easily maintained by the physical characteristics of skin and the immune system. Wounds, in contrast, have a damaged surface layer in contact with commensal microorganisms and often have an impaired vascular supply such that healing is suboptimal and infection control is inefficient. Infections that develop in chronic wounds such as ulcers are notoriously difficult to eradicate. Under these circumstances, the general objective of antiseptic therapy is to maintain control of microbial activity by preventing microbial overgrowth; complete sterilization is neither necessary nor practical. The goal of wound treatment with antiseptics is to decrease the rate and severity of wound infections while maintaining optimal wound healing. The two processes are clearly linked, as infected wounds commonly exhibit impaired healing due to the deleterious effects of bacterial products on regenerating tissue.

Antiseptics are chemical agents that oppose suppuration and decay of living tissue by killing offending microbes. Their clinical utility comes from their differential activity in killing microorganisms without greatly damaging host cells. The general properties desired in an antiseptic include broad, potent antimicrobial action with an ability to effectively penetrate necrotic tissue and eschar. Antiseptic therapy should not result in antimicrobial resistance, should not sensitize skin, and should be nontoxic locally and systemically. A vast array of efficacious antimicrobial compounds can exhibit some or all of these characteristics and has been used clinically. These agents broadly include alcohols, boric and other acids, carbolic acid and other phenols, mercury and other heavy metals, furans, hydrogen peroxide and other oxidizing agents, surfactants, and halogenated compounds containing chlorine and iodine. Of these, iodine-containing solutions in particular have retained popularity because of their high degree of efficacy while causing minimal pain, stinging, or irritation [2,3].

Tincture of iodine was developed as an antiseptic early in the nineteenth century and was employed to treat wounds during the Civil War. However, this form of iodine was found to be quite unstable and overly reactive. Early problems with elemental iodine, including stinging and irritation, allergic/anaphylactic reactions, low solubility, volatility, and poor absorption into tissue, were subsequently improved by complexing iodine with solubilizing agents.

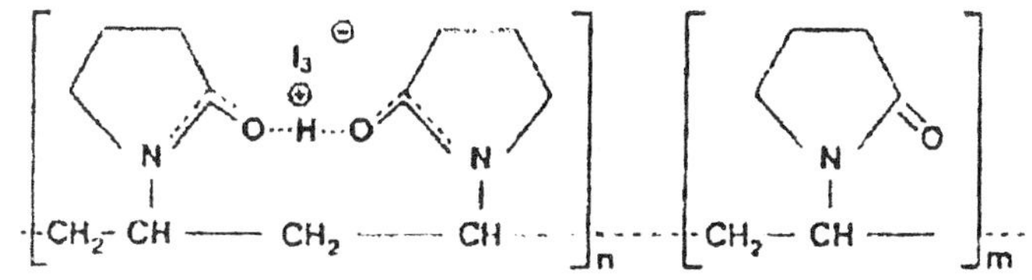

Figure 1 The structure of povidone-iodine.

II. THE BIOCHEMISTRY OF POVIDONE-IODINE

Povidone-iodine (Betadine® preparation, PVP-I) is a noncovalent complex (Fig. 1), formed by heating, between iodine and polyvinylpyrrolidone (povidone), a synthetic high molecular weight polymer used previously as a plasma expander, suspending agent, drug vehicle, and tableting aid. Povidone is a polymer that binds iodine fairly tightly, acting as an iodine-solubilizing carrier that gradually liberates free inorganic iodine in solution to skin and mucous membranes. The microbicidal effect in this complex is due to iodine; polyvinylpyrrolidone alone has no antibacterial activity [4]. Several buffered aqueous solutions of PVP-I have been developed for a variety of indications. Preparations containing 10% PVP-I typically yield 1% available iodine. Due to iodine complexing with povi-

Table 1 Properties of PVP-I as an Antimicrobial

Desired property	PVP-I characteristics
Broad antimicrobial efficacy	Broad-spectrum microbicide: kills bacteria including antibiotic-resistant strains, fungi, mycobacteria, viruses, spores, and protozoa
Rapid and long-lasting activity	Kills nearly all bacteria within seconds, more resistant microorganisms within minutes; antimicrobial activity persists on skin for hours
Rare development of microbial resistance	Intrinsic resistance to PVP-I is extremely rare; no evidence that prolonged use generates resistance
Good local tolerability	Iodine in the form of PVP-I is not painful or irritating, and does not sting; wound healing generally is not adversely affected
Well tolerated systemically	Although well tolerated, elevated blood iodine levels have been reported

done, PVP-I retains high antimicrobial capacity yet is comparatively more stable and less reactive than tincture of iodine, allowing iodine to be successfully used on skin surfaces without irritation or sensitization and on mucosal surfaces such as the oral cavity [5]. PVP-I is used in medical settings as a prophylactic and therapeutic antimicrobial, as well as a disinfectant [6]. Due to its effectiveness and general safety, PVP-I has been classified as category I (generally recognized as safe and effective and not misbranded) by the U.S. Food and Drug Administration (FDA) for uses both as a first aid antiseptic and also as a topical antiseptic for use in surgical hand scrub, patient preoperative skin preparation, and antiseptic handwash of healthcare professionals [7]. A summary of the properties of PVP-I in relation to desired characteristics for a topical antiseptic is presented in Table 1; each topic is discussed in depth in this chapter.

III. MECHANISM OF ACTION OF PVP-I AS AN ANTIMICROBIAL

Elemental iodine is rapidly lethal (microbicidal) to bacteria, fungi, viruses, and protozoa. In the absence of other organic matter that could compete for iodine binding, iodine solution can kill most microorganisms within seconds and rare, partially resistant organisms within minutes. PVP-I maintains a long duration of antimicrobial action because of slow, continuous iodine release [8]. Binding to organic matter is for the most part noncovalent, such that iodine later becomes released and antimicrobial activity, though delayed, is not greatly diminished. Iodine in commercial products is in great excess as an antimicrobial, thus protecting efficacy by means of mass action. In direct contrast to treatment with antibiotics, microorganisms are not known to develop tolerance or resistance to PVP-I.

Once released from PVP, elemental iodine has several forms in aqueous solution, with the most effective microbicidal forms being molecular I_2 and hypoiodous acid (HOI) [9]. In these forms, iodine is highly reactive with surrounding organisms via its strong oxidizing effects on functional groups of amino acids, nucleotides, and fatty acids [10]. Particularly susceptible are -NH groups of basic amino acids (lysine, arginine, histidine) and nucleotides, -SH groups of the amino acid cysteine, phenol groups of tyrosine, and double bonds of unsaturated fatty acids. Interaction of iodine with these groups in a cell results in rapid partitioning and disintegration of the cytoplasm, enzyme denaturation, pronounced coagulation of chromosomal material leading to nuclear denaturation, membrane pore forming and other loss of integrity in the bacterial cell membrane and fungal cell wall, and widespread loss of cytosolic material. Despite cell wall structural damage, most cells do not undergo lysis or rupture. The physicochemical rather than a biological mechanism of action may explain why PVP-I does not generate resistance in microorganisms [11]. PVP-I damages isolated cells more efficiently than cell clusters, in part explaining the high microbicidal activity of PVP-1 in the absence of significant host tissue injury [12].

IV. FORMULATIONS OF PVP-I

PVP-I (Betadine® preparation, Purdue Frederick Company; Massengill Medicated Douche®, SmithKline Beecham; Persist®, Becton Dickinson; DuraPrep®, 3M; CURITY®, Kendall) is an aqueous solution that comes formulated with a variety of detergents, emulsifiers, surfactants, and moisturizers, depending upon the clinical use. PVP-I can be applied via short soaks or repeated applications as free solutions. Available formulations of PVP-1 include those listed in Table 2.

V. SPECTRUM OF ANTIMICROBIAL ACTIVITY

The broad categories of microorganisms that are effectively killed by PVP-I include bacteria, fungi, viruses, and protozoa (Table 3) [13–22]. PVP-I kills both growing species and their frequently inactive spores. PVP-I is active against the broad diversity of microbial species that can contaminate surgical and chronic wounds (it is similarly effective against gram-negative and gram-positive and aerobic and anaerobic bacteria). Ranging from ''bothersome'' organisms such as fungi associated with athlete's foot to potentially lethal organisms such as hemolytic bacteria, *Mycobacterium tuberculosis*, and HIV, the spectrum of PVP-I activity is that of a broad-spectrum antimicrobial. As described in more detail below, iodine-containing solutions such as PVP-I have efficacy against clinically and epidemiologically significant new pathogens such as the methicillin-resistant and vancomycin-resistant bacteria that are common in the hospital setting. Compared to bacteriostatic antimicrobials, PVP-I is typically much more rapidly active (kills most organisms within seconds) and has a much wider activity profile. PVP-I is potent and results in relatively prolonged decontamination of skin due to residual activity lasting several hours [8,9].

VI. CLINICAL USES OF PVP-I, GENERAL CONSIDERATIONS

PVP-I is approved by FDA for hospital use in antiseptic handwashing of health-care personnel, preoperative skin decontamination, and as a surgical hand scrub. PVP-I is also approved for consumer use as a first aid preparation to reduce the

Table 2 Formulations of PVP-I (Betadine® Preparations)

Formulation	Clinical use	Active ingredients and important excipients
PVP-I solution (Betadine® preparation)	Preoperative site prepping, wound antisepsis, burn and laceration treatment, catheter site disinfection, catheter care	10% PVP-I, surfactant (Nonoxynol-9)
PVP-I surgical scrub (Betadine® preparation)	Surgical hand scrub, preoperative skin preparation	7.5% PVP-I, surfactant (ammonium nonoxynol-4 sulfate), lauramide (detergent and foam booster)
PVP-I skin cleanser (Betadine® preparation)	Antiseptic handwash	7.5% PVP-I
PVP-I ointment (Betadine® preparation)	Wound antisepsis, burn and laceration treatment, infection treatment	10% PVP-I
PVP-I swab stick (Betadine® preparation)	Preoperative skin preparation	10% PVP-I
PVP-I swabaid (Betadine® preparation)	Catheter site disinfection	10% PVP-I
PVP-I aerosol spray (Betadine® preparation)	Preoperative site prepping, wound antisepsis, burn and laceration treatment, catheter site disinfection, catheter care	5% PVP-I, surfactant (nonoxynol-9)
PVP-I mouthwash/gargle (Betadine® preparation)	Mouth/throat antisepsis	0.5% PVP-I
PVP-I douche (Betadine® preparation)	Vaginal antisepsis	0.3% PVP-I, surfactant (nonoxyl-9)
PVP-I perineal wash concentrate (Betadine® preparation)	Anogenital area antisepsis	10% PVP-I, surfactant (nonoxyl-10)

Table 3 Pathogens with Demonstrated Sensitivity to PVP-I

Gram-negative bacteria	Gram-positive bacteria	Mycobacteria	Fungi	Viruses	Other
Acinetobacter sp.	*Bacillus* sp.	*M. tuberculosis*	*Aspergillus* sp.	APC virus, type III, NIH	*Entamoeba histolytica*
Aerobacter aerogenes	*Clostridium* sp.	*M. chelonae*	*Blastomyces dermatitidis*	Cytomegalovirus	*Trichomonas vaginalis*
Aeromonas liquefaciens	*Corynebacterium* sp.	*M. fortuitum*	*Candida* sp.	Herpes simplex types I and 2	*Treponema pallidum*
Bacteroides sp.	Diphtheroids		*Cladosporium* sp.	HIV	Spores of *Bacillus* sp.
Burkholderia cepacia	*Lactobacillus acidophilus*		*Cryptococcus neoformans*	Influenza	Spores of *Clostridium* sp.
Campylobacter jejuni	*Enterococcus* sp.		*Debaryomyces* ap.	Papillomavirus	Spores of *Aspergillus* sp.
Chlamydia trachomatis	Vancomycin-resistant *Enterococcus* sp.		*Epidermophyton floccosum*	Polio	Spores of *Penicillium* sp.
Citrobacter sp.	*Micrococcus* sp.		*Microsporum audouinii*	Rabies	
Edwardsiella sp.	*Peptostreptococcus*		*Nocardia* sp.	Rubella	
Enterobacter aerogenes	*Propionibacterium acnes*		*Penicillium*	Vaccinia	
Escherichia sp.	*Sarcina lutea*		*Piedraia* sp.		
Gardnerella vaginalis	*Staphylococcus aureus*		*Pityrosporum ovale*		
Haemophilus vaginalis	*Staphylococcus epidermidis*		*Streptomyces* sp.		
Herella sp.	Methicillin-resistant *Staph.*		*Torulopsis glabrata*		
Klebsiella pneumoniae	*Streptococcus pneumoniae*		*Sacchyaromyces carlsbergenesis*		
Mima polymorpha	*Streptococcus faecalis*		*Trichophyton* sp.		
Morganella morganii	*Streptococcus pyogenes*				
Mycoplasma hominis					
Neisseria sp.					
Proteus sp.					
Providencia sp.					
Pseudomonas sp.					
Salmonella sp.					
Serratia sp.					
Shigella sp.					
Ureaplasma urealyticum					
Vibrio sp.					

chance of infection in minor cuts, scrapes, and burns. Other extended uses for Betadine® PVP-I preparations are described in Table 2. Off-label uses are specifically not recommended because of either lack of documented efficacy or unacceptable side effects. Use of PVP-I for irrigation of the gastrointestinal tract or joint spaces is not recommended because efficacy in these settings has not been demonstrated. Use of PVP-I in diluted form is contraindicated so that broad-spectrum, maximum antimicrobial activity is ensured. Finally, PVP-I should not be allowed to pool because prolonged contact can produce skin irritation.

An extensive literature describing the clinical utility of PVP-I presents weight-of-the-evidence support for this antiseptic. Animal studies provided early clear-cut evidence of efficacy because carefully controlled lesions could be created in order to obtain comparable pretreatment groups. PVP-I was shown to produce bacteriological control of rat wounds (decreased colony counts) with low-, moderate- and high-virulence organisms [23]. In mice, bacteriological control was associated with an improved rate of neovascularization [24]; reepithelialization, wound tensile strength, and complete healing time were either normal [12,15,25] or minimally retarded [24] in a number of animal studies (for review, see Ref. 26).

In contrast to well-controlled animal experimentation, human randomized clinical studies typically suffer from problems such as variable inclusion/exclusion criteria and poor comparability of treatment groups in terms of demographic characteristics, underlying pathologies, treatment interventions, and outcome measures. Nevertheless, PVP-I has been used for decades in surgical and chronic care settings. Its longevity reflects a confidence in its universal spectrum of potent antimicrobial activity, a mechanism of action other than those of antibiotics, and its generally excellent safety profile. While the earlier clinical literature (1950s–1980s) established the role of PVP-I for general infection control, the more recent literature stresses the special uses of this antiseptic for control of the developing opportunistic pathogens [27]. This is well illustrated by a recent supplement publication of *Dermatology* (vol. 195, suppl. 2, 1997) that includes several articles supportive of the antimicrobial activities and clinical utilities of PVP-I in treating nosocomial infections, antibiotic-resistant bacteria, and pathogenic viruses. FDA has formally approved, through a new drug application (NDA) process, specific products containing antimicrobials indicated for the following usage: preoperative prepping and postoperative preventive cleansing of surgical sites; disinfection of wounds, decubitus and stasis ulcers, lacerations and abrasions, second- and third-degree burns, and other skin and mouth infections; and use as a general infection-control agent for hospital personnel in contact with at-risk patients. Use of PVP-I in these types of indications is well described in case study reports.

The use of topical antiseptics makes clear intuitive sense based on our understanding of microbial pathogenesis of human disease. It is generally accepted that spontaneous healing of wounds, acceptance of skin grafts, and avoidance of systemic toxicity occurs best when bacterial counts are maintained at less than 10^5 per gram of tissue [1]. When this value is exceeded, there is impaired leukocyte function, deficient reepithelialization, and slowed wound contraction. While this balance is easily obtained by normal host defenses in healthy individuals without damaged skin lesions, bacterial overgrowth is common following surgical injury to skin and in the setting of chronic cutaneous wounds. Bacteremia associated with infected decubitus ulcers, a situation in which host defense mechanisms are particularly impaired due to poor circulation in the area of the ulcer and the often debilitated state of the patient, can result in sepsis conditions with as high as 55% mortality [28]. Chronicity of infection can also lead to additional infectious complications that can be equally difficult to treat, such as osteomyelitis. Whereas wound colonization (without true infection) is not considered an indication for antimicrobial treatment, the presence of substantial inflammation and frank infection presents a real risk of developing systemic toxicity that can prevent successful wound healing and not infrequently be lethal. Wound infections are commonly polymicrobial, and therefore broad-spectrum empiric control, as with PVP-I, becomes an advantage. As discussed in the following paragraphs, the general categories of antiseptic-based therapy include preoperative decontamination, prophylactic disinfection of acute skin lesions, and control of chronic wound infections.

VII. PVP-I FOR PROPHYLACTIC PREOPERATIVE SKIN DECONTAMINATION

PVP-I is most commonly used in the surgical setting to ensure preoperative decontamination, thus reducing the risk of surgical wound infection. The risk of operative infection is minimized by decontamination of patient skin surfaces in the area of the surgical procedure. The intent of such decontamination is to prevent introduction of skin flora into normally sterile sites, e.g., through invasive surgery, vascular catheter placement, or cerebrospinal space puncture. PVP-I is able to remove resident bacterial flora and leaves an antimicrobial residue on the skin capable of removing later transient contaminants. For these indications, PVP-I surgical scrub with sudsing capacity provided by a detergent and foam booster produces optimal preoperative decontamination. It has been well established that PVP-I surgical scrub (two 2.5-min Scrubs) produces nearly complete decontamination of surgical sites (98–100%) [29,31].

The recommended procedure for preoperative decontamination with PVP-I, based on the use of Betadine® Surgical Scrub preparation, is as follows:

Wet area with water
Apply PVP-I scrub (1 mL is sufficient to cover an area of 20–30 square inches)
Develop lather and scrub thoroughly for 5 minutes

Rinse off with sterile gauze saturated with water
Paint with PVP-I solution, or spray area with PVP-I aerosol
Allow area to dry for approximately 2 minutes

As a consequence of pre- and perioperative skin decontamination with PVP-I, the risk of postoperative infection is diminished. PVP-I decreases bacterial contamination of surgical sites such that incisions predominantly have no bacterial growth [32] and decreases the frequency of positive intraoperative wound cultures [33]. Postoperative infection rates are low following scrub or spray with PVP-I [4,34,35]. Surgical procedures involving the gastrointestinal tract are particularly susceptible to postoperative wound infection because of difficulty in eradicating gram-negative enteric bacteria even after use of preoperative oral antibiotics aimed at producing gut sterilization. PVP-I has been shown to be effective in preventing surgical wound infections in this setting [36] and has been shown to result in a much shorter postoperative stay (5 fewer days on average), indirectly suggesting a decreased severity of infection [37]. Finally, PVP-I mouthwash prior to dental extraction has been shown to reduce the positivity of gingival cultures and to decrease the extent of bacteremia [38] effects, that can be of critical importance, especially for patients at risk of endocarditis.

VIII. PVP-I FOR HANDWASHING

PVP-I is extensively used in the surgical setting to provide decontamination of the hands of operative personnel (surgeons, nurses, and others). PVP-I surgical scrub produces nearly complete decontamination of hands (98–100%) [29,30,39–42]. This effect persists for up to 6 hours after PVP-I washing, and an initial reduction in bacterial counts of 2–3 logs after the first washing can be increased to an even greater cumulative reduction (>4 logs) after three daily treatments [42,43]. The recommended procedure for preoperative handwashing with PVP-I, based on the use of Betadine® Surgical Scrub preparation, is as follows:

Wet area with water
Pour 5 mL PVP-I on the palm of the hand and spread over both hands
Rub scrub thoroughly over all areas for 5 minutes; clean thoroughly under fingernails; add water to develop copious suds
Rinse thoroughly under running water
Repeat above steps for another 5 minutes

IX. PVP-I AS A PROPHYLACTIC DISINFECTANT OF SUPERFICIAL ABRASIONS AND LACERATIONS

PVP-I has been used for decades as a disinfectant for superficial skin lesions such as abrasions and lacerations [2]. Due to the low incidence of quantifiable infectious complications in these situations, relatively large studies are required to objectively determine the true benefit of such prophylactic treatment. In two such large studies, PVP-I resulted in fewer infections in sutured wounds compared to saline alone [44,45] and compared to topical antibiotics [46]. Combining smaller prospective study results similarly has demonstrated the efficacy of PVP-I [47]. The general clinical acceptance of such treatment also is illustrated by numerous studies that fail to include an untreated control group when comparing PVP-I with other topical antiseptics. Except in the rare instance of iodine sensitivity, PVP-I is nonstinging and nonirritating, characteristics that have greatly contributed to its use in prophylactic infection control. In addition to prophylactic use on skin, PVP-I also is used in preventing infection in the eye. PVP-I was effective in preventing neonatal conjunctivitis (ophthalmia neonatorum) in one study, with superior efficacy and less ocular toxicity than alternative prophylactic treatments with antibiotics or silver nitrate [48].

X. PVP-I FOR MICROBIAL CONTAMINATION CONTROL OF ULCERS AND BURNS

Wounds such as burns, decubitus ulcers, and stasis ulcers have impaired vascular supply, localized ischemia, and devitalized tissue, all of which predispose to secondary infection. Localized wound infection causes pain and discomfort, prolongs spontaneous healing time, can prevent satisfactory surgical closures, and can lead to systemic infection. Bacterial contamination, in particular for decubitus ulcers, is likely to involve more than simple skin flora as these wounds are often contaminated by urine and/or feces, especially in the elderly and bedridden. As complete sterilization in these disorders is neither necessary nor practical, the goal of antiseptic treatment is to avoid frank infection and risk of sepsis. PVP-I therapy to exposed granulation tissue after initial debridement of necrotic tissue dramatically diminishes bacterial colony counts (by several logs) and produces excellent results for subsequent skin grafting [49]. Time to complete healing in experimental human wounds is faster in PVP-I–treated wounds than in untreated wounds, and PVP-I has proven as effective as triple antibiotic ointment in shortening deep abrasion reepithelialization time by 25% compared to nontreated controls [17]. Clinical improvement has been demonstrated in a wide variety of types of ulcers, including venous stasis, arterial, decubitus, and inflammatory (pemphigoid) [14,50]. PVP-I–treated ulcers have less leukocytoclastic vasculitis [51], and patients require amputation for septic complications less often [52].

Treatment of burn patients, similar to the situation with ulcer treatment in debilitated patients, is particularly difficult because of underlying immune impairment and nutritional deficiencies. Such patients easily make the transition from lesion colonization to infection and have reduced capacity to reverse sepsis once it occurs. A review of the experience of over 1500 burn patients treated with

PVP-I at frequent intervals has demonstrated overall excellent results [53]. Bacterial counts from burn sites in 100 PVP-I–treated patients in this study showed 45% sterility and no cultures above the 10^5 per gram tissue ''danger limit.'' Repeated, frequent application of PVP-I ointment, which was considered of great importance in this patient population, was possible because PVP-I was extremely well tolerated, could penetrate eschar, and minimized drying out of the burns. This finding has been quantitatively confirmed by the demonstration that successful control of bacterial growth and infection of burns depends directly upon the frequency of PVP-I application (qid > bid or tid > qd or qod) [54]. Several studies have shown that PVP-I effectively treats both superficial and deep burns, maintaining low systemic infection rates, preserving intact eschar, and promoting timely healing [54–58].

XI. COMPARISON OF PVP-I TO OTHER GENERAL ANTISEPTICS

PVP-I exhibits a broad array of desirable attributes, including a broad spectrum of microbicidal action, rapid activity (seconds for full kill), prolonged maintenance of skin sterilization, and minimal skin irritation or sensitization. A variety of other antiseptics exhibit some, but typically not all, of these attributes. As a result, PVP-I microbicides are the leading antiseptics in U.S. hospitals today. In vitro sensitivity studies have established that among antiseptics only PVP-I is capable of killing all classes of pathogens responsible for nosocomial infections: gram-positive and gram-negative bacteria, including antibiotic-resistant strains and some spores (both bacterial and fungal), as well as viruses, fungi, mycobacteria, and protozoa.

Today, gram-negative strains comprise over one third of bacteria isolated from hospital-acquired infections, and some commonly used antiseptics are ineffective against these organisms. With regard to in vivo comparative clinical efficacy, most clinical studies with PVP-I have involved either silver compounds, quaternary ammonium compounds, or chlorhexidine. PVP-I in many comparative studies shows superior decontamination of hands and surgical sites [29,30,34]. Fingertip cultures taken both before (short-term effect) and after (long-term effect) surgical procedures demonstrate superior hand sterility attained by PVP-I [34], greater decreases in bacterial contamination of ulcers [12], superior efficacy against antibiotic-resistant bacteria such as methicillin-resistant *Staphylococcus aureus* [59] and other gram-positive bacteria [42], and greater control of fungal contaminants such as *Candida albicans* [12] than the alternative antiseptic(s).

PVP-I also has been reported to give better penetration of eschar [12,60] and to maintain germicidal activity in the presence of blood, serum, or pus that is superior to that of several other topical antimicrobials [30]. With regard to the potential for microorganisms to develop resistance to antiseptics, a variety of gram-negative bacteria passaged in vitro and tested at each passage maintained full susceptibility to PVP-I, whereas they developed resistance to other antimicrobials [61]. Very rare bacteria may have intrinsic iodine resistance [62], although this does not appear to be a substantial problem given the continued evidence of widespread efficacy over half a century of clinical use. Development of tolerance with long-term clinical use (e.g., peritoneal dialysis) has not been observed [63]. Finally, subjective, visual, and objective (stratus corneum hydration measurement) tests show PVP-I to be nonirritating in contrast to other antiseptics [3].

XII. AVOIDING ANTIBIOTICS WITH PVP-I

Although it is well accepted that a frankly infected wound often requires systemic antibiotic therapy to cure the infection, avoid sepsis, and promote healing, it is also recognized that antibiotic usage is to be avoided when possible. First, antibiotics often do not work well with many types of wounds, especially for chronic granulating wounds. Due to poor circulation, especially in areas of ulceration, systemic antibiotics may not reach infected tissue and can often be of limited value in decreasing bacterial counts. Commonly, topical rather than systemic antisepsis is the initial attempt at infection control in such situations. PVP-I and other topical antiseptics do not require a good vascular bed and sufficient perfusion for their activity as do systemic antibiotics. Nonantibiotic topical antiseptics are equivalent or superior to topical antibiotics in minimizing infectious overgrowth of wounds [64].

Second, unnecessary antibiotic usage promotes the growing epidemic of antimicrobial resistance that presently afflicts hospital and chronic care settings. Antimicrobial resistance is a large and growing problem, both for the individual patient and as a public health issue. For example, organisms such as *Pseudomonas* and *Serratia*, which colonize wounds but rarely result in disease in the preantibiotic era, have evolved in response to widespread antibiotic usage into major pathogens [65]. Attempts at controlling infection often require treatments with one antibiotic after another in order to find a therapy to which the infectious organism is sensitive. This process is time-consuming, expensive, and often prolongs treatment within the hospital. Recent studies of antibiotic resistance in bacterial isolates from ulcers demonstrate that 25–50% of isolates are resistant to standard first- and second-line drugs [66,67]. In many instances, an antibiotic-resistant organism is simultaneously resistant to multiple antibiotics of different mechanistic classes [66]. Some of the most important current examples of antibiotic resistance are methicillin-resistant *S. aureus* (MRSA), *Enterococcus*, *Escherichia coli*, *Pseudomonas aeruginosa*, and vancomycin-resistant Enterococci (VRE). MRSA in particular is a leading nosocomial pathogen causing wound infections and often bacteremia. These bacteria are almost completely resistant as well to all other antibiotics in the therapeutic arsenal yet are fully sensitive to PVP-I [6,11,15,27,59,68,69]. Topical antisepsis may therefore be the only practical way to treat some wounds infected with multiple antibiotic-resistant organisms [27]. Whereas there can be cross-resistance by pathogens to types of antiseptics other than PVP-I [70], in most studies there has been no such demonstration with iodine solutions [11,61]. However, it has been reported that rare antibiotic-resistant bacteria may not be fully sensitive to PVP-I when used at less than the recommended concentration [27].

XIII. RISKS AND CONCERNS WITH PVP-I

Although PVP-I is well accepted and utilized by healthcare personnel and consumers, there are some known side effects as well as some additional concerns that are incompletely understood and currently unresolved. Known adverse effects of PVP-I include skin redness, swelling, pain and irritation, and rare instances of allergic hypersensitivity. When these reactions occur, it is recommended that use of the product be discontinued. Three unresolved concerns about PVP-I are principally related to questions of efficacy and potential adverse effects in the chronic setting and reflect the existence of either incomplete or contradictory data. Regarding efficacy, it is possible that PVP-I, because of its high nonspecific reactivity with proteins and lipids, may not function well in the presence of pus, blood, or necrotic tissue. A second potential risk is that PVP-I could delay wound healing: because of its nonspecific cytotoxic effects shown in vitro, it is possible that useful inflammatory cells and reparative connective tissues might be inadvertently damaged by PVP-I. Finally, a concern related to potential adverse effects is for localized iodine hypersensitivity and systemic toxicity, in particular, thyroid effects in the treated patient or, alternatively, hypothyroidism in the fetus or newborn of a treated patient. Each of these issues is explored below.

A. PVP-I Function in the Presence of Organic Substances

Chronic wounds such as burns and ulcers often contain substantial quantities of organic substances that are capable of binding iodine. At issue is whether such binding results in significant inactivation of PVP-I solutions. Early in vitro binding studies demonstrated that erythrocytes and free hemoglobin from lysed erythrocytes (but not plasma) could bind appreciable iodine from PVP-I solution, and it was suggested that PVP-I activity would be optimal in blood-free situations [71]. Subsequent work confirmed the in vitro inhibition of PVP-I by organic substances that bind iodine (necrotic tissue, pus, blood, fat, and glove powder)

[72]. For example, bacterial killing in vitro in the presence of organic compounds may require a considerably greater PVP-I concentration or longer contact time to achieve bacterial killing equivalent to that observed in the absence of such substances [73]. However, the clinical importance of this observation is uncertain since PVP-I is used therapeutically with great excess. For example, although pus in a 1:4 ratio with antiseptic decreases PVP-I potency by several logs, there is still >99% immediate bacterial killing, and at a 1:1 ratio, there is 96–99% killing with prolonged contact time (10–120 min). Other in vitro work similarly has shown adequate activity of PVP-I in the presence of 5–10% blood or serum [74], and in clinical use PVP-I has been reported to maintain acceptable germicidal activity in the presence of these organic substances without requiring precleansing of skin surfaces. Similarly, PVP-I was shown to be totally effective against vaginal bacteria even in the presence of serosanguinous, mucoid, or pus-like discharge [75].

B. PVP-I and Wound Repair

There is considerable current debate concerning the risk that PVP-I would inhibit wound healing. Risks of all topical antiseptics include the potential for impaired regeneration of healing tissue and impaired wound contracture and closure. Wound treatment requires removing damaged tissue, reestablishing blood supply, replacing destroyed tissue, and closing large open spaces to allow for reepithelialization. Expressed concerns with PVP-I include cytotoxicity to useful inflammatory cells (granulocytes and monocytes) and inhibition of angiogenesis, fibroblast proliferation, fibroplasia, scar remodeling, and epithelial cell migration to close the wound. These concerns derive primarily from a preclinical literature demonstrating that PVP-I is cytotoxic to cultured fibroblasts and to leukocytes in vitro (76–79). These findings may not be specific for PVP-I since other common topical antiseptics (hydrogen peroxide, acetic acid) appear to be equally cytotoxic to both bacteria and fibroblasts. Cytotoxicity, decreased wound healing rates, and impaired tensile strength of wounds have been demonstrated in vitro and in rats with experimental wounds treated with PVP-I [80]. PVP-I has been shown to inhibit the rate of wound reepithelialization in animals, although rates of neovascularization were improved [24].

However, other studies have failed to identify such effects, and it is well recognized that the relevance of in vitro wound healing studies to in vivo use in humans may be very limited [26,81]. At low PVP-I concentrations such as would be typical for retained iodine on and below skin outer surfaces, cell viability is well maintained [77]. It is possible that in vitro toxicity is largely affected by culture conditions. Culturing of skin from operating room specimens (e.g., for skin grafting) prepped with PVP-I yielded highly successful epithelial cell viability and growth while ensuring very low loss of tissue due to microbial contamination [82]. Compared to saline, PVP-I results in a smaller wound surface area that remains open and no difference in days to full wound closure in those wounds that completely heal [83]. Skin graft donor site wounds treated with PVP-I have been shown to exhibit normal or improved healing [84]. In a recent review article [26], it was concluded that PVP-I does not pose a significant risk to wound healing, although it was acknowledged that human in vivo experience is somewhat mixed [80,83,84]. Most studies show either no long-term adverse effects (closure rates and reepithelialization times unaffected by PVP-I) or improved rates of healing compared to other traditional treatment regimens [84–86].

The weight of human clinical evidence supports the conclusion that healing of human wounds is not adversely affected by therapeutic administration of PVP-I [12,87,88]. FDA has concluded, after much debate and careful consideration, that PVP-I should be considered ''generally recognized as safe and effective'' and that there is no evidence of delayed wound healing when PVP-I does not contain surfactant [7]. Perhaps the best condensation of the limited high-quality clinical literature relating to wound healing in the presence of iodine and other topical antiseptics has been presented in a systematic review of debridement methods conducted by the University of York, United Kingdom [89]. This systematic review included studies in which there was documentation of chronic nonhealing wounds, treatment groups were well matched, and all treatments were randomized and controlled. Cadexomer iodine, an antiseptic related to PVP-I, was shown, in contrast to a variety of other traditional wound care treatments, to produce superior rates of wound area reduction and/or complete wound healing, indicating a lack of adverse effects by iodine on wound repair. However, the human clinical literature regarding effects of PVP-1 on wound healing cannot be considered conclusive, and the overriding conclusion of the Consensus Conference for the Care of Chronic Wounds (presented at the 1999 Wound Healing Society Educational Symposium, Minneapolis, MN) was that there were insufficient clinical data from well-controlled ulcer healing trials to generate any definitive clinical care guidelines and recommendations.

C. Local Irritation and Systemic Toxicity

Iodine solution in its earlier form without povidone (tincture of iodine) proved suboptimum for chronic treatment of open wounds because of staining of the skin, local irritation, and sensitization of skin. Experimentally, such problems are avoided with PVP-I as solubility increases and chemical reactivity decreases due to binding with povidone. A review of the safety literature evaluating PVP-I concludes there is a lack of irritation as demonstrated by skin patch tests, sensitization assays, phototoxicity and allergenicity assays in humans, and eye and dermal Draize irritation assays in rabbits [17].

Absorption of iodine from common treatment of cuts, scrapes, and burns has been considered by the FDA. After evaluating PVP-I study data, FDA has concluded that transient increases in iodine blood levels do not adversely affect thyroid function [90]. However, there remains a risk regarding possible systemic toxicity in the situation in which PVP-I would be applied repeatedly, over a prolonged period, to a large surface area. Such conditions exist when treating a major burn with PVP-I; in fact, serum and urine iodine levels can become pronounced. Despite the risk that exposure to large quantities of iodine could cause hyperthyroidism or induce a thyrotoxic crisis in susceptible individuals, thyroid hormone abnormalities are not a major problem, and there is no evidence for significant thyroid dysfunction [12,53,57]. It is believed that abnormal thyroid measurements (low T3 and T4, increased TSH) in this setting are more likely attributable to the stress of the underlying condition than PVP-I therapy [12]. Another clinical situation of concern is that PVP-I treatment of pregnant or lactating mothers could induce transient hypothyroidism in the fetus or in the newborn. Although the literature is sparse and nondefinitive, it has been recommended that PVP-I be avoided in such situations [12,91]. Transient neonatal hypothyroidism also has been noted as a consequence of PVP-I use in the neonatal period for insertion of intravenous lines and blood gas measurements [92].

Rare case reports have been published of metabolic abnormalities associated with the use of PVP-I, mostly in the setting of treatment of burns or large volume irrigations [93–95]. In most instances these abnormalities include metabolic acidosis and hyperosmolarity [12,93,94] and are seen in the setting of acute renal failure. However, in general, systemic metabolic toxicity due to PVP-I absorption is believed to be extremely unusual [81]. Allergic reactions to PVP-I have been noted in rare case reports, although cutaneous sensitization to povidone-iodine as assessed by a patch test is extremely low [2,88]. There also have been reports (product complaints) that strong solutions of iodine can result in thermal burns or skin irritation secondary to pooling.

XIV. CONCLUSION

PVP-I is a broad-spectrum microbicide with activity against a wide range of bacterial, fungal, viral, and other pathogens. PVP-I activity is rapid and long-lasting, and there is no evidence that microorganisms develop resistance to it over time. PVP-I has excellent local tolerability and an absence of systemic toxicity. The literature supports the conclusion that PVP-I has clinical efficacy for preoperative prepping and postoperative cleansing of surgical sites, disinfection of lacerations, abrasions, burns and chronic ulcers, and as a general agent for infection control.

REFERENCES

1. RB Evans. An update on wound management. Hand Clin 7:409–432, 1991.
2. HA Shelanski, MY Shelanski. PVP-Iodine: history, toxicity and therapeutic uses. J Int Coll Surg 25:727–734, 1956.
3. RA Tupker, J Schuur, PJ Coenraads. Irritancy of antiseptics tested by repeated open exposure on the human skin, evaluated by non-invasive methods. Contact Derm 37: 213–217, 1997.
4. AL Games, E Davidson, LE Taylor, AJ Felix, BA Shidlovsky, A Prigot. Clinical evaluation of povidone-iodine aerosol spray in surgical practice. Am J Surg 97:49–53, 1959.
5. S Siggia. The chemistry of polyvinylpyrrolidone-iodine. J Am Pharm Assoc 44: 210–204, 1957.
6. W Fleischer, K Reimer. Povidone-iodine in antiseptic—state of the art. Dermatology 195 (suppl 12):3–9, 1997.
7. Federal Register, FDA, Department of Health and Human Services, June 17, 1994, pp. 31417–31423.
8. MS Marr, B Saggers. A skin disinfectant. Povidone-iodine (Betadine). Nurs Times 60:758–759, 1964.
9. W Gottardi. The influence of the chemical behavior of iodine on the germicidal action of disinfectant solutions containing iodine. J Hosp Infect 6 (suppl): 1–11, 1985.
10. K Reimer, H Schreier, G Erdos, B Konig, W Konig, W Fleischer. Molecular effects of a microbicidal substance on relevant microorganisms: electron-microscopic and biochemical studies on povidone-iodine. Z Hyg Umweltmed 200:423- 434, 1998.
11. ET Houang, OI Gilmore, C Reid, EI Shaw. Absence of bacterial resistance to povidone iodine. J Clin Pathol 29:752- 755, 1976.
12. M Steen. Review of the use of povidone-iodine (PVP-I) in the treatment of burns. Postgrad Med J 69 (suppl 13):584–592, 1993.
13. M Asanaka, T Kurimura. Inactivation of human immunodeficiency virus (HIV) by povidone-iodine. Yonago Acta Med 30:89–92, 1987.
14. A Gilgore. The use of povidone-iodine in the treatment of infected cutaneous ulcers. Curr Ther Res 24:843–848, 1978.
15. D Goldenheim. An appraisal of povidone-iodine and wound healing. Postgrad Med J 69:597–105, 1993.
16. IC Kaplan, DC Crawford, AG Dumo, RT Schooley. Inactivation of human immunodeficiency virus by Betadine. Infect Control 8:412–414,1987.
17. B Oshlack, RP Grandy, C Chelle, TM Ast, PD Goldenheim. Povidone iodine cream for antisepsis. In: S Selwyn, ed. Proceedings of the First Asian/Pacific Congress on Antisepsis. Royal Soc Med Services Intemat Congress and Symp Series No. 129, 1988, pp. 73–80.
18. SA Plotkin. The effect of povidone-iodine on several viruses. In: HC Polk, NJ Ehrenkranz, eds., Medical & Surgical Antisepsis with Betadine Microbicides. New York: The Purdue Frederick Co., 1972, 9–16.
19. RR Rinaldi, ML Sabia. A double-blind controlled study: Betadine solution treatment of athlete's foot. Arch Pod Med Foot Surg 3:1–9, 1976.

20. SA Sattar, VS Springthorpe. Survival and disinfectant inactivation of the human immunodeficiency virus: a critical review. Rev Infect Dis 13:430–447, 1991.
21. DD Scherr, TA Dodd. In vitro bacteriological evaluation of the effectiveness of antimicrobial irrigating solutions. J Bone Joint Surg 58:119–122, 1976.
22. DC Sokal, PL Hermonat. Inactivation of papillomavirus by low concentrations of povidone-iodine. Sex Transm Dis 22:22–24, 1995.
23. MC Robson, RHM Schaerf, TJ Krizek. Evaluation of topical povidone-iodine ointment in experimental burn wound sepsis. Plast Reconstruct Surg 54:328–334, 1974.
24. D Kjolseth, JM Frank, JH Barker, GL Anderson, AI Rosenthal, RD Acland, D Schuschke, FR Campbell, GR Tobin, LJ Weiner. Comparison of the effects of commonly used wound agents on epithelialization and neovascularization. J Am Coll Surg 179:305–312, 1994.
25. RG Geronemus, PM Mertz, WH Eaglstein. Wound healing. The effects of topical antimicrobial agents. Arch Dermatol 15:1311–1314, 1979.
26. DA Mayer, MJ Tsapogas. Povidone-iodine and wound healing: a critical review. Wounds 5:14–23, 1993.
27. DN Payne, SA Gibson, R Lewis. Antiseptics: a forgotten weapon in the control of antibiotic resistant bacteria in hospital and community settings? J R Soc Health 118: 18–22, 1998.
28. CS Bryan, CE Dew, KL Reynolds. Bacteremia associated with decubitus ulcers. Arch Intern Med 143:2093–2095, 1983.
29. P Dineen. Hand-washing degerming: a comparison of povidone-iodine and chlorhexidine. Clin Pharmacol Ther 23:63–67, 1978.
30. SM Joress. A study of disinfection of the skin: a comparison of povidone-iodine with other agents used for surgical scrubs. Ann Surg 155:296–304, 1962.
31. The Medical Department, The Purdue Frederick Company. Efficacy of Betadine surgical scrub in reducing bacterial counts on the hands after two consecutive 2.5-minute scrub periods. Data on file, The Purdue Frederick Company, Stamford, CT.
32. VH Crowder, IS Welsh, GH Bomside, I Cohn. Bacteriological comparison of hexachlorophene and polyvinylpyrrolidone-iodine surgical scrub soaps. Am Surg 33: 906–911, 1967.
33. RA Garibaldi. Prevention of intraoperative wound contamination with chlorhexidine shower and scrub. J Hosp Infect 11 (Suppl B):5–9, 1988.
34. AS Close, BF Stengel, HH Love, ML Koch, MB Smith. Preoperative skin preparation with povidone-iodine. Am J Surg 108:398–401, 1964.
35. OI Gilmore, C Reid, A Strokon. A study of the effect of povidone-iodine on wound healing. Postgrad Med 153:122–125, 1977.
36. IG Gray, MIR Lee. The effect of topical povidone iodine on wound infection following abdominal surgery. Br J Surg 68:310–313,1981.
37. IA Walsh, ImcK Watts, PI McDonald, II Finlay-Jones. The effects of topical Betadine microbicides on the incidence of infection in surgical wounds. In: WA Altemeier, ed. World Congress/Antiseptics Proceedings. New York: HP Publishing Co., 1980, pp. 78–81.
38. IW Scopp, LD Orvieto. Gingival degerming by povidone-iodine irrigation: bacteremia reduction in extraction procedures. J Am Dent Assoc 83:1294–1296, 1971.
39. C Poon, DI Morgan, F Pond, I Kane, BR Tulloh. Studies of the surgical scrub. Aust NZ J Surg 68:65–67, 1998.

40. V Hingst, I luditzki, P Heeg, H-O Sonntag. Evaluation of the efficacy of surgical hand disinfection following a reduced application time of 3 instead of 5 minutes. J Hosp Infect 20:79–86, 1992.
41. EIL Lowbury, HA Lilly, IP Bull. Disinfection of hands: removal of transient organisms. Br Med J 2:230–233, 1964.
42. The Medical Department, The Purdue Frederick Company. Microbial effectiveness of Betadine surgical scrub (povidone-iodine) versus hexachlorophene. From studies by VanderWyk RW, 1971. Data on file, The Purdue Frederick Company, Stamford, CT.
43. I Faoagali, I Fong, N George, P Mahoney, V O'Rourke. Comparison of the immediate, residual, and cumulative antibacterial effects of Novaderm R, Novascrub R, Betadine Surgical Scrub, Hibiclens, and liquid soap. AMJ Infect Control 23:337–343, 1995.
44. A Gravett, S Sterner, IE Clinton, E Ruiz. A trial of povidone-iodine in the prevention of infection in sutured lacerations. Ann Emerg Med 16:167–171,1987.
45. WI Morgan. The effect of povidone-iodine (Betadine) aerosol spray on superficial wounds. Br J Clin Pract 33:109–110, 1979.
46. TCN Morgan, R Firmin, B Mason, V Monks, D Caro. Prophylactic povidone iodine in minor wounds. Injury 12:104–105, 1980.
47. AHN Roberts, FEY Roberts, RI Hall, IH Thomas. A prospective trial of prophylactic povidone iodine in lacerations of the hand. J Hand Surg 10:370–374, 1985.
48. SJ Isenberg, L Apt, M Wood. A controlled trial of povidone-iodine as prophylaxis against ophthalmia neonatorum. N Engl J Med 332:562–566, 1995.
49. JF Connell, LM Rousselot. Povidone-iodine. Extensive surgical evaluation of a new antiseptic agent. Am J Surg 108:849–855, 1964.
50. AG McKnight. A clinical trial of povidone-iodine in the treatment of chronic leg ulcers. Practitioner 195:230–234, 1965.
51. C Pierard-Franchimont, P Paquet, JE Arrese, GE Pierard. Healing rate and bacterial necrotizing vasculitis in venous leg ulcers. Dermatology 194:383–387, 1997.
52. SM Eweda, NI Ahmed, TA El-Wahab. Comparative study to determine the effectiveness of utilization different dressing techniques and solutions in dressing of septic diabetic foot ulcers. J Med Res Inst 15:44–51, 1994.
53. PR Zellner, S Bugyi. Povidone-iodine in the treatment of burn patients. J Hosp Infect 6(suppl):139–146, 1985.
54. NG Georgiade, WA Harris. Open and closed treatment of burns with povidone-iodine. Plast Reconstr Surg 52:640–644, 1973.
55. M De Kock. Topical burn therapy comparing povidone-iodine ointment or cream plus aserbine, and povidone-iodine cream. J Hosp Infect 6 (suppl): 127–132, 1985.
56. T-Z Li, H-G Guo, X-L Lin, B-J Qin, J-Y Ahu, L-T Li, A-Y Su, W-P Wang, A-Y Guo, W-W Peng. Comparative trial between 10% Betadine ointment and silver sulpha diazine in the treatment of burns. In: S Selwyn, ed. Proceedings of the First Asian/Pacific Congress on Antisepsis. Royal Soc Med Services Intemat Congress and Symp Series No. 129, 1988, pp. 83–91.
57. G Meszaros, L Menesi, Z Kopcsanyi. Treatment of thermally injured patients with Betadine7 solution and cream. Ther Hung 41:132–136, 1993.
58. R Sinha, RK Agarwal, M Agarwal. Povidone iodine plus neosporin in superficial burns—a continuing study. Burns 23:626–628, 1997.

59. AR McLure, J Gordon. In-vitro evaluation of povidone-iodine and chlorhexidine against methicillin-resistant *Staphylococcus aureus*. J Hosp Infect 21:291–299, 1992.
60. S Kominos, CE Copeland, B Grosiak. Penetration of topical antiseptics through eschar. Recent Antisepsis Techniques in the Management of the Burn Wound. Proceedings of a symposium sponsored by the Department of Surgery, Division of Plastic, Maxillofacial and Oral Surgery, Duke University Medical Center, Durham, NC, 1974, pp. 21–23.
61. HN Prince, WS Nonemaker, RC Norgard, DL Prince. Drug resistance studies with topical antiseptics. J Pharm Sci 67:1629–1631, 1978.
62. RL Anderson. Iodophor antiseptics: intrinsic microbial contamination with resistant bacteria. Infect Control Hosp Epidemiol 10:443–446, 1989.
63. L Klossner, HR Widmer, F Frey. Nondevelopment of resistance by bacteria during hospital use of povidone-iodine. Dermatology 195 (suppl 12):10–13, 1997.
64. CM Stahl-Bayliss, RP Grandy, RD Fitzmartin, C Chelle, B Oshlack, P Goldenheim. The comparative efficacy and safety of 5% povidone-iodine cream for topical antisepsis. Ostomy Wound Manage 31:40–49, 1990.
65. L Weinstein. Gram-negative bacterial infections: a look at the past, a view of the present, and a glance at the future. Rev Infect Dis 7 (suppl 14):S538–544, 1985.
66. AS Colsky, RS Kirsner, FA Kerdel. Analysis of antibiotic susceptibilities of skin wound flora in hospitalized dermatology patients. The crisis of antibiotic resistance has come to the surface. Arch Dermatol 34:1006–1009, 1998.
67. P Teng, V Falanga, FA Kerdel. The microbiological evaluation of leg ulcers and infected dermatoses in patients requiring hospitalization. Wounds 5:133–136, 1993.
68. D Michel, GA Zach. Antiseptic efficacy of disinfecting solutions in suspension test in vitro against methicillin-resistant *Staphylococcus aureus*, *Pseudomonas aeruginosa* and *Escherichia coli* in pressure sore wounds after spinal cord injury. Dermatology 195 (suppl 12):36–41, 1997.
69. CE Haley, M Marling-Cason, JW Smith, JP Luby, PA Mackowiak. Bactericidal activity of antiseptics against methicillin-resistant *Staphylococcus aureus*. J Clin Microbiol 21:991–992, 1985.
70. T Kunisada, K Yamada, S Oda, P Hara. Investigation on the efficacy of povidone-iodine against antiseptic-resistant species. Dermatology 195 (suppl 12):14–18, 1997.
71. RW Lacy. Antibacterial activity of povidone iodine towards non-sporing bacteria. J Appl Bacteriol 46:443–449, 1979.
72. JL Zamora, MP Prince, P Chuang, LO Gentry. Inhibition of povidone-iodine's bactericidal activity by common organic substances: an experimental study. Surgery 98: 25–29, 1985.
73. W Sheikh. Comparative antibacterial efficacy of Hibiclens7 and Betadine7 in the presence of pus derived from human wounds. Cuff Ther Res 40:1096–1102, 1986.
74. L Gershenfeld. Povidone-iodine as a topical antiseptic. Am J Surg 94:938–939, 1957.
75. H Vorheff, UP Vorheff, P Mehta, JA Ulrich, RH Messer. Antimicrobial effect of chlorhexidine and povidone-iodine on vaginal bacteria. J Infect 8:195–199, 1984.
76. J Custer, RF Edlich, M Prusak, J Madden, P Panek, OH Wangensteen. Studies in

the management of the contaminated wound. V. An assessment of the effectiveness of phisophex and betadine surgical scrub solutions. A J Surg 121:572–575, 1971.
77. PJ Van den Broek, LFM Buys, RV Furth. Interaction of povidone-iodine compounds, phagocytic cells, and microorganisms. Antimicrob Agents Chemother 22: 593–597, 1982.
78. AR Johnson, AC White, B McAnalley. Comparison of common topical agents for wound treatment: cytotoxicity for human fibroblasts in culture. Wounds 1:186–192, 1989.
79. O Damour, SZ Hua, F Lasne, M Villain, P Rouselle, C Collombel. Cytotoxicity evaluation of antiseptics and antibiotics of cultured human fibroblasts and keratinocytes. Burns 18:479–485, 1992.
80. W Lineaweaver, R Howard, D Soucy, S McMorris, J Freeman, C Crain, J Robertson, T Rumley. Topical antimicrobial toxicity. Arch Surg 120:267–270, 1985.
81. RI Burks. Povidone-iodine solution in wound treatment. Phys Ther 78:212–218, 1998.
82. N Georgiade, A Eiring, R Georgiade, F Richard, K Pickrell. The effects of various skin sterilization techniques on the viability of skin. Plast Reconstr Surg 21:479–482, 1958.
83. D Dennis, A Luterman, ML Ramenofsky, JJ Lefante, PW Curreri. Does PVP-iodine interfere with wound healing? Infect Surg 2:371–374, 1983.
84. RP Gruber, L Vistnes, R Pardoe. The effect of commonly used antiseptics on wound healing. Plast Reconstr Surg 55:472–476, 1975.
85. By Lee, FS Trainor, WR Thoden. Topical application of povidone-iodine in the management of decubitus and stasis ulcers. J Am Geriatr Soc 27:302–306, 1979.
86. J Michael. Topical use of PVP-I (Betadine) preparations in patients with spinal cord injury. Drugs Exp Clin Res XI:107–109, 1985.
87. D Mlangeni, F Daschner. Povidone-iodine. Evaluation of povidone-iodine as an antiseptic. Antiinfect Drugs Chemother 13:161–167, 1995.
88. R Niedner. Cytotoxicity and sensitization of povidone-iodine and other frequently used anti-infective agents. Dermatology 195 (suppl 2):89–92, 1997.
89. M Bradley, N Cullum, T Sheldon. The debridement of chronic wounds: a systematic review. Health Technology Assessment 3 (no. 17 , part 1), 1999.
90. FDA Consumer Magazine. OTC Options: Help for Cuts, Scrapes, Burns. Department of Health and Human Services, Rockville, Maryland, 1996.
91. S Ito. Drug therapy: drug therapy for breast-feeding women. N Engl J Med 343: 118–126, 2000.
92. P Smerdely, A Lim, SC Boyages, K Waite, D Wu, V Roberts, G Leslie, J Arnold, E John, CJ Eastman. Topical iodine-containing antiseptics and neonatal hypothyroidism in very-low-birth weight infants. Lancet 2(8664):661–664, 1989.
93. M Ryan, ZA Al-Sammak, D Phelan. Povidone-iodine mediastinal irrigation: a cause of acute renal failure. J Cardiothor Vasc Anesth 13:729–731, 1999.
94. KJ Lavelle, DI Doedens, SA Kleit, RE Forney. Iodine absorption in burn patients treated topically with povidine-iodine. Clin Pharmacol Ther 17:355–362, 1975.
95. C Scoggin, IR McClellan, IM Cary. Hypernatraemia and acidosis in association with topical treatment of burns. Lancet 1(8018):959, 1977.

5

Quaternary Ammonium Compounds

Edward B. Walker
Weber State University, Ogden, Utah

I. INTRODUCTION

Quaternary ammonium chlorides (QACs) occupy a unique niche in the world of antimicrobial compounds. Rather than being a single, well-defined substance, as is the case for many such active ingredients, QACs are composed of a diverse, eclectic collection of substances that share a common chemical motif, namely a molecular structure containing a positively charged nitrogen atom covalently bonded to four carbon atoms. This quaternary ammonium group is responsible for the name of these antimicrobial compounds and also plays a dominant role in determining their chemical behavior.

The first reports of quaternary ammonium compounds with biocidal activity appeared in 1916 [1]. Since that time, QACs have grown in popularity and been utilized extensively as active ingredients in many types of products, including household cleaners, institutional disinfectants, skin and hair care formulations, sanitizers, sterilizing solutions for medical instruments, preservatives in eye drops and nasal sprays, mouthwashes, and even in paper processing and wood preservatives. As a group, QACs are effective across a broad spectrum of microorganisms, including bacteria, certain molds and fungi, and viruses. However, the specific activity of QACs is as diverse as their range of chemical structures. QAC antimicrobial effectiveness is highly formulation and packaging dependent, since many ingredients affect QAC activity. Some components reduce the QAC efficiency, while others may synergize their activity or expand the spectrum of affected microorganisms. This fact has led to some confusion and apparent contradiction in the published literature as to the actual effectiveness of QACs in their role as antimicrobials.

In addition to their antimicrobial activity, QACs also behave as surfactants, assisting with foam development and cleansing action. They are also attracted to the skin and hair, where small amounts remain bound after rinsing. This contributes to a soft, powdery afterfeel on skin, unique hair-conditioning effects, and long-lasting, persistent activity against microorganisms. These various attributes and multifunctional roles of QACs appeal to formulators and are responsible for their incorporation into many consumer products.

The purpose of this chapter is to introduce the reader to the variety of QACs available for use in topical antimicrobials and examine their efficacy in this role. The intent is to help the reader gain an appreciation for the usefulness of QACs and to introduce some of the unique attributes that should be considered for their successful incorporation into skin care products.

II. CHEMICAL DIVERSITY OF QACS

It should be noted that quaternary amines represent a very large and diverse group of chemical compounds, most of which are not considered in this chapter. Quaternary nitrogen atoms are found in the chemical structures of many naturally occurring substances in living cells, such as choline and others. These substances are generally not toxic to microorganisms. The synthetic antimicrobial QACs discussed here represent a relatively small subgroup of this larger class of chemical substances and do not occur naturally.

The first antimicrobial quaternary amines synthesized were rather simple, with three methyl groups and one linear, saturated alkyl group attached to a single, positively charged nitrogen atom (alkyltrimethylammonium chloride) (Fig. 1). The nonpolar ''tail'' and polar ''head'' regions make them similar to soap molecules. However, since soap molecules contain anionic head groups, these cationic molecules were dubbed ''invert soaps.'' Since that time, chloride is the anion most often used to balance the nitrogen's ionic charge; however, a plethora of other counter anions are available today, including bromide, iodide, acetate, methosulfate, ethosulfate, lactate, and benzoate, to name a few [2]. The most common alkyl chain lengths in commercially available raw materials vary from C8 to C22, with an even number of carbon atoms, due to the natural sources from whence they are synthesized.

In 1935, almost 20 years after the introduction of these relatively simple aliphatic QACs, Dormagk reported that the substitution of a benzyl group on the quaternary nitrogen increased antibacterial activity [3]. This second type of QAC, namely alkyldimethylbenzylammonium chlorides (ADBACs), were the first class of QAC to find acceptance and utilization as disinfectants in medical applications.

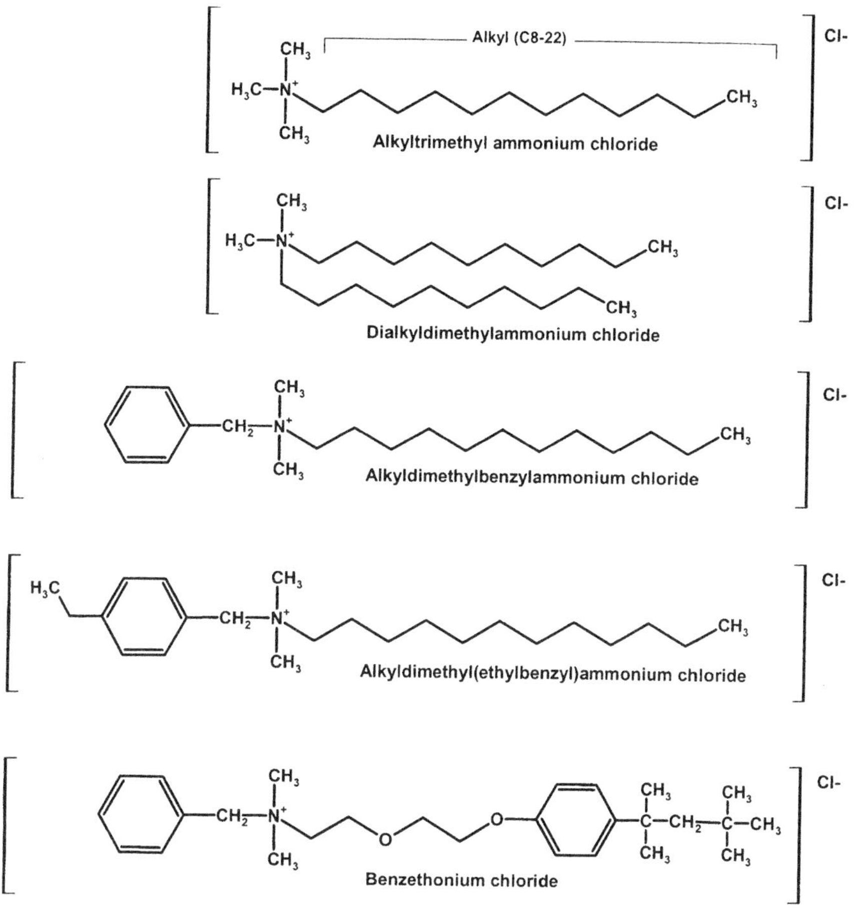

Figure 1 Generalized chemical structures of quaternary ammonium chloride compounds with antimicrobial activity.

Soon thereafter, chemists added an ethyl group to the aromatic ring, further increasing detergency action and antimicrobial activity.

A third class of QAC emerged 30 years later in 1965, when new synthetic techniques were developed to allow the economically feasible formation of twin chain or dialkyldimethylquaternary ammonium chlorides (DDACs) [4]. These QACs exhibit even higher biocidal efficacy than their predecessors.

In 1938, a creative departure from the relatively simple alkyl groups substituted on the quaternary nitrogen resulted in an entirely different quaternary amine, benzethonium chloride (BEC) [5]. The distinguishing feature of this rather complex group is best described by BEC's IUPAC name, *N,N*-dimethyl-*N*-[2-[2-[4-(1,1,3,3-tetramethylbutyl)phenoxy]ethoxy]ethyl]benzene-methanaminium chloride [6]. As seen in Figure 1, this complex nitrogen substituent contains branched hydrocarbons, phenoxy, and ethoxy components. Ethoxylation increases detergency, and the complex nature of this group enhances the spectrum of activity when compared to more traditional QACs. The unique chemical character of BEC establishes it as a very different fourth class of QACs.

The chemical literature continues to describe new QACs with ever-increasing diversity [7]. However, most of these have not yet found broad acceptance or applications in topical antimicrobials and therefore will not be discussed here.

III. CHEMICAL PROPERTIES OF QACS

QACs are actually ammonium salts. Like acid salts, these quaternary ammonium salts are colorless, odorless, crystalline solids that have high melting points. Historically, QACs are synthesized from a reaction between tertiary amines and alkyl halides, resulting in an ammonium salt of the corresponding halide (normally chloride). Since the four organic substituents are covalently bound to the nitrogen atom, they do not readily dissociate in acid-base reactions as do simple amines. Hence, QACs do not lose their positive ionic charge at elevated pH. Quaternary amine functional groups are also recognized for their lack of reactivity and are extremely stable at the relatively moderate temperatures and chemical environments associated with the manufacture of topical skin care products.

As seen in Figure 1, QACs have polar "heads" and nonpolar "tails," a general characteristic of all detergent molecules. The degree of solubility in various solvents and detergency of QACs are both influenced by the nature of the nonpolar tails. The shorter chains cause higher foaming and decrease solubility in nonpolar solvents. Longer chains increase solubility in nonpolar solvents and act as excellent emulsification agents. All QACs show excellent solubility in water, alcohol, and acetone. Solubility in more nonpolar liquids such oils and esters depends on the nature of the QAC. This variable solubility becomes more important as QACs are incorporated into emulsions, where the size of their hydrophobic substituents determines their partition coefficient and hence their distribution between the water and oil phases. Their effects on surface tension are also dependent upon their substituents, which can play an important role at the oil water emulsion interface.

QACs' normally high solubility in water can be dramatically decreased by anionic substances. For example, QACs are not compatible with traditional anionic fatty acids found in soaps. The ionic attraction between these positively and negatively charged groups causes precipitation of the two substances. Similarly, divalent and trivalent anionic salts and negatively charged polymers such as polyacrylates form strong salt linkages with multiple QAC molecules, causing precipitation or at the very least altering the concentration of total available, unbound QAC. QACs also adhere to a variety of surfaces. This behavior is manifested in their residual, persistent antimicrobial activity following application to the skin or to various inanimate surfaces. Their affinity for protein, negative charges, and nonpolar surfaces has been suggested by many as a basis for their purported mechanisms of antimicrobial action. However, from a consumer standpoint, more noticeable results of these surface phenomena include silky, conditioned hair with reduced flyaway (static) and skin that feels soft and powdery following the application of QACs.

IV. BENZALKONIUM CHLORIDE

Benzalkonium chloride, (BAK) is the most well known of all the QACs. It is actually a mixture of alkybenzyldimethylammonium chlorides, differing only in the length of their alkyl groups. While chain length of the various analogs includes C_8 to C_{18}, the actual distribution is clearly defined: not less than 40% C_{12} and 20% C_{14}, with the sum of these two not to be less than 70% of the total [7]. Benzalkonium chloride is most often commercially available dissolved in water or alcohol or mixtures thereof at either 50% or 80% active concentrations.

V. BENZETHONIUM CHLORIDE

Benzethonium chloride (BEC) is supplied as a white crystalline, odorless material. Early in BEC's history, a methylated derivative, methylbenzethonium chloride, was also available, but it has all but disappeared from use in the last two decades. Both compounds are soluble in polar solvents such as water, low molecular weight alcohols, and glycols and also in nonpolar solvents such as carbon tetrachloride, tetrachloroethane, and benzene [8]. Aqueous solutions of benzethonium chloride are stable within a pH range of 4–7 [9–11]. BEC offers better detergency action than BAK and exhibits different antimicrobial activity.

VI. SAFETY OF QACS

Assessment of QAC safety has been the subject of numerous studies. Almost without exception, QACs have been determined to be safe for topical skin care products. They rarely sensitize the skin or cause allergic reactions [12]. The Cosmetic, Toiletries, and Fragrance Association (CTFA) has published safety reviews on both benzalkonium and benzethonium chlorides [8,13]. In 1989 their safety panel concluded that benzalkonium chloride was safe and did not sensitize normal human skin at concentrations up to 0.1%. They also reported that it can be safely used as an antimicrobial agent at concentrations up to 0.1%. Two years later, the FDA published a tentative final monograph (21CFR333) recognizing benzalkonium chloride as safe and effective at levels of 0.10–0.13% in first aid antiseptics [14]. Benzalkonium chloride is not absorbed through the skin or mucosal linings. Cosmetic products containing benzalkonium chloride may be applied to the skin, hair, and vaginal mucosa and may come in contact with the nasal mucosa and eyes. Products containing benzalkonium chloride may be applied several times daily over a period of several years [13].

The CTFA Final Report on the Safety Assessment of benzethonium chloride, published in 1985, concludes that this QAC is safe at levels of 0.5% in cosmetic products applied to the skin [8]. It was noted that benzethonium chloride produced mild skin irritation at 5% but not at lower concentrations. The report also limits the maximum concentration of 0.02% of this ingredient for cosmetics used in the eye area. The FDA's tentative final monograph (21CFR333) for first aid antiseptics specifies that benzethonium chloride's safe and effective concentrations are 0.1–0.2% for these products [14]. Aqueous and alcohol solutions of benzethonium chloride are not absorbed through the skin [15]. Oral LD_{50} values of 368–654 mg/kg in rats have been reported [16]. Benzethonium chloride has also been shown to be nonmutagenic in at least two different types of tests [17,18].

VII. CHEMICAL STRUCTURE AND MECHANISM OF QAC ANTIMICROBIAL ACTIVITY

Antimicrobial activity of QACs extends across a broad spectrum of microbes, including bacteria, yeasts, molds, and viruses. This varies between organisms and efficacy for each organism also depends upon the chemical nature of each QAC.

Many investigators have offered explanations for QAC mechanisms of antimicrobial action. Most theories center on the interactions between QACs and the membranes of their target cells. The surfactant nature of these agents suggests that they act as other membrane-active agents, perturbing homeostasis. However, the manner in which they interact with the membrane itself is not always clear. Disruption of membrane integrity has been observed by monitoring the release of intercellular materials following treatment by QACs [19,20]. Following exposure to one QAC, gram-positive cells released cellular components more rapidly than gram-negative bacteria. Potassium ions were released first, followed by phosphates, then by larger molecular weight molecules.

The exact target of attack at the membrane surface is still contested. The negatively charged phospholipids offer a natural target, due to electrostatic attraction. However, the detergent action of QACs works extremely well on nonpolar oils and lipids as well as anionic lipids. QACs also have a strong affinity for proteins. Subsequent denaturation of proteins or inhibition of important transport proteins may be a primary or complementary mechanism of biocidal activity [8].

The activity of QACs against viruses may depend on both the affinity of the QAC for the proteinaceous coat and the lipid nature of the viral envelope. QACs are quite active against lipophilic viruses such as HIV, influenza, and herpes simplex 1 & 2. However, their activity is lower against nonlipophilic viruses such as adenovirus or parvovirus [21].

In one study, the absorption of three QACs onto the surface of *Candida albicans* blastospores at room temperature revealed that binding required from 30 seconds to 5 minutes, depending on the structure of the QAC. The Langmuirian adsorption of all three agents resulted in a concentration-dependent formation of drug-monolayers on the surface of the blastospore. Dissimilarities in total QAC binding and binding kinetics between the different QACs were attributed to differences in the orientations of both the cationic nitrogen atom and the accompanying lipophilic portions of each QAC at the blastospore surface [22]. This selective interaction may eventually help understand the differential fungicidal activities of commercial QAC-based disinfectants [23].

More evidence for the important role of the lipophilic portions of QACs in determining biocidal activity is that the alkyl chain length of can dramatically affect activity. The killing of *C. albicans* by a series of QACs with different hydrocarbon chain lengths was reported to be closely related to the binding of the compounds to the cells and damage of the cell membranes [24]. In addition, binding to the cell surface was related to the critical micelle concentration for each QAC. The investigators in this study proposed that interfacial micelle-like aggregates form at the cell surface as a step in the binding process. The investigators also showed fundamental differences between binding to the yeast and binding to gram-negative bacteria.

Further experiments with bacteria reveal the dependence of activity on chain length of alkyldimethylbenzylammonium chlorides [25]. Table 1 summarizes these results, (Fig. 2) showing that as chain length increases from C_8 to C_{14}, activity increases. However, as chain length increases beyond this point, activity decreases.

Table 1 Effect of Chain Length of Alkyldimethylbenzylammonium Chlorides on Minimal Inhibitory Concentration

Alkyl chain length (carbon #)	Minimal inhibitory concentration (ppm) *Staphylococcus aureus*	*Salmonella typhosa*	*Pseudomonas aeruginosa*
8	3000	4500	6000
9	800	1400	2500
10	450	300	1200
11	160	130	400
12	45	40	120
13	25	20	50
14	15	12	40
15	25	20	70
16	30	25	200
17	170	15	360
18	450	60	1000
19	330	90	1300

QACs dissolved in deionized water; 10-minute contact time.
Source: Ref. 25.

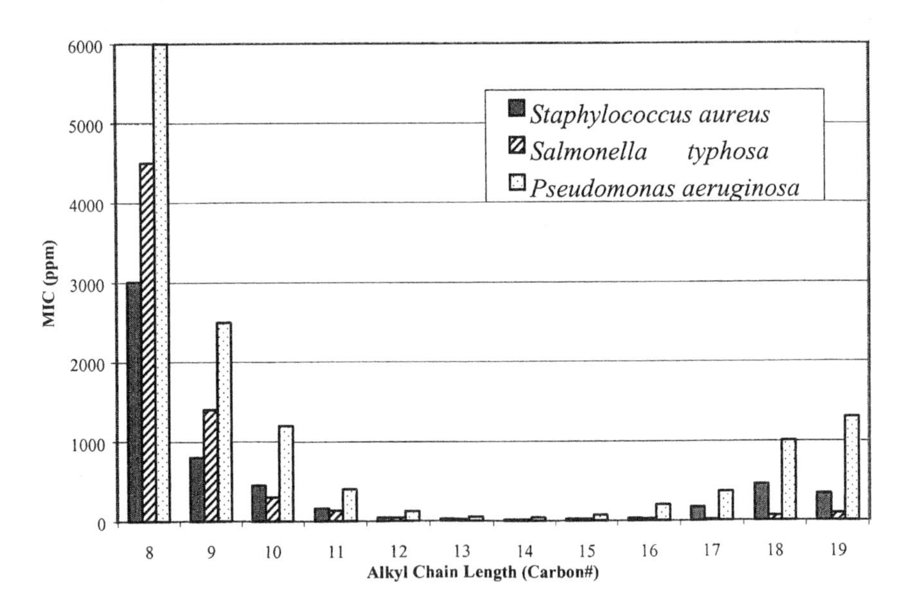

Figure 2 Influenza of alkyl chain length on MIC: graphic representation of data from Table 1.

A relatively new category of amphiphilic hydrolyzable QACs, alkanoylcholines with hydrocarbon chains of 10–14 carbon atoms, were also comparable to the activities of more traditional, nonhydrolyzable QACs of corresponding chain length and increased with an increasing number of carbon atoms [26].

Temperature also affects the activity of QACs against *Pseudomonas aeruginosa* and *Escherichia coli*. Tests of 18 proprietary disinfectant products used in the food industry showed that their activity decreased as the temperature was decreased from 20° to 10°C [27]. This may be related to temperature-induced changes in QAC critical micelle concentration. Alternatively, changes in microbial membrane fluidity may be responsible for the observed changes. More such data with well-defined QAC composition and extended temperature ranges are needed to understand the correlation between temperature and antimicrobial activity of QACs.

VIII. EFFECTS OF FORMULA COMPOSITION AND QAC ACTIVITY

One of the most challenging aspects of incorporating QACs into antimicrobial products is avoiding activity-depleting effects of the matrix. Excipient ingredients such as surfactants, emulsifiers, oils, preservatives, and even impurities in raw materials can exert negative effects on activity. Alternatively, if formulated correctly, synergistic combinations with other ingredients can enhance activity. Comparisons between QAC activity in deionized water and formulas involving added ingredients must be interpreted cautiously. Of course, it also demands that final product formulations be extensively tested to determine if QAC activity has been adversely affected.

For example, anionic surfactants will bind to cationic QACs via electrostatic attractions, inactivating or even precipitating the QACs. Divalent inorganic anions such as polyprotic acids or even sulfates will attract QACs, often forming insoluble complexes. Bactericidal activity of QACs often increases at higher pH values [28]. Some surfactants and emollients can capture QACs when present above their critical micelle concentration. For example, lecithin and polysorbates are well-known neutralizers of QACs (as well as parabens and phenols). These two ingredients are listed as QAC neutralizers of choice for the required testing method in the FDA's Topical Antimicrobial Drug Product Tentative Final Monograph [14]. They should obviously be avoided in formulating QAC-based antimicrobial products. High levels of a 12-mole nonylphenol ethoxylate negatively affect the microbiological efficacy of certain QACs, such as dialkyldimethylammonium chlorides [21]. EDTA and other metal-sequestering agents enhance QAC activity, as hard water can adversely affect activity of many QACs. An excellent review of 49 combinations of various active ingredients with preservatives and excipients reveals that some ingredients enhance activity while some destroy QAC action [29].

IX. SPECTRUM OF QAC ACTIVITY

QACs are efficacious, broad-spectrum antimicrobials used for a wide variety of applications. In general, QACs exhibit excellent activity against bacteria, with higher efficacy against gram-positive than gram-negative organisms [19]. It should be remembered that such relative activity measurement is most often linked to minimal inhibitory concentrations (MICs). Since commercially available products usually have QAC concentrations approximately 10–100 times the MIC, these types of comparisons are often misleading and should be considered carefully when comparing the activity of various QACs and their respective products.

Vancomycin-resistant enterococci (VRE) often contaminate the hospital environment. QACs were found to be highly effective against eight strains of VRE, using a quantitative suspension test method [30]. They are also effective against *Listeria* spp. [31].

Disinfectant solutions containing QACs have been used for decades as hard-surface cleaners. This is particularly interesting since microorganisms in suspension are easier to kill than microbes that are dried on the surface. Russell reported that it requires as much as 50 times more disinfectant to kill dried bacteria than bacteria in a liquid suspension [32]. As hard-surface cleaners, QACs are known for their activity against a host of viruses. The U.S. Environmental Protection Agency (EPA) allows for the validation of antiviral activity and for the label claim describing such. However, the U.S. Food and Drug Administration (FDA) does not currently have any published monographs providing for antiviral testing or any associated claims for topical antimicrobials. This causes some confusion among consumers who do not understand the labeling and associated claims allowed by two different regulatory agencies. The key to proper label claims in the United States is the site where the product is used: on hard, inanimate surfaces antiviral claims are allowed; on skin, viral inactivation claims are not allowed.

This regulatory differentiation does not exist in third-world countries, especially where viral infections such as HIV represent a huge percentage of infections among their populations. Appropriate, nonmisleading antiviral label claims for topical antimicrobial products are allowed in many countries outside the United States. QACs are active against many viruses, such as HIV. During 1 minute of exposure, HIV is inactivated by QACs [33]. QACs and other active ingredients were assessed against cell-free HIV in culture medium and cell-associated HIV suspended in medium or whole human blood. QACs completely inactivated cell-free HIV following a 1-minute exposure. However, cell-associated HIV was more resilient, requiring exposure of 5 minutes or more for some disinfectants. Other viruses are also inactivated by QACs, including various animal viruses [34].

If the target microbes are covered or encapsulated under a protective layer of organic soil, such as serum, blood, food residues, or fecal matter, they are more difficult to kill. If the organic soil adsorbs the QAC, it reduces the amount available for attacking the microorganism(s). Test procedures have been described to evaluate activity in the presence of organic soil loads [35].

Occasionally, the clean surface itself may impede the action of active ingredients. QACs may be adsorbed and thus inactivated by various fabrics, sponges, and plastics [8]. One surface that could raise a serious concern is latex, used in the manufacture of gloves worn in laboratories and hospitals. It has been shown that QACs exhibit fast-acting antibacterial activity on surgical glove latex material and on human skin [36].

X. EXAMPLE OF BROAD-SPECTRUM ANTIMICROBIAL ACTIVITY FOR A BEC TOPICAL PRODUCT

Such a review of QACs in antimicrobial products would be incomplete without examination of a representative commercially available product. As an example of one such product, antimicrobial test data for Fresh Cleanse Lotion-Handwash [37] is presented below. These antibacterial tests are required by the FDA to support the claims for products labeled in the United States as either ''First Aid Antiseptic'' or ''Healthcare Professional Handwash'' (HCPHW) [14,38]. This lotion-handwash is a QAC-based formulation containing benzethonium chloride as the active ingredient at a concentration of 0.2%. Since the formulation was designed to function as either a leave-on antiseptic or as a rinse-off handwash, testing for both label claims was required. The resulting data represents an excellent study of how a specific QAC such as benzethonium chloride may be formulated into an oil-in-water emulsion with numerous excipients so as to not impede its activity and to illuminate the excellent antimicrobial activities available from these versatile active ingredients. (Testing was conducted at BioScience Laboratories, Inc.)

In vitro time-kill studies against 50 strains of bacteria (including both laboratory and clinical isolates) are required for certification of products labeled as ''Health Care Personnel Handwash.'' (Representative data are presented in Table 2.) Only single exposure times (usually 15 seconds) are shown due to space constraints. For compliance, at lease a 3-log reduction in bacteria is required, but, it may be seen that in many cases that 5,6, or even 7 log reductions are observed for this short exposure time. This broad spectrum of activity for benzethonium chloride amply demonstrates QAC utility and efficacy in topical antimicrobial products.

In addition to the in vitro testing of healthcare personnel handwash prod-

Table 2 Time Kill Data from In Vitro Testing of 0.2% Benzethonium Chloride Handwash Product

		In vitro time-kill			In vitro MIC,
Microorganism species	ATCC#	Exposure time (sec)	Log_{10}	Percent reduction	product dilution
Acinetobacter baumannii	19606	15	6.73	99.9990	1:256
Acinetobacter baumannii	CI	15	6.52	99.9993	1:256
Bacteroides fragilis	25285	15	7.31	99.9986	1:512
Bacteroides fragilis	CI	15	6.50	99.9980	1:512
Bacillus subtilis (vegetative cells)	6633	15	4.55	99.9972	
Campylobacter jejuni	29428	15	7.63	99.9999	
Candida albicans	10231	15	6.33	99.9889	1:256
Candida albicans	CI	15	6.56	99.9806	1:128
Candida tropicalis	750	15	6.25	99.9992	1:256
Candida tropicalis	CI	15	6.16	99.9590	1:1.024
Clostridium sporogenes	7955	30	3.57	99.9737	
Enterobacter cloacae	13047	15	6.13	99.9994	1:128
Enterobacter cloacae	CI	15	6.09	99.9993	1:128
Enterococcus faecalis	29212	15	6.60	99.9995	1:1.024
Enterococcus faecalis (VRE)	CI	15	6.74	99.9997	1:768
Enterococcus faecium	19434	15	6.46	99.9995	1:512
Enterococcus faecium (VRE)	CI	15	6.36	99.9981	1:512
Escherichia coli	11229	15	6.32	99.9999	1:128
Escherichia coli	CI	15	5.76	99.9992	1:128
Escherichia coli	25922	15	6.06	99.9996	1:128
Escherichia coli (0157:H7)	CI	15	6.08	99.9992	1:128
Escherichia coli (0157:H7)	35150	15	5.90	99.9999	
Haemophilus influenzae	19418	15	4.77	99.9981	1:2.048
Haemophilus influenzae	CI	15	5.18	99.9600	1:1.024
Klebsiella oxytoca	43165	15	5.84	99.9996	1:128
Klebsiella oxytoca	CI	15	6.12	99.9990	1:128
Klebsiella pneumoniae	11296	15	5.61	99.9992	1:512

Kiebsiella pneumoniae	CI	15	5.79	99.9990	1:64
Listeria monocytogenes	7644	15	6.84	99.9999	
Micrococcus luteus	7468	15	5.81	99.9672	1:3.072
Micrococcus species	CI	15	5.97	99.8764	1:4.096
Proteus mirabilis	7002	15	7.18	99.9997	1:64
Proteus mirabilis	CI	15	5.99	99.9997	1:64
Pseudomonas aeruginosa	15442	15	6.49	99.9993	1:64
Pseudomonas aeruginosa	CI	15	6.36	99.9990	1:32
Pseudomonas aeruginosa	27853	15	6.20	99.9983	1:64
Pseudomonas aeruginosa (MDR)	CI	15	6.08	99.9986	1:32
Salmonella typhi	6539	15	6.12	99.9999	
Serratia marcescens	14756	15	6.10	99.9997	1:64
Serratia marcescens	CI	15	6.17	99.9900	1:128
Shigella dysenteriae	13313	15	7.17	99.9999	
Staphylococcus aureus	6538	15	3.60	99.9999	1:4.096
Staphylococcus aureus	CI	15	4.77	99.9959	1:4.096
Staphylococcus aureus	13565	15	6.02	99.9999	
Staphylococcus aureus	29213	15	4.34	99.8650	1:2.048
Staphylococcus aureus MRSA	CI	15	3.35	99.9993	1:2.048
Staphylococcus epidermidis	12228	15	5.80	99.9997	1:4.096
Staphylococcus epidermidis	CI	15	5.98	99.9994	1:5.120
Staphylococcus haemolyticus	29970	15	5.53	99.9991	1:4.096
Staphylococcus haemolyticus	CI	15	5.61	99.9995	1:512
Staphylococcus hominis	27844	15	5.23	99.9767	1:6.144
Staphylococcus hominis	CI	15	5.64	99.9941	1:512
Staphylococcus saprophyticus	15305	15	5.66	99.9997	1:2.048
Staphylococcus saprophyticus	CI	15	5.60	99.9986	1:2.048
Streptococcus pneumoniae	6303	15	3.13	99.9999	1:2.048
Streptococcus pneumoniae	CI	15	5.57	99.9999	1:2.048
Streptococcus pyogenes	19615	15	6.75	99.9828	1:40.960
Streptococcus pyogenes	CI	15	4.71	99.9998	1:40.960
Vibrio vulnificus	27562	15	4.92	99.9988	
Yersinia enterocolitica	9610	15	6.32	99.9999	

Table 3 Healthcare Personnel Handwash Glove Juice Test of 0.2% Benzethonium Chloride Product with *Serratia marcescens* (ATCC#14756)

Sample	Sample size	Mean	Standard deviation	95.0% Confidence interval	Log reduction from baseline	Percent reduction from baseline
Baseline	30	9.04	0.20	8.97–9.12	N/A	N/A
Wash 1	30	6.64	0.39	6.50–6.79	2.40	99.60
Wash 3	30	6.34	0.46	6.16–6.51	2.70	99.80
Wash 7	30	6.02	0.44	5.85–6.19	3.35	99.96
Wash 10	30	6.01	0.38	5.87–6.16	3.27	99.95

Table 4 In vitro Inactivation of Viruses by 0.2% Benzethonium Chloride Handwash Product

Virus	Type	Exposure time (sec)	Percent reduction
Herpes simplex	Type 1	15	99.70
		30	99.98
		60	99.80
Human immunodeficiency virus (HIV)	Type 1	15	99.90
		30	99.94
		60	99.94
Influenza virus	Type A2	15	99.94
		30	99.94
		60	99.98

Table 5 Transepidermal Moisture Gain/(Loss) Comparative Test

Percentages	Day 1	Day 2	Day 3	Day 4	Day 5
0.2 Benzethonium chloride	4.4	2.4	7.2	5.0	6.7
7.5 Iodine	(9.9)	1.1	3.4	1.5	0.8
0.75 CHG	(1.4)	(6.4)	(0.1)	(1.9)	(1.1)
1 PCMX	(1.9)	(1.9)	(3.3)	(3.3)	(2.6)
4 CHG	(4.1)	(6.2)	(4.7)	(6.1)	(3.6)
0.3 Triclosan	(2.6)	(17.9)	(19.9)	(16.9)	(21.1)

100 handwashes/day conducted over a 5-day period.
Positive numbers indicate moisture content increase from baseline.
Negative numbers (shown in parentheses) indicate moisture content decreases from baseline.

Table 6 Skin Irritation Comparative Test of Fresh Cleanse Lotion-Handwash

0.2% Benzethonium Chloride	0.75% CHG	0.3% Triclosan	1% PCMX	4% CHG	4% CHG
0.0	1.0	1.0	1.0	2.0	3.5

100 handwashes/day conducted over a 5-day period.
Test conducted at conclusion of the 5-day study.
Baseline (''perfect skin''), shown as 0.0. The higher the number, the higher the irritation.

ucts, in vivo glove juice tests are also required. Following inoculation of the hands with 5.0 mL of *Serratia marcescens* (ATCC#14756) with a concentration of 10^5 CFU/mL, the number of bacteria must drop by 3 logs within 10 washings to pass this test. This performance criterion was successfully met before the seventh handwash cycle see Table 3.

Also of interest is the activity of this formulation against viruses. In vitro inactivation of selected viruses demonstrates that QACs, especially benzethonium chloride, may retain their virus-inactivating activity even in complicated matrices (Table 4).

Often the perception of antibacterial handwashes used in a professional setting is that they are harsh and irritating to the skin (often a well-earned reputation). Professional health care givers are required to wash their hands as many as 100 times daily! If a product that is highly efficacious is also irritating to the skin, it is likely that high-frequency washing compliance will suffer. Therefore, it is important that such products be gentle, moisturizing, and not irritate the skin. Tables 5 and 6 show data that compare these cosmetic aspects of Fresh Cleanse Lotion-Handwash to randomly selected HCPHW products with non-QAC active ingredients. While many such cosmetic aspects are formulation dependent, the data do show that QAC products can be formulated with many desirable skin-care attributes beyond antibacterial activity.

XI. CONCLUSIONS

Quaternary ammonium compounds represent a class of highly effective, versatile substances for use as active ingredients in topical antimicrobial products. They exhibit a broad spectrum of activity against bacteria, fungi, mold, and viruses. The extent of their activity depends on the nature of the quaternary ammonium compound, especially with respect to their alkyl substituents. Special care must be exercised when formulating with this class of antimicrobials, since their activity is affected by many of the ingredients found in skin-care products. Their added attributes of foaming, emulsification, stability, and gentleness to the skin are helpful to formulation chemists and offer many possibilities for incorporation into topical antimicrobial products.

REFERENCES

1. WA Jacobs. The bactericidal properties of the quaternary salts of hexamethyltetramine. I. The problem of the chemotherapy of experimental bacterial infections. J Exp Med 23:563–568, 1916.
2. In: The International Cosmetic Ingredient Dictionary and Handbook. 8th ed. Washington, DC: Cosmetic, Toiletry, and Fragrance Association, 2000.

3. G Dormagk. Eine neue Klasse von Disinfektionsmitteln. Dtsch Med Wochschr 61: 828–832, 1935.
4. RD Ditoro. New generation of biologically active quaternaries. Soap Chem Specialties: 47–52, 86–88, 91–92, 1969.
5. U.S. Patent 2,115,250 (1938).
6. S Budavari, ed. The Merck Index. An Encyclopedia of Chemicals, Drugs, and Biologicals. 11th ed. Rahway, NJ: Merck and Co., Inc., 1989.
7. USP 24/NF19, p. 2419.
8. Final report on the safety assessment of benzethonium chloride and methylbenzethonium chloride. J Am Coll Toxicol 4(5): 65–106, 1985.
9. DA Joslyn, K Yaw, AL Rawlins. Germicidal efficacy of phemerol. J Am Pharm Assoc Sci Ed 32:49–51, 1943.
10. GG Hawley. Condensed Chemical Dictionary. 8th ed. New York: Van Nostrand Reinhold, 1971.
11. M Windholz, ed. The Merck Index: An Encyclopedia of Chemicals, Drugs, and Biologicals. 9th ed. Rahway, NJ: Merck and Co., Inc., 1976.
12. A Ancona, A Arevalo, E Macotela. Contact dermatitis in hospital patients. Dermatol Clin 8(1):95–105, 1990.
13. Final report on the safety assessment of benzalkonium chloride. J Am Coll Toxic 8(4):589–625, 1989.
14. Topical antimicrobial drug products for over-the-counter human use; tentative final monograph for first aid antiseptic drug products. Fed Reg 56(140): 33664–33680, 1991.
15. MW Counts. Lonza Corporation (personal communication).
16. HC Hodge, H Sterner. Tabulation of toxicity classes. Am Indust Hyg Assoc Quart 10:93–96, 1949.
17. S DeFlora. Study of 106 Organic and Inorganic Compounds in the Salmonella/microsome test. Carcinogenesis (London) 2(4):283–298, 1981.
18. NTP. Technical Bulletin. Dept. of Health and human Services, Public Health Service, 9, April 1983.
19. AD Russell, I Chopra. Understanding Antibacterial Action and Resistance, Chichester, West Sussex, England: Ellis Horwood Limited, 1990.
20. MRJ Salton. Lytic agents, cell permeability and monolayer penetrability. J Gen Physiol 52:2275–2528, 1968.
21. MW Counts, L Young-Bandala, & LK Hall. Hard surface disinfection using optimized quaternary ammonium compounds. Unpublished data.
22. LJ Schep, DS Jones, MG Shepherd. Primary interactions of three quaternary ammonium compounds with blastospores of *Candida albicans* (MEN strain). Pharm Res 12(5):649–652, 1995.
23. B Terleckyj, DA Axler. Efficacy of disinfectants against fungi isolated from skin and nail infections. J Am Podiatr Med Assoc 83(7):386–393, 1993.
24. B Ahlstrom, M Chelminska-Bertilsson, RA Thompson, L Edebo. Submicellar complexes may initiate the fungicidal effects of cationic amphiphilic compounds on *Candida albicans*. Antimicrob Agents Chemother 41(3):544–550, 1997.
25. JJ Merianos. Methods for Testing Disinfectants. In: SS Block, ed. Disinfection, Sterilization and Preservation. 4th ed. 1991, pp. 1009–1027.

26. B Ahlstrom, M Chelminska-Bertilsson, RA Thompson, L Edebo. Long-chain alkanoylcholines, a new category of soft antimicrobial agents that are enzymatically degradable. Antimicrob, Agents Chemother 39(1):50–55, 1995.
27. JH Taylor, SJ Rogers, JT Holah. A comparison of the bactericidal efficacy of 18 disinfectants used in the food industry against *Escherichia coli* O157:H7 and *Pseudomonas aeruginosa* at 10 and 20 degrees C. J Appl Microbiol 87(5):718–725, 1999.
28. T Furuta. Effect of alkaline builders and surfactants on the bactericidal activity of didecyldimethylammonium chloride. J Antibact Antifung Agents 20(12):617–622, 1992.
29. A Ancona, A Arevalo, E Macotela. Contact dermatitis In hospital patients. Dermatol Clin 8(1):95–105, 1990.
30. G Saurina, D Landman, JM Quale. Activity of disinfectants against vancomycin-resistant *Enterococcus faecium*. Infect Control Hosp Epidemiol 18(5):345–347, 1997.
31. A Van de Weyer, MJ Devleeschouwer, J Dony. Bactericidal activity of disinfectants on *Listeria*. J Appl Bacteriol 74(4):480–483, 1993.
32. AD Russell. Factors influencing the efficacy of antimicrobial agents. In: AD Russell, B Hugo, GAJ Ayliffe, eds. Principles and Practice of Disinfection, Preservation and Sterilization. 2nd ed. London: Blackwell Science, 1992, pp. 89–113.
33. JD Druce, D Jardine, SA Locarnini, CJ Birch. Susceptibility of HIV to inactivation by disinfectants and ultraviolet Light. J Hosp Infect 30(3): 167–180, 1995.
34. MA Kennedy, VS Mellon, G Caldwell, LN Potgieter. Virucidal efficacy of the newer quaternary ammonium compounds. J Am Anim Hosp Assoc 31(3):254–258, 1995.
35. G Reybrouck. Evaluation of the Antibacterial and Antifungal Activity of Disinfectants. In: AD Russell, B Hugo, GAJ Ayliffe, eds. Principles and Practice of Disinfection, Preservation and Sterilization. 2nd ed. London: Blackwell Science, 1992, pp. 114–133.
36. SW Newsom, M Shaw. Tests for the Antibacterial activity of surgical glove material. J Hosp Infect 26(4): 279–286, 1994.
37. Fresh Cleanse™ Lotion-Handwash Product was obtained from First Scientific Corp., 1877W 2800S, Suite 200, Ogden, Utah 84401.
38. Topical antimicrobial drug products for over-the-counter human use. Tentative final monograph for health-care antiseptic drug products. Fed Reg 59(116):31402–31452, 1994.

6
Chlorhexidine Gluconate

Daryl S. Paulson
BioScience Laboratories, Inc., Bozeman, Montana

I. CHLORHEXIDINE GLUCONATE

Chlorhexidine gluconate (CHG) was first synthesized in 1950 by ICI Pharmaceutical in England [1]. CHG was found to have high levels of antimicrobial activity, but relatively low levels of toxicity to mammalian cells [1,2]. Additionally, CHG has a strong affinity for skin and mucous membranes. As a result, it has been used quite effectively as a topical antimicrobial for wounds, skin prepping, and mucous membranes (especially in dentistry), where it provides, by virtue of its proclivity for binding to the tissues, extended antimicrobial properties. CHG also has value as a product preservative, including for ophthalmic solutions, and as a disinfectant of medical instruments and hard surfaces.

CHG is a cationic molecule that is generally compatible with other cationic molecules, such as the quaternary ammonium compounds [2]. Some nonionic substances such as detergents, although not directly incompatible with CHG, may inactivate the antimicrobial properties of CHG, depending upon the compound and concentration levels. CHG is ''incompatible'' with inorganic anions, except in very dilute concentrations, and may also be incompatible with organic anions present in soaps containing sodium lauryl sulfate, and with a number of pharmaceutical dyes [3,4].

The antimicrobial activity of CHG is pH dependent, with an optimal use range of 5.5–7.0, a nice consistent match with the body's usual range of pH. However, the relationship between antimicrobial effectiveness of CHG and pH varies with the microorganism. For example, CHG's antimicrobial activity against *Staphylococcus aureus* and *Escherichia coli* increases with an increase in pH, but the reverse is true for *Pseudomonas aeruginosa*.

The antimicrobial activity of CHG against vegetative forms of both gram-positive and gram-negative bacteria is broad and pronounced [5]. It is generally inactive relative to bacterial spores, except when they are exposed to CHG at elevated temperatures. Mycobacteria reportedly are inhibited, but not killed by CHG in aqueous solutions. A variety of lipophilic viruses (e.g., herpes virus, HIV, influenza virus, many respiratory viruses, and cytomegalovirus) are rapidly inactivated by exposure to CHG. CHG, like many other topical antiseptics, does not have significant antimicrobial activity against small protein-coated viruses, such as many enteric viruses, poliomyelitis, and papilloma virus [5]. Finally, many important pathogenic fungi, particularly those in the yeast phase, are sensitive to CHG.

A. Antimicrobial Action

At relatively low concentration levels, CHG exerts bacteriostatic effects on many bacterial species, both gram-negative and gram-positive. At higher concentration levels, CHG generally demonstrates rapid bactericidal effects. However, the precise effectiveness in terms of time varies from species to species and as a function of concentration of CHG.

The antimicrobicidal effects against microorganisms occur in a series of steps related to both cytological and physiological changes, culminating in the death of the bacterial cell. Chemically, CHG is attracted to bacterial cell walls and is absorbed into certain phosphate-containing cell wall compounds, thereby penetrating the bacterial cell wall, even in the presence of cell wall molecular exclusion mechanisms. Once cell wall penetration has occurred, the CHG is attracted to the cytoplasmic membrane. Upon penetration of the cytoplasmic membrane, low molecular weight cellular components (e.g., potassium ions) leak out of the membrane, and membrane-bound enzymes such as adenosyl triphosphatase (ATPase) are inhibited. Finally, the cell's cytoplasm precipitates, forming molecular complexes with phosphated compounds, including ATP and nucleic acids [3].

As a rule of thumb, bacterial cells carry a total negative surface charge. It has been observed that, at sufficient CHG concentration levels, the bacterial total surface charge rapidly becomes neutral and then positive. The degree of change in the bacterial surface change is directly related to the concentration levels of CHG, but generally reaches a steady-state equilibrium within about 5 minutes of exposure. The rapid electrostatic attraction between the cationic CHG molecules and the negatively charged bacterial cell surface contributes to the rapid reaction rate, i.e., the rapid bactericidal effects exerted by CHG.

CHG at antiseptic concentrations (0.5–4%) demonstrates a high degree of antimicrobial activity, both -static and -cidal, on vegetative phases of gram-positive and gram-negative bacteria, but it has little sporicidal activity. While there have been concerns that prolonged use of CHG can lead to reduced sensitivity and, ultimately, the development of resistant strains of bacteria, this concern has not been verified, even after prolonged and extensive use. There is no evidence that plasmid-mediated antimicrobial resistance, particularly common in gram-negative bacteria, has developed. This has been borne out in studies of common indicator species such as *E. coli*, *P. aeruginosa*, *Serratia marcescens*, and *Proteus mirabilis*. Although several researchers have reported a reduced sensitivity to CHG among certain methicillin-resistant strains of *S. aureus* (MRSA), at clinical-use concentrations, this concern has not been substantiated. MRSA strains appear to be as susceptible to CHG, as are non-MRSA strains.

In general, CHG is a highly effective antimicrobial in its immediate, persistent, and residual properties [6]. Concentrations of CHG [4% to as low as 0.5%] provide excellent immediate and persistent action, with the added benefit (when used repeatedly over time) of good residual effects.

CHG, by virtue of its residual antimicrobial properties, may be clinically useful in reducing the probability of surgical infections when used in full or partial bodywashes prior to elective surgery [7]. If the product is used once a day as a shower soap or washed directly onto the proposed surgical site over the course of 3–5 days, the resident microbial population levels are reduced dramatically. Logically, then, when a person undergoes surgery, the remaining microbial populations residing on the proposed surgical site, having been significantly reduced, result in far fewer microorganisms to be eliminated by preoperative prepping procedures and, therefore, available for infecting the surgical site.

Currently, there is much interest in alcohol tinctures of CHG. These alcohol/CHG products may prove to be highly effective for use as fast-acting, preoperative skin preps, as well as preinjection and prearterial/venous catheterization preps. Preparations of alcohol/CHG in combination have reported high immediate antimicrobial properties (presumably due to the alcohol) with the benefit of persistent antimicrobial properties (due to the CHG) to provide a clinical performance superior to either alcohol or CHG alone.

II. MEDICAL APPLICATIONS

A. Surgical Scrub Formulations

Four percent chlorhexidine gluconate (CHG) forms the basis for many common presurgical scrubs in use today. These demonstrate high immediate degerming properties and high persistent antimicrobial properties and show good residual antimicrobial effects used both with and without a surgical scrub procedure.*

* A product's immediate antimicrobial efficacy is a quantitative measurement of both the mechanical removal of microorganisms by washing and the product's ability to rapidly inactivate microorgan-

Recently, certain manufacturers have developed effective 2% CHG surgical scrub products.

In our evaluations with 4% CHG, adequate $\log_{10}$ reductions to meet FDA requirements have been observed in wash procedures without the use of a scrub brush. Several manufacturers have combined alcohol [>60%] with 1, 2, 3, and 4% CHG, also satisfying the current FDA requirements for a presurgical scrub formulation without the use of a scrub brush.

CHG provides good protection against transmission of resident microbial flora from the hands of surgical staff to a surgical wound, if a puncture or tear to the gloves occurs. This is a challenge for any antimicrobial 6 hours after the surgical gloves were donned, because that occlusive condition accentuates the growth of normal skin bacteria, particularly *Staphylococcus epidermidis*. However, studies show that CHG effectively prevents significant regrowth [6]. Additionally, CHG has demonstrated remarkable broad-spectrum properties against the bacteria listed in Table 1, as based on our laboratory results:

B. Preoperative Skin Preparation

Initially, CHG-based preoperative skin preparations experienced much resistance from surgical staff, in that they were accustomed to observing a "yellow stain" on the proposed surgical skin site after prepping with an iodophor; CHG dried clear. However, it is now common practice to add a highly visible dye to stain the skin. And the product has the same broad antimicrobial spectrum as noted in the 4% surgical scrub section.

A novel application of CHG is its use as a preoperative antimicrobial wash prior to elective surgeries. In this endeavor, the patient is provided a CHG-impregnated applicator to apply at the proposed operative site prior to undergoing a surgical procedure. The purpose of this application is to capitalize on the residual antimicrobial effects of the product. So, over the course of a 2- to 3-day application period, the baseline counts at the proposed operative site are dramatically reduced. This theoretically will pose a far less microbially populated area with which the preoperative skin preparation must contend.

Additionally, certain manufacturers are exploring the potential benefit of continuing the CHG application procedure after the surgery has been completed. This interest has been stimulated by the positive results of using CHG-impregnated wound dressings.

---

isms residing on the skin surface. The persistent antimicrobial effect is the product's ability to prevent microbial recolonization of the skin surfaces, either by microbial inhibition or lethality after application of the product. The residual efficacy is a measurement of the product's cumulative antimicrobial properties after it has been used repeatedly over time.

Table 1 Bacteria

Gram-negative bacteria
Acinetobacter species; various clinical isolates and ATCC#19606
Bacteroides species; various clinical isolates, including *B. fragilis*, and ATCC#25285
Candida species, clinical isolates, including *C. albicans* and ATCC#10231
Enterobacter species; various clinical isolates and ATCC#13048
Escherichia coli; various clinical isolates, including serotype 0157:H7 and ATCC#11229 and #25922
Haemophilus influenzae; various clinical isolates and ATCC#49247
Klebsiella species; various clinical isolates of *Klebsiella pneumoniae* and ATCC#11296
Proteus mirabilis; various clinical isolates and ATCC#7002
Pseudomonas aeruginosa; various clinical isolates and ATCC#15442 and #27853
Serratia marcescens; ATCC#14756
Gram-positive bacteria
Enterococcus faecalis; various clinical isolates and ATCC#29212
Enterococcus faecium; various clinical isolates and ATCC#51559
Micrococcus luteus; various clinical isolates and ATCC#7468
Staphylococcus aureus; various clinical isolates and ATCC#6538 and #29213
Staphylococcus epidermidis; various clinical isolates and ATCC#12228
Staphylococcus hominis; various clinical isolates and ATCC#27844
Staphylococcus haemolyticus; various clinical isolates and ATCC#29970
Staphylococcus pneumoniae; various clinical isolates and ATCC#33400
Streptococcus pyogenes; various clinical isolates and ATCC#19615
Staphylococcus saprophyticus; various clinical isolates and ATCC#15305

C. Healthcare Personnel Handwash

Recall that a healthcare personnel handwash is intended for use by healthcare personnel between patient examinations, x-rays, CAT scans, phlebotomy, etc., so that they do not serve as a disease vector, transmitting patient A's infectious microorganisms to patient B. Healthcare personnel handwashes, like surgical scrub products, have two primary ways of degerming the hands: physical removal by the mechanical action of handwashing and the antimicrobial product's lethal effect on the contaminating microorganisms. In this respect, the healthcare personnel handwash need not be as antimicrobially active as a surgical scrub—that is, it need not remove at least 3 logs of the normal skin flora, as well as the transient microorganisms. The product, as tested according to FDA standards, need only remove 3 logs of transient microorganisms. This, however, is not that easy to do. Additionally, the product must have very broad antimicrobial properties, for healthcare personnel potentially will come into contact with a much wider assortment of contaminative bacteria, fungi, and viruses than do surgical staff.

In some very important ways, then, it is even more desirable that susceptibility among a wide range of microorganisms be demonstrated for the healthcare personnel handwash.

D. Precatheter Insertion Skin Preps

We have worked with various tinctures of CHG as precatheter insertion skin preps, and they seem to provide both very rapid antimicrobial action and long-term antimicrobial persistence (>48 hours) with one product application. This can be very important, because if the normal flora (e.g., *Staphylococcus epidermidis*) counts are kept to low levels, this should serve to reduce the risk of nosocomial infection. Additionally, because the catheter need not be reinserted in a new site as often, this could reduce not only the increased probability of infection, but physical trauma to a vein as well.

E. Food Service

There has been interest shown in using CHG washes for food service personnel. At this point, no real applications have been seen. However, in our work with CHG, when *E. coli* are aliquotted into low-fat hamburger meat, we have seen an apparent reduction in antimicrobial effectiveness of CHG, suggesting that the fat load in hamburger interferes with its antimicrobial efficacy.

In summary, there appears to be much interest in developing CHG into specific niche categories in the medical field. It will be interesting to see these events unfold.

REFERENCES

1. SF Bloomfield. Chlorhexidine and iodine formulations. In: JM Ascenzi, ed. Handbook of Disinfectants and Antiseptics. New York: Marcel Dekker, 1996, pp. 133–158.
2. GW Denton. In: SS Block, ed. Chlorhexidine, Disinfection, Sterilization, and Preservation. 5th ed. Philadelphia: Lippincott, Williams, and Wilkins, 2001, pp. 321–336.
3. RC Weast, ed. CRC Handbook of Chemistry and Physics. 4th ed. Boca Raton, FL: CRC Press, 1984.
4. B Budavari. The Merck Index. 12th ed. Merck & Co., Whitehorse Station, NJ: Merck & Co., 1996.
5. NS Ranganathan. Chlorhexidine. In: JM Ascenzi, ed. Handbook of Disinfectants and Antiseptics. New York: Marcel Dekker, 1996, pp. 235–264.
6. DS Paulson. Topical Antimicrobial Testing and Evaluation. New York: Marcel Dekker, 1999.
7. DS Paulson. Efficacy evaluation of a 4% chlorhexidine gluconate solution as a full body shower wash. Am J Infect Control 21:4, 205–209, 1993.
8. Tentative Final Monograph for Healthcare Antiseptic Drug Products, Proposed Rule. Federal Register, Food and Drug Administration, June 17, 1994.

7
Alcohols

Daryl S. Paulson
BioScience Laboratories, Inc., Bozeman, Montana

I. ALCOHOLS

Alcohols have become ever more accepted as the primary skin antiseptic for healthcare personnel hand sanitizer applications in both North America and Europe [1,2]. Additionally, alcohol-based ''hand sanitizers'' are well accepted for use in degerming the hands of food servers/handlers and consumers at large [3]. Use of alcohols for sanitation is not new, and those containing 60–70% ethanol or isopropanol have been used for years as hand antiseptics in both the medical and food-handling environments.

Generally, alcohols are of limited value as sporicidals but have broad-spectrum antimicrobial properties relative to both gram-positive and gram-negative vegetative bacteria, as well as yeasts, fungi, and some viruses [4]. The antimicrobial effects of alcohol are dependent on specfic concentrations, exposure, and condition. Like many other topical antimicrobial compounds, alcohols are considered to be nonspecific antimicrobials because they are active through a multiplicity of toxic effect mechanisms [3,4]. This has important implications for the spectrum, speed, and, ultimately, overall effectiveness of alcohols as disinfectants [5]. The predominant mode of action appears to stem from protein coagulation/denaturation. Additionally, they disrupt the cell wall and/or cytoplasmic membrane of a cell, leading to cell lysis [6]. Relatively low concentrations, about twice the minimum inhibitory concentration (MIC), can accomplish lysis.

Protein coagulation has also been reported occurring in microorganisms at alcohol concentrations of 60–70%. In the absence of water, proteins are not denaturated as readily as when water is present. This is likely why absolute ethanol is less bactericidal than are mixtures of alcohol and water. Coagulation of cellular proteins leads to loss of cellular functions.

The physicochemical properties of alcohols are associated with their chemical structure [3–5]. Factors such as water miscibility, solvency, surface tension, vapor pressure, and protein denaturancy vary with chemical structure and help explain corresponding variations in biological activity. Microbial resistance and the development of alcohol-resistant bacterial strains is not a significant issue with alcohols, especially at use-level concentrations employed for antisepsis and disinfection [5].

II. ANTISEPTIC PROPERTIES OF ALCOHOLS

Alcohol is and has been widely used for the destruction of the vegetative forms of microorganisms preceding such procedures as venipunctures, hypodermic injections, finger pricks, and other procedures that break the intact skin [4]. It is also widely used in some European countries and has been increasingly promoted in the United States for use as a surgical hand scrub [5].

The antimicrobial action of isopropanol has been reported to be greater than that of ethanol [4,5]. For example, by making counts of surviving bacteria after 30 seconds of exposure to varying concentrations of alcohol, isopropanol has been shown to be slightly more bactericidal than either ethanol or methanol for *Escherichia coli* and *Staphylococcus aureus.*

III. VIRUSES

There is no general agreement in the literature regarding antiviral activity of alcohols. It is well established that lipophilic, enveloped viruses are easier to inactivate by alcohols and other general disinfectants than are "naked" viruses [4,5]. Enveloped viruses extensively studied include vaccinia virus, togavirus, influenza A virus, and rabies virus. Naked viruses also have been investigated quite extensively, including picornaviruses, poliovirus, coxsackievirus, and echovirus. Enteroviruses, such as hepatitis A and B, and rotaviruses have been studied for their resistance to chemical and physical influences, including the effectiveness of alcohols.

IV. ALCOHOLS FOR SKIN DEGERMING

While it is not possible, nor probably desirable to sterilize the skin surfaces of humans, the number of microorganisms residing on the skin can be greatly reduced. Skin bacteria are generally classified as "transient" and "resident" flora [6]. Transient bacteria are those that do not normally live on the skin surface and are often loosely attached to the skin by lipid substances. Their removal is relatively easy compared with removal of the resident, or ''normal,'' bacteria. Studies of skin antiseptics generally are divided into those that assess effects on transient flora and those that assess effects on resident flora. In the former, transient flora are provided by contamination with a surrogate bacterium in studies of healthcare personnel handwashes, general use handwashes, and foodhandler handwashes [6,7]. Studies that assess effects on resident flora include those for preoperative skin preparations, surgical scrubs, preinfection skin preparations, and precatheter insertion skin preparations [6,7].

V. TRANSIENT MICROORGANISMS

In 1977, Rotter et al. developed a test method for the evaluation of healthcare personnel handwash products [3,4], and a modification of this method has been adopted by the German and Austrian governments [5]. The method includes artificial contamination of the hands with *E. coli* and decontamination with two 30-second applications of 3 mL of 60% (by volume) isopropanol, as a standard for comparison with any test product applied according to manufacturer's directions. European researchers have reported 4 or greater $\log_{10}$ reductions with all alcohols tested, in comparison with 2–3 $\log_{10}$ reductions with soaps containing phenolic antimicrobials, chlorhexidine gluconate (CHG), or povidone-iodine [8–10].

Marples and Towers developed a model to assess the transfer of organisms by contact [11]. They contaminated a fabric-covered bottle with *Staphylococcus saprophyticus*. Subjects grasped the contaminated bottle, then a sterile cloth-covered bottle, and the microbial population transferred was determined. Washing human subjects' hands with nonmedicated hand soap reduced transfer by 95%, whereas washing in 500 mL of 70% ethanol for 30 seconds reduced transfer by 99.9%. Fifteen minutes after testing, the contaminating organism was still present on the hands of subjects who washed with soap, but it was apparently undetectable on alcohol-treated hands. Use of an alcohol-impregnated towelette and application of a small volume (0.2 mL) of 80% ethanol resulted in reductions of a lesser magnitude, 80% and 93%, respectively.

Reportedly, a bland soap handwash was ineffective in preventing transfer by hand of gram-negative bacteria to catheters following brief contact with a heavily contaminated patient source; an alcohol hand rinse was generally effective. Other studies have compared the effectiveness of 10% povidone-iodine, 70% isopropanol, and 2% aqueous CHG for the prevention of infection associated with central venous and arterial catheters [5]. CHG resulted in the lowest incidence of local catheter-related infection (2.3% vs. 7.1% and 9.3% for isopropanol and povidone-iodine, respectively) and catheter-related bacteremia (0.5% vs. 2.3% and 2.6% for isopropanol and povidone-iodine, respectively [5].

Additionally, in a survey overview on the effectiveness of handwashing and hand disinfection for the removal of nosocomial pathogens from heavily contaminated hands, 0.5% CHG in 70% isopropanol was most effective, followed by 70% ethanol and, to a lesser extent, 40% isopropanol [5]. Comparison of the efficacy of 62% ethanol, 70% isopropanol, and benzylalkonium chloride-based hand sanitizers using the healthcare personnel handwash protocol indicated that all three had equivalent efficacy at greater than 2-log reduction after the first wash, whereas the benzylalkonium chloride hand sanitizer demonstrated residual efficacy.

VI. RESIDENT SKIN MICROORGANISMS

A number of investigators have evaluated the effectiveness of alcohols as hand rinses after short contact times. Morrison et al. compared three alcohol-based hand rinses—including 70% isopropanol, 0.5% CHG in 70% isopropanol, and a 60% isopropanol formulation containing evaporative retardant—in 14 subjects [12]. The 60% isopropanol with evaporative retardant was associated with significantly greater reductions after each of four consecutive handwashes. Similarly, Larson has long reported the benefits of alcohol in the medical arena [5]. She reported significant reductions from baseline counts among subjects using the evaporation-retarded formulation of isopropanol after a single 15-second application. After using this formulation 15 times per day for 5 consecutive days, subjects using either one of two alcoholic hand rinses, 70% isopropanol or 4% CHG in detergent, had significant reductions in their colonizing flora, but the two alcoholic hand rinses continued to be associated with the greatest reductions. There was no significant change in bacterial counts among subjects using a nonmedicated control soap. In another study, Larson demonstrated a significant dose response with two alcohol hand rinses: subjects using a 3 mL hand rinse had significantly greater reductions in bacterial flora counts than did subjects using a 1 mL rinse [13]. Based on these data, alcohols deserve serious consideration for potential use as surgical scrubs and hand-degerming.

The sequential use of a chlorhexidine gluconate–containing detergent followed by an alcoholic disinfectant reduced the release of resident skin bacteria significantly better than did a sequence of unmedicated soap and alcohol used for the same periods. Paulson compared five surgical hand-scrub preparations (4% CHG brush, 2% CHG solution, povidone-iodine brush, parachlorometaxylenol brush, and alcohol-impregnated brush) [14]. Only the CHG products demonstrated antimicrobial effectiveness in all three parameters (immediate, persistent, and residual). A comparison also was made between a 5-minute povidone-iodine scrub and a 1-minute povidone-iodine scrub, followed by alcohol foam [15]. The total number of colonies was less after the 1-minute scrub with alcohol foam than after the standard 5-minute scrub for both 1-hour and 2-hour groups.

Larson and coworkers investigated both the effect of blood and the effect of a protective foam on the antimicrobial activity of alcohol and other active agents [16,17]. The effect of blood on the efficacy of 70% isopropanol, 0.5% CHG in 70% isopropanol, 7.5% povidone-iodine in a detergent base, 4% CHG in a detergent base, and a nonantimicrobial soap was evaluated in 71 subjects. In the presence of blood, the two alcohol-containing products resulted in significantly greater reductions in the number of colony-forming units (CFUs) than the other products. In the absence of blood, 70% isopropanol was associated with significantly greater reductions; soap resulted in a significantly lower reduction [16]. The effects of a skin protectant on glove integrity and the efficacy of surgical scrubs with 70% isopropanol, 4% CHG in a detergent base, 7.5% povidone-iodine in a detergent base, and a nonantimicrobial soap (control) were determined. No significant differences were found in CFUs on hands with or without protectant immediately after scrubbing or at 2 hours after scrub on gloved or ungloved hands [17]. The efficacy of alcohol hand rubs with two different kinds of handwashing machines was studied in vivo [18]. It was concluded that an alcohol-based solution containing an effective antimicrobial detergent preceded by a soap wash is necessary to reduce hand-surface bacteria to a satisfactory degree using these techniques.

Contrary to popular opinion, alcoholic products seem to be quite acceptable to users [19]. Newer formulations containing emollients eliminate the drying effects of alcohol on skin and significantly increase acceptability [3]. Intermittent use of an alcohol hand gel containing emollients reduced the soap-induced skin irritation of healthcare personnel, improved their skin condition (cracking, scaling, and redness), and maintained normal skin hydration [3].

Although all methods were initially comparable, with bacterial reductions of greater than 99%, recolonization of a test site was significantly reduced after 60 minutes when prepared with an alcohol and iodophor drape, compared with the other methods. Jeng and Severin investigated the performance of a povidone-iodine gel alcohol (5% povidone-iodine and 62% ethanol in gel form) as a 30-second, one-time application preoperative skin preparation [20]. The povidone-iodine gel alcohol formulation delivered rapid and persistent antimicrobial activity against a broad spectrum of bacteria, both in vitro and in vivo, and was found to be an effective skin preparation formulation for use in a single-step 30-second application.

For persons who were allergic to iodine, CHG and isopropanol were reported more efficacious than ''green'' soap and isopropanol. Further studies are certainly required to determine whether CHG is superior to povidone-iodone or tincture of iodine for this procedure.

VII. ALCOHOLS USED TO PREVENT INFECTIONS

Murie and MacPherson compared postoperative wound-infection rates associated with two hand-scrub techniques in the operating room [21]. They alternated each month for 6 months between 0.5% chlorhexidine in 95% methanol and 4% chlorhexidine in detergent. Among 226 patients—117 in one group and 109 in the other—no difference in infection rates was found. Additionally, the alcoholic preparation was calculated to be five times less expensive, less time-consuming, and more acceptable to users. Dorif et al. compared the use of triple dye versus alcohol for umbilical cord care in neonates in terms of impact on staphylococcal infections in a newborn nursery over a 1-year period [22]. There was no significant difference in infection rates (0.4% for newborns receiving triple dye and 0.6% for those receiving alcohol treatments). The alcohol-treated infants, however, had fewer cord complications and better healing of umbilical stumps.

A study was carried out to determine the comparative efficacies of 4% CHG in a detergent base and 60% isopropanol hand rinse (with optional use of a bland soap) in reducing nosocomial infections in intensive care units (ICUs) [23]. The authors concluded that the CHG product reduced the nosocomial infection rate more effectively than did use of alcohol and soap and attributed the results, at least in part, to better handwashing compliance when the CHG product was used. Alcohol preparations kill bacteria rapidly, theoretically permitting briefer washing time. Voss and Widmer stated that alcoholic hand disinfection, with its rapid activity, superior efficacy, and minimal time commitment, allows 100% healthcare worker compliance without interfering with the quality of patient care [24]. Alcohol preparations are inexpensive and can be used without a sink, when sinks are unavailable, or when tap water is contaminated—a major advantage in developing countries.

A randomized clinical trial comparing the effectiveness of an alcoholic solution with that of the standard handwashing procedure used in clinical wards and ICUs was carried out in a large public university hospital [25]. The alcoholic solution resulted in significantly fewer CFUs recovered after the cleansing procedure, and overall acceptance was rated as good by 72% of the healthcare workers. These investigators concluded that the use of alcoholic solutions is effective and safe and deserves more attention, especially in situations in which the compliance rate is hampered by architectural problems or nursing work overload.

Recently, alcohols have been combined with CHG to form a tincture of CHG to synthesize the benefits of both. Theoretically, the tincture should be much faster-acting than CHG by itself and provide a longer-lasting persistent antimicrobial effect than alcohol alone. The merits of this concept have not been reported in challenge use with normal subject volunteers or in the actual clinical setting as yet. The panoply of uses for alcohol, however, as a primary topical antimicrobial has not been totally explored, and new applications undoubtedly will be witnessed in the future.

REFERENCES

1. WC Beck. Benefits of alcohol rediscovered. AORN J 1984; 40:172–176.
2. EL Larson. Antiseptics. In: RN Olmstad, et al., eds. Infection Control and Applied Epidemiology Principles and Practices. St. Louis, MO: Mosby, 1996:1–197, G1-G7.
3. EJ Fendler. Personal communication. GoJo Industries, 2001.
4. ML Rotter. Alcohols for antiseptics of hands and skin. In: JM Ascenzi, ed. Handbook of Disinfectants and Antiseptics. New York: Marcel Dekker, 1996:177–233.
5. Y Ali, MJ Dolan, EJ Fendler, EL Larson. Alcohols. In: SS Block, ed. Disinfection, Sterilization, and Preservation. 5th ed. Philadelphia: Lippincott, Williams and Wilkins, 2001:229–253.
6. AJ Isquith, WR Chesbro. Pools, confluxes and transport of amino acids in *Streptococcus faecium*. Biochem Biophys Acta 1963; 74:642–646.
7. DS Paulson. Topical Antimicrobial Testing and Evaluation. New York: Marcel Dekker, 1999.
8. ML Rotter. Povidone-iodine and chlorhexidine gluconate containing detergents for disinfection of hands [letter]. J Hosp Infect 1981; 2:275–280.
9. ML Rotter. Händedesinfektion. In: W Weuffen, Spiegelbergeer, eds. Handbuch der Desinfektion and Sterilisation. Berlin: Springer-Verlag, 1984; 62–79.
10. ML Rotter, W Koller, G Wewalka, et al. Evaluation of procedures for hygienic hand-disinfection: controlled parallel experiments on the Vienna test model. J Hygiene Cambridge 1986; 96:27–37.
11. RR Marples, AG Towers. A laboratory model for the investigation of contact transfer microorganisms. J Hygiene Cambridge 1979; 82:237–249.
12. AJ Morrison, J Gratz, I Casbezudo, et al. The efficacy of several new handwashing agents for removing non-transient bacterial flora from hands. Infect Control 1986; 7:268–272.
13. EL Larson, PI Eke, MP Wilder, et al. Quantity of soap as a variable in handwashing. Infect Control Hosp Epidemiol 1987; 8:271–375.
14. DS Paulson. Comparative evaluation of five surgical hand scrub preparations. AORN J 1994; 60:246–256.
15. N Deshmukh, JW Kramer. A comparison of 5-minute povidone-iodine scrub and 1-minute povidone-iodine scrub followed by alcohol foam. Mil Med 1998; 163:145–147.
16. E Larson, L Bobo. Effective hand degerming in the presence of blood. J Emerg Med 1992; 10:7–11.
17. E Larson, JK Anderson, L Baxendale, et al. Effects of a protective foam on scrubbing and gloving. Am J Infect Control 1993; 21:297–301.
18. S Namura, S Nishijima, K Mitsuya, et al. Study of the efficacy of antiseptic handrub lotions with hand washing machines. J Dermatol 1994; 21:405–410.

19. LJ Pereira, GM Lee, KJ Wade. An evaluation of five protocols for surgical handwashing in relation to skin condition and microbial counts. J Hosp Infect 1997; 36: 49–65.
20. DK Jeng, JE Severin. Povidone iodine gel alcohol: a 30-second, one-time application preoperative skin preparation. Am J Infect Control 1998; 26:488–494.
21. JA Murie, G MacPherson. Chlorhexidine in methanol for the preoperative cleansing of surgeons' hands: a clinical trial. Scott Med J 1980; 25:309–311.
22. C Dorif, D Warshauer, M Roth, et al. Use of triple dye vs. alcohol in newborn and care for prevention of *Staphylococcus aureus* infections. Proceedings of the 1985 Interscience Conference on Antimicrobial Agents and Chemotherapy. Washington, DC: American Society for Microbiology, 1985:185.
23. BN Doebbeling, GL Stanley, CT Sheetz, et al. Comparative efficacy of alternative hand-washing agents in reducing nosocomial infections in intensive care units. N Engl J Med 1992; 327:88–93.
24. A Voss, AF Widmer. No time for handwashing? Handwashing versus alcoholic rub: can we afford 100% compliance? Infect Control Hosp Epidemiol 1997; 18:205–208.
25. M Zaragoza, M Saliés, J Gomez, et al. Handwashing with soap or alcoholic solutions? A randomized clinical trial of its effectiveness. Am J Infect Control 1999; 27:258–261.

Part III
Product Applications

Topical antimicrobials provide an important first line of defense against surgical infections. Chapter 8 discusses microorganisms and disease, whereas Chapter 9 describes in some depth nosocomial infections. The following chapter discusses nosocomial infections as they relate to postsurgical infections. Since staphylococcal species, particularly coagulate-negative varieties, are so commonly associated with postsurgical infections, certain aspects of them are covered in Chapter 12. In the medical arena, research and development is ongoing in finding new ways to prevent or at least reduce the incidence of postsurgical infections. One interesting approach is to utilize chlordexidine gluconate or a preoperative wash prior to elective surgeries as presented in Chapter 11. Theoretically, this should reduce the normal resident microorganisms populations at the proposed surgical site. However, the preoperative skin preparation performed just prior to surgery will be used on very clean skin and should then sterilize the skin. Some argument exists as to whether a chlorhexidine-based product can be useful in this endeavor. Chapter 13 explores this aspect. Chapter 14 is a review section written by one of the original scientists, Mary Bruch, who developed the glove juice sampling procedure. Chapter 15 provides some suggestions for developing healthcare personnel handwash products.

8
Microorganisms and Disease

Daryl S. Paulson
BioScience Laboratories, Inc., Bozeman, Montana

I. INTRODUCTION

The earth's biosphere contains untold numbers and kinds of microorganisms. Humans have an interaction with microorganisms that is dynamic and modified by many factors. From the beginning of microbiology, with Van Leeuwenhoek's application of the microscope, small ''animalcules''—microbes—were suspected of causing disease [1]. It was not until Koch published his classic paper on anthrax that these suspicions were proven, through rigorous experimentation as per Koch's four postulates [2]. Although Koch's postulates were for the determination of causation attributable to a single microbial species, the disease process is often more complex, involving multiple etiologies [3].

Secondary infections often occur when a primary pathogen reduces the immunological resistance of the patient, so that other microorganisms of limited pathogenicity produce disease [4]. For example, a primary viral infection may lead to a secondary bacterial infection. Staphylococcal pneumonia is rarely a primary disease, but is often observed as a sequel to viral influenza. This predisposition to a secondary infection is generally due to a stress disruption of both cellular and humoral immunity and a suppression of phagocytosis due to a distinct drop in the opsonization of antigen [5,6].

There are dual infections where a disease is the result of a synergistic relationship between two microorganism species. Such relationships are both complex and more common than generally thought. A classic example of this is the human disease of diphtheria [6]. While the disease is caused by *Corynebacterium diphtheriae*, the toxin produced by this bacterium that accounts for its pathogenicity is due to lysogenic conversion of the *C. diphtheriae* by a specific bacteriophage. Only the bacteriophage-infected bacteria are capable of producing the profound disease symptomology of diphtheria [3,7].

Microbial pathogenicity refers to the potential capacity of a microorganism to cause disease, as applied to a species or group of microorganisms. *Virulence* refers to the degree of pathogenicity within a species or group of microorganisms, usually related to the minimum number of microorganisms required to elicit a disease (infectious dose). Often avirulent and virulent strains will occur in the same species.

While microorganisms may be introduced directly into a host's tissue via a cut, wound, puncture, burn, or insect inoculation, most infectious disease is due to a microorganism penetrating the mucous membrane barrier of the respiratory tract, the gastrointestinal tract, or the urogenital tract. With very few exceptions, pathogenic microorganisms possess the capacity to adhere to and colonize those mucous membranes and, in most cases, to penetrate them. The ability to cause disease is further dependent upon the microorganism's ability to evade and resist the host's immunological defense system. The pathogenicity of a microorganism group is rooted in a combination of multiple microorganism and host factors.

II. INFECTION

For an infectious disease to manifest, at least five events generally occur [6]:

1. Encounter: The host must be physically exposed to the microorganism.
2. Entry: The microorganism must enter the host.
3. Spread: The microorganism must spread from the site of physical entry.
4. Multiplication: The microorganism must multiply in the host.
5. Damage: The microorganism and its metabolites and/or the host's immune response cause tissue damage.

The host is exposed to the microorganism in a variety of situations relevant to topical antimicrobial compounds. The scenarios tending to infection described below relate to the healthcare setting, but much the same sets of conditions are involved in transmission of foodborne disease via the environmentally or fecally contaminated hands of food handlers. Generally, there is (1) the exposure of patients to pathogenic microorganisms via a healthcare worker's contamination by a different patient, worker, or environmental source; (2) the exposure of a patient's surgical site or a worker's wound site to contaminated hands; and (3) the contamination of a surgical, catheter, or injection site with the patient's own normal microbial flora [8,9].

In situation 1, the patient is exposed to pathogenic microorganisms from a different patient via the contaminated hands of healthcare personnel. That is, the healthcare provider serves as a disease vector, transferring patient A's pathogenic microorganisms to patient B. It should be noted that the microorganisms transferred are transient pathogens, not normal microorganisms of the skin.

Situation 2 represents the exposure of a patient's surgical site by the contaminated hands of surgical personnel. This generally occurs when a surgical glove is torn or nicked, and the normal resident microorganisms of the caretaker's skin enter the patient at the surgical site [10]. The degree of infectious disease is dependent upon the surgical site, the number of opportunistic microorganisms transferred to the patient, and the immunological competence of the patient.

Situation 3 occurs as a result of the contamination of a surgical site, venous/arterial catheter site, or an injection site with the patient's normal skin microorganisms. Contaminative (nosocomial) infections of surgical and catheter sites are not uncommon and are dangerous because of entry of microorganisms into the blood stream, potentially resulting in a bacteremia and septicemia. Patients who are immunocompromised are at a greater risk of septicemia [4]. Injection site infections are generally localized but can result in a septicemic condition.

III. SPREAD

Spread of pathogenic microorganisms from patient to patient via healthcare personnel hand contact can take a variety of forms [10]. It can be hand-to-hand contact, such as shaking hands. Often, the contaminating bacteria are respiratory or gastrointestinal in origin, but not exclusively. Most anatomical sites subsequently examined, particularly if they are anatomically or physiologically compromised, can become a spread site.

General infection resulting from direct contact with the hands of surgical personnel or introduction of normal skin microorganisms through catheters is facilitated by the blood and the lymphatic systems [6,11,12]. Spreading can be passive or active, depending upon a microorganism's motility and/or manufacture of extracellular hydrolytic enzymes that allow them to break through ''walling off'' mechanisms of the inflammatory response. For example, streptococci produce a protease that breaks up fibrin, a hyaluronidase that hydrolyzes hyaluronic acid (an important component of connective tissue), and a deoxyribonuclease that causes the release of DNA from lysed white cells and reduces the viscosity of pus.

IV. MULTIPLICATION

Once the microorganisms have spread from the entry site, they frequently multiply to cause a systemic infection. Generally, there is a time lapse between the

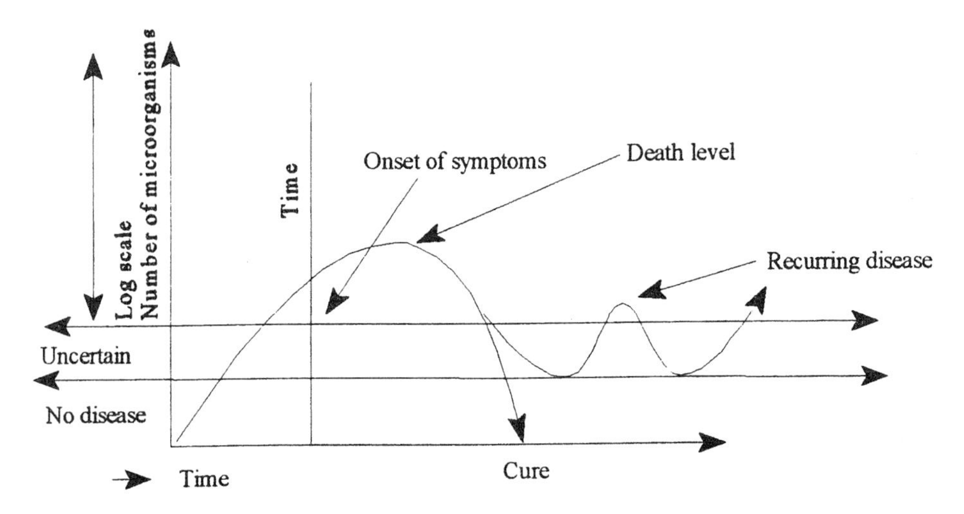

Figure 1 Incubation period of infectious microorganisms.

exposure and the manifestation of symptoms, referred to as the incubation period (Fig. 1) [10].

V. IMMUNE SYSTEM

The ability of pathogenic microorganisms to multiply in the body is influenced by the body's immunological system. In normally functioning immune systems, the body is able to generate a variety of ''immune system'' cells and antibody molecules that are capable of detecting (''recognizing'') and eliminating an apparently infinite variety of foreign substrates and microbial forms, which include viruses, bacteria, and fungi.

Functionally, the immune system can be divided into two interrelated activities: recognition and response [5]. Immunological recognition is highly specific in that it can distinguish one foreign macromolecule or pathogenic microorganism from another. The immune system is also able to discriminate between ''self'' and ''not self'' at the molecular level of cell structure. Once a foreign microorganism or molecular form is recognized, the immune system mounts an immune (''effector'') response through the participation of various immunological cells, phagocytic cells, and certain molecular substrates [7]. The effector response then may or may not effectively eliminate or neutralize the infectious microorganism species or foreign molecular form. The immune system is able to translate the recognition established at initial exposure to a foreign microorganism or molecule to subsequent exposures, because certain immunoactive cells retain "memory" of them. Upon a subsequent exposure, a heightened and more rapid immune response is generated, which serves to eliminate the invader quickly and even prevent a disease [8].

Often, however, an individual who is sick or a patient who has undergone a surgical procedure is, to some degree, immunologically compromised. Hence, the potential for morbidity and mortality through infectious disease is increased.

VI. MICROBIAL NUTRITION

At first glance, it would appear that the body offers a variety of rich mediums for the support of microorganisms. Body fluids such as plasma contain sugars, vitamins, minerals, and other substances on which bacteria and fungi can subsist [7]. However, for microorganisms other than normal flora, life in or on the body is not so simple. For example, if fresh blood plasma is incubated with challenge microorganisms, microbial growth is generally nonexistent or sparse. This is because antimicrobial substances such as lysozyme and molecular constituents of the complement system are inhibitory [5].

Bacteria require free iron for the synthesis of their cytochromes and other enzymes. This also appears to be a limiting factor for growth of bacteria in the body. Plasma, as well as a variety of other body fluids, contains very little free iron, probably due to its avid binding to a wide range of proteins. In fact, the body actually sequesters iron to defend itself against bacterial multiplication. When a sufficient number of microorganisms has been detected within the body, iron-binding proteins literally pour into plasma and other tissue fluids, as the body strives to limit the free iron available to bacteria.

The range of nutritional requirements of microorganisms found as normal microbial flora is a reflection of their ecological niches. For example, *Staphylococcus epidermidis*, the predominant skin surface–colonizing bacterium, requires several amino acids and vitamins that are common on the skin surfaces. However, microorganisms common in both soil and water are much less fastidious. They can achieve their organic requirements from simple carbon compound sources widely available in the body, as well as in the natural environment. Both *Escherichia coli* and *Pseudomonas* spp. are examples of bacteria that can thrive on very nutrient-minimal media [10].

VII. PHYSICAL FACTORS

Physical factors also affect microbial multiplication within the body. Important physical factors include temperature range of an anatomical region of the body, osmotic pressure of fluids, and humidity. Microorganisms that are normal inhabitants of the body, or those that can only live on the body, tend to have much more limited tolerance to physical changes than ones also found commonly in the environment, like *Pseudomonas* spp. [10].

VIII. ENDOGENOUS "OPPORTUNISTIC" INFECTION

Normally, endogenous commensal microorganisms cause no disease in human hosts, because a dynamic balance exists between the host's immunological defense and the microorganism's ability to cause disease [6]. Yet, if that balance is upset, endogenous microorganisms take advantage of the situation. Thus, in major or minor surgical wounds, staphylococci species such as *S. aureus* and *S. epidermidis* can establish an infection [13].

Opportunistic infectious disease is increasingly prevalent, due particularly to medical advances that prolong life in patients suffering disease (e.g., cancer, diabetes, etc.) that would have killed them in earlier times or treatments that, themselves (e.g., cancer therapy), compromise the immune system [6,8]. Opportunistic infections are also common following invasive diagnostic procedures and treatments.

IX. ADHERENCE

In order for infectious disease to occur, microbial adherence to tissues and membranes is necessary. Mucous membranes are exposed constantly to microorganisms from the environment—from food, water, the air, dust, and other sources [10]. Most of the microorganisms are bound and/or mechanically removed via discharge of mucus from the upper respiratory tract, or via the urine, saliva, and tears, or by entrapment and elimination in the intestinal contents before they have the opportunity to multiply. Some adhere to inert bodily structures. For example, *Streptococcus mutans* attaches to the enamel surface of the teeth (tartar, plaque) in order to elicit dental caries [13].

Microorganisms attach to host cells via an often unique and specific process. Microbial surface structures called adhesins react and combine with complementary receptor sites on host cells. The specific interaction is how individual microbial strains demonstrate a predilection for a particular host body site [3].

Because the net charge for microorganisms and human cells is negative, the repulsion generated, one would think, would prevent adhesion [3,8]. The overall net charge between these cells, however, does not preclude localization—attractive forces between them—particularly in areas where hydrophobic forces override the repellent forces. The relatively weak hydrophobic forces can permit attachment, but firm binding requires an interaction between ligands on the microorganism and complementary receptor sites on the host cells. The mutual participation of a large number of these specific interactions results in firm attachment between the microorganisms and host eukaryote cells.

Bacterial surface components that are, or contain, the adherence ligands include fimbriae, fibrillae, surface polysaccharides, and specialized terminal structures found, for example, on mycoplasma cells. These components of the bacterial cell are referred to as adhesins. Let us look at them in a little greater detail.

The surfaces of some gram-negative bacteria have structures called fimbriae, also known as pili. Many of these structures have been identified as attachment vehicles. Pili are, for the most part, composed of protein, so they tend to react with specific proteinaceous sites based on host cells—amino acids, sequencing, and peptide configurations.

A number of gram-positive bacteria, as well as gram-negative, contain cell surface hair-like projections termed fibrillae. In *Streptococcus pyogenes* bacteria, these structures are composed of protein and lipoteichoic acid complexes. The lipid portion of the lipoteichoic acid complex appears to be the ligand structure that binds to the complementary protein or glycoprotein host cell sites. *Mycoplasma* spp. and some filamentous bacteria are able to attach by their terminal of attachment structures to host cells, specifically epithelial cells [6,8].

Not all attachment and adherence factors are positive to survival. Phagocytosis of fimbriated bacteria is more efficient than of non–pili-containing cells, apparently because phagocytic attachment is easier, a condition required for phagocytosis [10]. Fimbrial adhesins, lectins (carbohydrate-binding proteins), also mediate a tight bonding adherence between bacteria and host cells. Similar adhesins exist in fungi, viruses, and protozoa. Generally, however, such host advantages are more than offset by the increased virulence such attachment mechanisms provide bacteria.

X. SPREAD

Many microorganisms, including normal flora, remain on the epithelial cell surface without invading underlying tissue. This type of colonization is generally harmless. Most others do not gain access to deeper tissues unless injury is sustained to the area [3].

Once microbial organisms gain entry and adhere within a human host, they encounter host defenses in both humoral and cellular arms of the immune system [6]. In order for the microorganism to spread, it must minimize the effectiveness of the host's immune system, given it is in normal working order.

Some bacteria produce nontoxic substances that inhibit both the humoral and cellular components of the host's immune system. Generally, microorganisms first encounter host immunological phenomena at the epithelial tissues. A common immunoglobin frequently present at these sites is IgA, a secretory antibody. It does not activate, complement, or enhance phagocytosis via opsonization, but it does inhibit bacterial adherence and is antiviral in many instances. It is also reactive with certain bacteria to prevent their absorption into epithelial tissue. However, some bacteria are capable of biochemically clearing and, therefore, inactivating the effects of IgA [3,10].

The human body contains a number of self-protection barriers, such as blood and tissue microbicidal agents, including complement, lysozyme, betalysins, and iron-binding host proteins [6]. However, because this is not the focus of this book, we will look at other aspects of interest in the topical anti-infective arena.

REFERENCES

1. PL Carpenter. Microbiology. 4th ed. Philadelphia: W. B. Saunders, 1977, pp. 24–39.
2. BA Freeman. Burrows Textbook of Microbiology. 22nd ed. Philadelphia: W. B. Saunders, 1985, pp. 1–17.
3. C Mims, N Dimmock, A Nash, J Stephen. Mim's Pathogenesis of Infectious Disease. 4th ed. New York: Academic Press, 1995.
4. VT Schuhart. Pathogenic Microbiology. Philadelphia: J. B. Lippincott Co., 1978.
5. MS Thaler, RD Kalusner, HJ Cohen. Medical Immunology. Philadelphia: J. B. Lippincott Co., 1977.
6. M Schaechter, G Medhoff, BI Eisenstein. Mechanisms of Microbial Disease. 2nd ed. Baltimore: Williams & Wilkins, 1993.
7. RF Boyd. Basic Medical Microbiology. 5th ed. Boston: Little, Brown & Co., 1995.
8. RP Gaynes, TC Horan. Surveillance of nosocomial infections. In: CG Mayhall, ed. Hospital Epidemiology and Infection Control. 2nd ed. Philadelphia: Lippincott, Williams & Wilkins, 1999, pp. 1285–1318.
9. DS Paulson. A broad-based approach to evaluating topical antimicrobial products. In: JM Ascenzi, ed. Handbook of Disinfectants and Antiseptics. New York: Marcel Dekker, 1996, pp. 23–42.
10. DS Paulson. Handbook of Topical Antimicrobial Testing and Evaluation. New York: Marcel Dekker, Inc., 1999.
11. DG Maki. Infections caused by intravascular devices used for infusion therapy: pathogenesis, prevention and management. In AL Bisno, FA Waldvogel, eds. Infections Associated with Indwelling Medical Devices. Washington, DC: ASM Press, 1994, pp. 155–212.
12. E Jawetz, JL Melnick, EA Adelberg. Review of Medical Microbiology. Los Angeles: Lange, 1982.
13. C Crossley, GL Archer. The Staphylococci in Human Disease. New York: Churchill Livingston, 1997.

9
Nosocomial Infection

Daryl S. Paulson
BioScience Laboratories, Inc., Bozeman, Montana

It has been estimated that every year between 5 and 10% of patients admitted to hospitals in the United States and Europe will acquire an infection that was not present before they were admitted to the hospital [1]. A number of these nosocomial (hospital-acquired) infections lead to the patient's death at one extreme or, at the least, require additional antimicrobial chemotherapy. Among critically ill patients, the prevalence of hospital-acquired infection can reach 50% in intensive care units, where patients remain for prolonged periods, often undergoing invasive therapeutic support, such as mechanical ventilation. Within hospitals, the surgical and medical wards usually have the highest infection rates, while pediatric and neonatal services have the lowest.

The Centers for Disease Control and Prevention (CDC) utilizes a subsidiary, the National Nosocomial Infection Surveillance (NNIS) program, to monitor the incidence of nosocomial infection in the United States. Interestingly, the size of the hospital or a medical school affiliation does not seem to matter; the rates of infection across them are equivalent [2]. In point of fact, large hospitals that treat many seriously ill patients, as well as large teaching hospitals, frequently have the highest incidence of nosocomial infectious disease.

The majority of nosocomial infections seem to be endogenous, that is, the infections are due to translocation of microorganisms from the patient's own normal flora. Examples include *Escherichia coli* infections from indwelling bladder catheterizations and *Staphylococcus epidermidis* infections from percutaneously inserted intravenous catheters [3]. The microorganisms migrate from their muscosal or skin habitat along the external surface of the catheters and across the meatal or skin barrier to gain access to the site of infection. Additionally, the environment supporting commensal microorganism populations changes during hospitalization because of mucosal modifications (pH changes, surface receptor changes, etc.) and the selective pressures of antibiotics, which enhance development of antibiotic-resistant strains.

There is also the exogenous route of nosocomial infection caused by microorganisms harbored in the hospital setting. The harbor includes the environment (bedding, sinks, toilets, walls, floors, etc.), other patients, and the hospital staff [4]. Microorganisms are frequently transferred by cross-infection due to hand contact between hospital staff and patients. Cross-infection has been estimated to account for 10–20% of hospital-acquired infections [5]. Hand-body contact, or contact spread, is a mode of nosocomial transmission that is particularly hard to control because the pathogens are environmentally robust and generally more resistant to antimicrobials.

Three principal factors play into the probability of a patient acquiring a nosocomial infection [6,7]:

1. Physiological (including emotional/mental states) susceptibility of the patient to the infectious opportunistic microorganism
2. The virulence of an infection-causing microorganism
3. The nature of the patient's exposure to the infection-causing microorganism

I. PHYSIOLOGICAL SUSCEPTIBILITY OF A PATIENT TO AN INFECTIOUS OPPORTUNISTIC MICROORGANISM

In general, hospital patients have an increased susceptibility to infection [7]. Common chemotherapeutic treatments (e.g., corticosteroids, antineoplastic agents, and antibiotics) individually and synergistically contribute to increased susceptibility [8]. This is due mainly to immune system suppression and/or altering the patient's normal flora, often with replacement by antibiotic-resistant hospital microorganisms, as previously discussed.

Susceptibility is further enhanced by a patient's emotional-mental state. Seriously ill, older patients, patients in a great deal of pain, those with serious chronic diseases, and those under emotional stresses of rejection by their peers, financial difficulties associated with their disease, depression and/or anxiety, and feeling not in control are all more susceptible to nosocomial infection [9].

Surgical procedures often result in nosocomial infections at the surgical site due to the patient's own normal resident, but opportunistic, skin microorganisms (e.g., *S. epidermidis*), which will readily colonize the traumatized tissues. Antibiotic-resistant microorganisms within the hospital environment may also contaminate the surgical wound [10]. Other common medical procedures, such as urinary and/or venous catheterization, as discussed previously, or endoscopy, biopsy procedures, etc., often increase the susceptibility to nosocomial infection

Table 1 Modes of Transmission of Nosocomial Pathogens

Mode of transmission	Reservoir/source	Examples of pathogens
Contact	Patients/healthcare workers, fomites, medical devices	*Staphylococcus aureus* *Enterococcus* spp. Enterobacteriaceae *Clostridium difficile* Respiratory syncytial virus Rotavirus Adenovirus *Candida* spp.
Droplet spread	Healthcare workers, patients	*Staphylococcus aureus* Respiratory syncytial virus Influenza virus
Device-related	Water/respiratory equipment, endoscopes	*Pseudomonas aeruginosa* *Acinetobacter* spp. *Stenotrophomonas maltophilia*
Medication-related	Water/IV fluids, disinfectants	*Burkholderia cepacia* *Acinetobacter* spp. *Serratia marcescens*
Transfusion, needlestick	Patients/blood	Hepatitis B virus Hepatitis C virus HIV, etc.
Transplantation	Patients/donor tissues	Cytomegalovirus *Toxoplasma gondii* Creutzfeld-Jacob agent
Airborne	Patients Hot water/showers Soil/dust	*Mycobacterium tuberculosis* *Legionella* spp. *Aspergillus* spp.
Foodborne	Animals/food products Water/enteral feeding	*Salmonella* spp. *Enterobacter* spp. *Pseudomonas aeruginosa*

by traumatizing tissues and/or breaching skin—or epithelial—barrier properties [11] (Table 1).

II. VIRULENCE OF AN INFECTION-CAUSING MICROORGANISM

Unfortunately, many of the hospital-acquired infections are caused by very virulent microorganisms. This is due, in part, to the very nature of a hospital, serving as a collection point for treating the more serious infectious diseases. Virulence is the ability of a microorganism to cause infectious disease in normal humans. This situation is often manifested in a hospital where a good portion of the patients are immune-compromised by medical/surgical conditions and treatment regimens, not to mention emotional stress.

III. NATURE OF PATIENT'S EXPOSURE

Patients are, for the most part, in a debilitated state in the hospital [12], and many of the nosocomial infectious agents, particularly bacteria, are antibiotic-resistant, even multiantibiotic-resistant. The microorganisms causing disease in patients who are admitted for diagnosed infectious disease are, by definition, virulent [13]. Hence, the spread of nosocomial infection in the hospital setting is a serious and ongoing problem.

Infectious disease can be spread from patient to patient via direct contact with healthcare personnel, food, water, fomites (medical instruments/devices), and airborne transmission. Considering the biological variables, nosocomial infection most probably will never be eliminated, but it can be kept at a minimum level by ongoing infection-control programs.

IV. COMMON NOSOCOMIAL INFECTIONS

As stated previously, one of the more common nosocomial infection sites is the urinary system. Not surprisingly, the gram-negative *E. coli* is primarily responsible for these infections. However, significant numbers of urinary infections are also caused by other gram-negative microorganisms, such as *Klebsiella* spp., *Proteus mirabilis* and other Enterobacteriaceae, and *Pseudomonas aeurginosa*, as well as gram-positive ones, such as *Staphylococcus* spp. and various enterococci. The utilization of Foley urinal catheters increases the risk of urinary tract infections. It is estimated that as many as 20% of all short-term urinary catheterization patients develop a urinary tract infection. In general, viruses are only minimally associated with urinary tract infections.

Most microorganisms that cause endogenous hospital infections have been shown to contaminate transiently the hands of healthcare workers and to be disseminated in that way. Vehicles of cross-contamination infection also include contaminated medical devices (thermometers, endoscopes, electrodes), infected blood products, or transplanted tissues/organs [13].

Individual-to-individual spread of pathogens also follows the fecal-oral route, secreted droplet paths, or airborne channels [6]. Pathogens routinely transmitted from person to person in the hospital setting include hospital-acquired infectious agents, such as viruses [varicella-zoster virus, respiratory syncytial virus (RSV), influenza virus, herpes simplex virus, hepatitis A virus, rotavirus, and adenovirus], bacteria (*Staphylococcus aureus*, *Neisseria meningitidis*, *Haemophilus influenzae*, and *Mycobacterium tuberculosis*), and parasites (*Cryptosporidium* spp.). The risk factors of disease in patients exposed to these agents include age, immune function status, emotional state, adherence to aseptic techniques by hospital staff, patient's condition, number of patients sharing rooms, handwashing practices, and effectiveness of antimicrobial products employed by the staff [12,14].

V. SURGICAL SITE INFECTIONS

Two main complications after surgery are infection and the systemic inflammatory response syndrome (SIRS), an activation of the inflammatory cascade. The SIRS, a response often but not always seen accompanying infection [15], is a systemswide inflammatory response to surgical trauma and injury and/or other noninfectious insults to the patient that result in the activation of pro-inflammatory cytokines and other biochemical mediators. Sepsis is defined as the systemic inflammatory response with microbial infection and is considered severe when organ dysfunction occurs [16]. The term *septic shock* is used when hemodynamic abnormalities are present (e.g., lysis of red blood cells). Multiple organ dysfunction syndrome (MODS) is a term describing malignant failures in organ function in acutely ill patients.

Sources of infection in surgery patients are primarily the surgical wounds themselves [12,15]. Wound contamination may derive from the patient's own bacteria (e.g., *S. epidermidis*) or from the surgical team performing the surgery, the hospital staff, or the environment, especially if prolonged support (antibiotics, IVs, intubation, etc.) is required [3]. The most common causes of postsurgical site infections are *S. aureus* and *S. epidermidis* [17]. There are also significant postsurgical infection rates due to gram-negative rods (e.g., *E. coli*) particularly in abdominal, bowel, or pelvic surgeries [7]. *Candida* spp. and enterococci are also seen in many postsurgical infections because they are, to a large part, dependent upon the source of transmission (punctured surgical gloves, patient's normal flora in close proximity to a surgical site, airborne microbial species, contamination to the site by healthcare personnel in charge of dressing, etc.).

Risk factors in surgical site infections include types of surgery (clean, clean-contaminated, contaminated, or dirty), advanced age of patient, obesity, malnutrition, extended hospitalization, diabetes, extended surgery time, immune compromise, and inappropriate antibiotics used prophylactically to prevent postsurgical infections [12].

VI. ROLES OF INFECTION

Incidence of complications due to infection after surgery is estimated to be 20–30% but is higher for patients requiring intensive care. It has been reported that among surgical patients with stays longer than 30 days, 63% developed pneumonia, 69% bacteremia, and 43% urinary infections [12]. The risk of developing surgical infections is increased by the extent of the surgical procedure, the degree of contamination at the surgical site, the psycho-physical state (organ function level, degree of chronic illness, stress level, affect state, and world view), blood loss/replacement during surgery, and the duration of the hypotensive state [12,18].

The role of infection as a postsurgical concern has been rated minimal by some and high by others, particularly as a cause of multiple organ failure (MODS). Both types of MODS—rapid (within 72 hours of a surgical procedure) and delayed (beyond 72 hours)—are associated with high infection rates—in the delayed form, about 90% [10].

Most patients undergoing major and prolonged surgery develop a systemic inflammatory response within a few days, which is a normal, adaptive phenomenon. Generally, SIRS at a moderate level is beneficial to the patient but, at high levels, is potentially harmful, even fatal [15,16]. Cardiac and vascular surgeries with aortic cross-clamping are procedures that predispose a patient to potentially severe SIRS, MODS, and acute respiratory distress syndrome (ARDS). The exact incidence of SIRS and MODS after other types of surgeries (gastroenteric, orthopedic, or urological) is not known.

Microbial translocation, particularly bacterial, is a possible cause of activation of SIRS. Surgical patients are at a high risk of bacterial translocation* secondary to enteric overgrowth, intestinal ischemia, bowel status, or hemorrhagic shock. However, significance of bacterial translocation in the development of MODS is unclear.

After the surgery and onset of SIRS, negative feedback systems tend to limit or neutralize the auto-destructive inflammatory process. While this is beneficial from a self-destructive perspective, it is often associated with major post-surgical infections [19].

VII. ANTIBIOTIC-RESISTANT MICROORGANISMS

It has long been known that strains of microorganisms responsible for nosocomial infections have evolved over the years in response to antibiotic therapies and

* Bacterial translocation is the passage of bacteria or their toxins from surgically involved tissues (e.g., intestinal lumen) to other anatomical sites (e.g., heart and circulatory system, liver, spleen, and lymph nodes).

have become increasingly resistant to them [20]. The risk of acquiring a nosocomial infection then increases with the length of hospital stay. After the massive utilization of penicillin and sulfa drugs in the 1940s and 1950s, the first markedly resistant microbial strain to appear was of *S. aureus*. With the narrow-spectrum antimicrobials—cephalosporins and aminoglycosides—gram-negative rods such as *Klebsiella*, *Serratia*, *Enterobacter*, and *Pseudomonas* are commonly resistant. Additionally, there are now methicillin-resistant *S. aureus* (MRSA) and vancomycin-resistant enterococci (VRE) strains with which to contend. It has also been observed that patients' normal microflora (e.g., viridans streptococci, saprophytic *Neisseria* spp., and diptheroids) are replaced by resistant microorganisms found in the hospital environment.

VIII. INFECTION-CONTROL PROGRAMS

Hospital infection-control programs, at least in theory, are designed to detect and control the spread of nosocomial infections. Recall that nosocomial infections are spread in at least four ways [12,13,21]:

1. Direct contact: food, contaminated instruments, or punctured surgical gloves
2. Indirect contact: hand contact between patients and healthcare personnel
3. Droplet contact: exposure to droplets from coughing, etc. (>5 μm)
4. Aerosolized droplets: inhalation of droplets (≤ 5 μm)

While handwashing, scrubbing, or prepping cannot prevent or influence all of these, they will be the focus of this book.

The most important aspects for topical antimicrobials include:

1. Healthcare workers washing their hands with effective (validated) antimicrobials and methods.
2. Surgical personnel performing adequate (validated) scrub procedures with effective antimicrobials.
3. Adequate prepping of patients with effective and relevant antimicrobials prior to injections, phlebotomy procedures, catheterization (venous, arterial, or urethral), and surgical procedures.

IX. PREOPERATIVE SKIN PREP

In order to be effective, preoperative skin preparation formulations must degerm an intended surgical site rapidly as well as provide a high level of bacterial inactivation and persistent antimicrobial activity—up to 6 hours—post–skin prepping

[22]. The preoperative skin preparations should demonstrate a broad spectrum of antimicrobial activity against both gram-negative and gram-positive bacteria, as well as fungi, by means of a *valid* time-kill study and a *valid* minimum inhibitory concentration study per each tested microorganism. Although *S. epidermidis* is the most prominent microorganism on the skin, it is by no means the most threatening.

Many microorganisms produce a slime (glycocalyx), or biofilm, that protects them from phagocytes, other host immunological reactions, and many antibiotics. *S. aureus*, a much more virulent potential pathogen than *S. epidermidis*, often ''normally'' colonizes the anterior nares, nasopharynx, perineal region, and skin. Once provided entry via a surgical wound site, it is able to produce alpha, beta, gamma, and delta toxins (hemolysins), which enhance its virulence by destruction of host cells, including red blood cells, monocytes, and platelets. The enzyme leukocidin, secreted by *S. aureus,* is a potent toxin to phagocytes—neutrophiles, monocytes, and macrophages. Its secretion of clumping factors, coagulase, and hyaluronidase enhance its survival in infected tissues [17]. Additionally, protein A, a surface component of most *S. aureus* strains, competes with neutrophils for the Fc portion of IgG antibodies and, thereby, is antiphagocytic. Staph infections are far worse in patients with diabetes, burn patients, and those who have undergone extensive surgical procedures [14,17].

S. epidermidis is a normal, opportunistic microbial skin resident and is not a threat until the skin is compromised by a surgical incision. Once the skin is compromised, the microorganisms, being opportunistic, can enter the patient, rapidly colonize the wound, and spread throughout the body. They release exotoxins, including hemolysins, and a variety of enzymes (7).

Invasive microorganisms enter cells in the vicinity of the patient's surgical site that are not naturally phagocytic [7]. Penetration of these cells is achieved by specific attachment to the host cells by the microorganism and induction of local rearrangement of the cytoskeleton through polymerization and depolymerization of actin [23].*

Ultimately, the microorganism is engulfed by a phagocyte, particularly once opsonization (coating of surface proteins by host antibodies, stimulating phagocytosis) has occurred. If the microorganism is resistant to phagocytosis, the microorganism may be transported across superficial epithelial tissue to be released in the subepithelial space—a process call transcytosis [7]. After transcytosis, the underlying tissues can be invaded. This process can continue and spread throughout the body.

* Actin is released through the tail of a nonmotile microorganism. The elongation process has sufficient force to move microorganisms through cellular tissue. Actin remains stationary, the microorganism moves. Microorganisms that are motile on the basis of actin include *Listeria monocytogenes*, *Shigella* spp., and various viruses.

Other microorganisms that can also be transient on the skin and cause postsurgical infection are *E. coli*, *Enterococcus* species, *P. aeruginosa*, *Enterobacter* species, *Candida albicans*, *Klebsiella pneumoniae*, and *Proteus mirabilis* [7,12].

Current FDA regulations require that preoperative prepping solutions be evaluated in challenge by using 25 ATCC species and 25 clinical isolates of 22 species of bacteria/yeast, quite a challenge to a preoperative preparation formulation. The ATCC microorganisms listed in the 1994 TFM are [24]:

Gram-negative: *Acinetobacter* spp., *Bacteroides fragilis*, *H. influenzae*, *Enterobacter* spp., *E. coli* (ATCC ##11229 and 25922), *Klebsiella* spp. (including *K. pneumoniae*), *P. aeruginosa* (ATCC ## 15442 and 27853), *P. mirabilis*, and *Serratia marcescens* (ATCC #14756).

Gram-positive: staphylococci [*S. aureus* (ATCC #6538 and 29213), *S. epidermidis* (ATCC #12228), *Staphylococcus hominis*, *Staphylococcus haemolyticus*, and *Staphylococcus saprophyticus*], *Micrococcus luteus* (ATCC #7468), and streptococci [*Streptococcus pyogenes*, *Enterococcus faecalis* (ATCC #29212), *Enterococcus faecium*, and *Streptococcus pneumoniae*].

Yeasts to be tested include *Candida* spp., particularly *C. albicans.*

Those products that have good biocidal action against these microorganisms can be considered effective. Yet, surprisingly, many manufacturers of over-the-counter (OTC) formulations do not have documented evidence of effectiveness from in vitro studies [time-kill and minimum inhibitory concentration (MIC)] to support their label claims for preoperative skin preps. In vivo studies performed on human volunteers are designed to assure that the products are antimicrobially effective within 10 minutes of application and are antimicrobially persistent for up to 6 hours postapplication.

The kinds of antimicrobial products used most commonly as preoperative preps are iodophors and chlorhexidine gluconate products.

X. PRECATHETERIZATION PREPS

The precatheterization (venous or arterial) skin preps are very similar to the preoperative skin preps in terms of efficacy requirements. And there is a definite thrust in the industry for long-term precatheterization insertion preps—those that remain effective for over 24 hours [22]. A substantial, yet underappreciated potential exists for bloodstream infections originating at the vascular access site [25]. More than one-half of all epidemics of nosocomial bacteremia or candidemia are caused by vascular access of some type, and these are, for the most part, preventable.

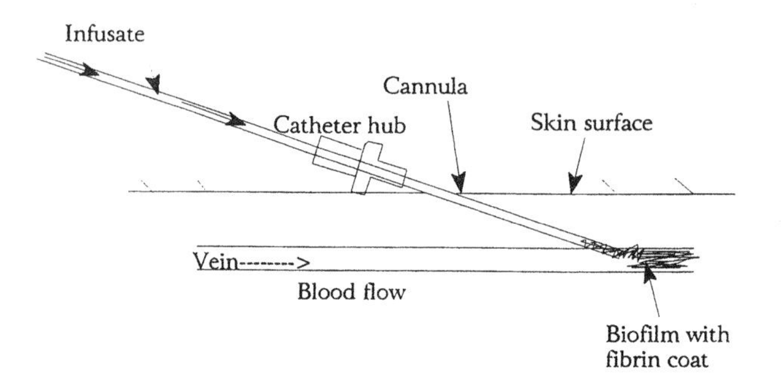

Figure 1 Intravascular portion of the cannula. (From Ref. 25.)

Nosocomial bloodborne infections in hospitalized patients are associated with a two- to threefold increase in mortality. The origin of infection is usually either cannula-related or infusate-related. Cannula-related infections include those derived from percutaneous devices used for vascular access (e.g., needles, hubs, and plastic catheters). Maki reports that between 5 and 25% of intravascular devices are microbially colonized at the time of vascular withdrawal [25]. Large numbers of microorganisms are observed on the intravascular portion of the cannula or its tip (Fig. 1).

Coagulase-negative *Staphylococcus* species, particularly *S. epidermidis*, are the most common causes of catheter-related bacteremia [25]. Heavy colonization of the skin-insertion site has been shown to be strongly correlated with catheter-related bacteremia. In hemodialysis patients, the risk of *S. aureus* bacteremia is six times greater than in nonhemodialysis patients. And numerous incidents of intravascular infection have been traced to microbially contaminated topical disinfectants.

One of the most serious types of infection occurs when the blood clot biofilm surrounding the intravascular portion of the cannula becomes microbially colonized, leading to phlebitis, a bloodstream infection that usually persists after the cannula has been removed. This is an extremely serious condition, particularly in burn and ICU patients. Before leaving this topic, let us discuss cannula-related infection in more detail.

XI. PERIPHERAL VENOUS CATHETERS

Small Teflon® or polyurethane catheters, as well as scalp-vein and butterfly needles inserted peripherally (e.g., anterior cubital region, including the forearm and wrist), are currently associated with very low incidence of infection—about one in 500.

XII. ARTERIAL CATHETERS

Arterial catheters are generally used to monitor blood O_2 and CO_2. The rate of catheter-related bacteremia is about 1%.

XIII. UMBILICAL CATHETERS

For newborns, catheterization of the umbilical artery or vein is used to provide vascular access. It is estimated that the bacteremia rate in this region is about 5%.

XIV. CENTRAL VENOUS CATHETERS

Central catheters have been shown to be the most important risk factor in nosocomial *Candida* infections, which rival in seriousness any underlying disease. Catheters inserted in the subclavian or internal jugular vein have an infection rate of 3–5%—in some hospitals, 7–10%. Percutaneous inserted, noncuffed venous catheters used in hemodialysis are associated with the highest infection rate, 10%.

XV. PREINSERTION TOPICAL PREPS

It is exceedingly important that the topical antimicrobial products used for these indications are fast-acting and effective [22]. Unlike preoperative skin preps, the invasive procedure is performed less than 10 minutes after prepping. So that the skin can be sufficiently degermed, many manufacturers have been developing combination drugs that are both fast-acting and have long-term antimicrobial persistence. Alcohols (70% IPA) are commonly used in preps to provide immediate degerming of the skin, but these lose all antimicrobial activity upon evaporating [22]. To counter this down side, manufacturers have been adding iodophors or chlorhexidine gluconate to ensure antimicrobial persistence—up to 48 hours post–skin prepping.

XVI. PREINJECTION/PHLEBOTOMY SKIN PREPS

For the most part, there is considerably less risk of nosocomial infection for these applications because the device is inside the patient's body so briefly. It appears that 70% alcohol alone is adequate for this application. However, other antimicrobials, such as povidone iodine and chlorhexidine gluconate, are also used.

XVII. HEALTHCARE PERSONNEL HANDWASH

Healthcare personnel handwash formulations need to remove and/or destroy quickly any transient microorganisms picked up on the hands of a healthcare provider from patient A and prevent their passage to patient B [22]. The product is intended to break this disease-transmission cycle at the level of the contaminated healthcare workers' hands by removing pathogenic microorganisms from their hands.

Importantly, the product must demonstrate low skin irritation potential upon repeated and prolonged usage—20–30 washes per day—over the course of 5 consecutive workdays, or personnel are likely to comply poorly with handwashing requirements. This mildness, however, is usually attained at the price of reduced antimicrobial efficacy [22].

The healthcare personnel handwash formulation must be antimicrobially active against a broad spectrum of microorganisms (gram-negative and gram-positive bacteria, fungi, and viruses) after only a 15- to 30-second exposure time, and usually at a diluted strength. In efficacy testing, it is highly desirable that the antimicrobial formulations be evaluated using the microorganisms that healthcare personnel are most likely to contact.

All of the microorganisms used to evaluate surgical scrubs should be evaluated at a use-dilution level for the healthcare personnel antimicrobial formulation. This should also be an ongoing process for healthcare facility infection control personnel. Clinical isolates should be periodically evaluated against the product formulations used [22].

Additionally, when healthcare workers are palpating an infected area or performing a procedure in a microbially contaminated anatomical region, latex exam gloves should be worn, generally a standard procedure. Thereby, the hands are never directly exposed to large populations of microorganisms. However, this in no way reduces the need for antimicrobial efficacy in healthcare personnel handwash formulations.

There are many actives used in healthcare personnel handwash formulations, including 70% ethyl or isopropyl alcohol, iodophors, quaternary ammonias, PCMX, triclosan, and chlorhexidine gluconate [22]. Recently, there has been much interest in waterless healthcare personnel handwash formulations. These are used by healthcare personnel when they do not have access to sinks and running water [22]. Most of these products are alcohol-based, with many being combinations of alcohol and chlorhexidine gluconate. It is very difficult for these nonrinse products to not only provide large immediate reductions in contaminating microorganisms but to demonstrate persistent antimicrobial effects, that is, upon recontamination with pathogenic microorganisms to continue effective degerming of the hands. All too often, with no water wash, the hands, upon being recontaminated, tend to experience a build-up of organisms, obviously an undesirable situation.

XVIII. PRESURGICAL SITE WASHES

It has long been debated whether or not a preoperative skin preparation is adequate in degerming the skin prior to making a surgical incision. The FDA has a 3 $\log_{10}$ reduction criterion, but in highly colonized areas of the skin (e.g., inguinal), most individuals have microbial populations of over 1.0×10^5 cm^2 [22]. Many surgeons recognize this problem, and, in order to promote greater degerming of the proposed surgical site, patients are requested to wash the area daily, or more often, prior to a surgical procedure. The intent is to reduce the normal microbial population at the presurgical site so that when the region is prepped prior to surgery, the organisms then, being few in number, are virtually totally removed from the skin.

A problem, however, is that not all antimicrobial compounds have residual properties. Residual antimicrobial properties are those properties achieved by the antimicrobial compound binding to the stratum corneum. Hence, as the product is used repeatedly, it increases on the skin and is not removed by perspiring or washing.

Of the various products containing iodophors, alcohols, parachlorometaxylenol, triclosan, quaternary ammonia, or chlorhexidine gluconate, only those with chlorhexidine gluconate have demonstrated this residual antimicrobial property. There have been arguments as to whether a CHG product will provide significant added protection against infection. According to the literature, CHG products must be applied to the presurgical site at least 2 days prior to being preoperatively prepped. Additionally, antimicrobial effects are noted using the CHG up to 5 days prior to a surgery. Use more than 5 days out appears not to provide an effect.

In summary, nosocomial infections are extremely important in the hospital setting, and the development and proper utilization of effective topical antimicrobials is a first-line and crucial aspect of preventing them.

REFERENCES

1. KJ Rothman, S Greenland. Measures of disease frequency. In: KJ Rothman, S Greenland, eds. Modern Epidemiology. 2nd ed. Philadelphia: Lippincott, Williams & Wilkins, 1998, pp. 29–46.
2. MJ Struelens. Hospital infection control. In: D Armstrong, J Cohen, eds. Infectious Diseases. Vol. 1. St. Louis: Mosby, 1999.

3. D Pittet, RP Wenzel. Nosocomial bloodstream infections: secular trends, rates, mortality, and contributions to total hospital deaths. Arch Intern Med 155:1177–1184, 1995.
4. PS Falk. Infection control and employee health services. In: CG Mayhall, ed. Hospital Epidemiology and Infection Control. 2nd ed. Philadelphia: Lippincott, Williams & Wilkins, 1999, pp. 1381–1386.
5. RW Haley, DH Culver, JW White. The efficacy of infection surveillance and control programs in U.S. hospitals. Am J Epidemiol 121:182–205, 1985.
6. CG Mayhall. Hospital Epidemiology and Infection Control. Philadelphia: Lippincott, Williams & Wilkins, 1999.
7. C Mims, N Dimmoch, A Nash, J Stephens. Mim's Pathogenesis of Infectious Disease. 4th ed. New York: Academic Press, 1995.
8. CM Mindorff, DJ Cook. Critical review of hospital epidemiology and infection control. In: CG Mayhall, ed. Hospital Epidemiology and Infection Control. 2nd ed. Philadelphia: Lippincott, Williams & Wilkins, 1999, pp. 1273–1282.
9. GS Dacher. Psychoneuroimmunology. New York: Paragon, 1991.
10. RG Dixon. Costs of nosocomial infections and benefits of infection control programs. In: RP Wenzel, ed. Prevention and Control of Nosocomial Infections. Baltimore: Williams & Wilkins, 1987, pp. 19–25.
11. LB Laxson, MJ Blaser, SM Parkhurst. Surveillance for the detection of nosocomial infections and the potential of nosocomial outbreaks. Am J Infect Control 12:318–1324, 1984.
12. ES Wong. Surgical site infections. In: CG Mayhall, ed. Hospital Epidemiology and Infection Control. 2nd ed. Philadelphia: Lippincott, Williams & Wilkens, 1999, pp. 189–210.
13. Association of Operating Room Nurses. Recommended practices: sterilization and disinfection. AORJ J 45:440–462, 1987.
14. JS Garner, WR Jarvis, TG Emori. CDC definitions for nosocomial infections. Am J Infect Control 16:128–140, 1988.
15. CA Dietch. Multiple organ failure: pathophysiology and potential future therapy. Am Surg 210:116–123, 1992.
16. J Carlet, M Garrouste-Orgeas, J Timsit, P Moine. Infective complications after trauma and surgery. In: D Armstrong, J Cohen, eds. Infectious Disease. Vol. 1. St. Louis: Mosby, 1999, pp. 178–203.
17. KB Crossley, GL Archer. The Staphylococci in Human Disease. New York: Churchill Livingston, 1997.
18. RP Gaynes, TC Horan. Surveillance of nosocomial infections. In: CG Mayhall, ed. Hospital Epidemiology and Infection Control. 2nd ed. Philadelphia: Lippincott, Williams & Wilkins, 1999, pp. 1285–1318.
19. A Kurz, DI Sessler, R Lenhardt. Perioperative normal thermia to reduce the incidence of surgical wound infections and shorten hospitalization. N Engl J Med 334: 1209–1215, 1996.
20. F Schmitz, AC Fluit. Mechanisms of resistance. In: D Armstrong, J Cohen, eds. Infection Diseases. Vol. II. St. Louis: Mosby, 1999, pp. 863–885.
21. G Pugliese, B Lambert, KA Kroc. Development and implementation of infection control policies and procedures. In: CG Mayhall, ed. Hospital Epidemiology and

Infection Control. 2nd ed. Philadelphia: Lippincott, Williams & Wilkins, 1999, pp. 1357–1366.
22. DS Paulson. Handbook of Topical Antimicrobial Testing and Evaluation. New York: Marcel Dekker, 1999.
23. J Verhoef, F Schmitz. Staphylococcis and other micrococcaceae. In: D Armstrong, J Cohen, eds. Infectious Diseases. Vol. II. St. Louis: Mosby, 1999, pp. 232–297.
24. Tentative Final Monograph for Antiseptic Drug Products, Part 21 CFR, Parts 333 and 369. Food and Drug Administration, Department of Health and Human Services, Part 3. Federal Register, Vol. 59, No. 116, Friday, June 17, 1994, pp. 31402–31452.
25. DG Maki. Infections caused by intravascular devices used for infusion therapy: pathogenesis, prevention, and management. In: AL Bisno, FA Waldvogel, eds. Infections Associated with Indwelling Medical Devices. 2nd ed. Washington, DC: ASM Press, 1994, pp. 155–212.

10
Postsurgery Infection

Daryl S. Paulson
BioScience Laboratories, Inc., Bozeman, Montana

Despite advances in surgical procedures, increased understanding of wound pathologies, widespread use of broad-spectrum antibiotics, and more effective topical antimicrobial skin preps and dressings, postsurgical infection remains a source of morbidity and mortality for surgical patients [1]. It has been estimated that 2–5% of all surgical patients per year (approximately 16 million) develop a postsurgical infection. Postsurgical infections account for about 24% of all nosocomial infections, the second most common—urinary infections being the first. Postsurgical site infections prolong the hospital stay by 7.4 days on average. The total added cost, including indirect costs, may exceed $10 billion per year [2].

The incidence of surgical site infections traditionally has been stratified by the National Nosocomial Infection Surveillance (NNIS) system. Participating hospitals are categorized by size and medical school affiliation. Surgical site wounds are classified by the degree of microbial contamination at the operative site. The classification scheme is as follows [1,3]:

Clean sites: surgical sites in which no inflammations were encountered, and the alimentary, genital, respiratory, or urinary systems were not operatively involved. Clean wounds are also closed wounds, and, if necessary, drainage is closed.

Clean-contaminated sites: alimentary, genital, respiratory, or urinary operative sites that are without unusual contamination. Specifically, surgery involving the appendix, biliary tract, vagina, or oropharynx are categorized in this group.

Contaminated sites: surgeries involving open, fresh, accidental trauma, or surgeries with major aseptic technique violation or much leakage of contents from the intestinal tract. Such sites include infected biliary tracts

with infected bile, infected urinary tracts, or those from general surgical procedures that are inflamed, but not purulent.
Direct infection sites: old traumatic wounds with devitalized tissue, foreign bodies, or fecal contamination, or sites that are also purulent.

In practice, there is disagreement as to the relevance of these categories, to correlating postsurgical infection rates with surgical site category. The general problem is that clean sites often progress to postsurgical infections. Other variables enter the picture, including degerming efficacy of the presurgical preps used, contamination originating from surgical personnel, and contamination of surgical instruments.

I. INFECTION

From a clinical perspective, a surgical site is infected when purulent drainage is present at the incision site. Generally, local swelling, erythema, tenderness, and elevated skin temperatures at the site are noted. Many hospitals require that purulent exudate from the incision site and/or drainage tube be cultured. This enables infection-control personnel to trend infection locale and microbial species as part of a program to control nosocomial infection rates (surveillance and epidemiology) and to track antibiotics to which the cultured microorganisms are sensitive. The microorganisms most commonly responsible for postsurgical infections are shown in Table 1 [1].

Table 1 Microorganisms Commonly Isolated from Surgical Site Infections (1990–1992)

Microorganism	Percentage
Escherichia coli	8
Coagulase-negative *Staphylococcus* spp. (e.g., *S. epidermidis*)	14
Enterococcus spp.	12
Pseudomonas aeruginosa	8
Staphylococcus aureus	19
Enterobacter spp.	7
Candida albicans	3
Klebsiella pneumoniae	3
Proteus mirabilis	3
Streptococcus spp.	3

Staphylococcus aureus and coagulase-negative *Staphylococcus* spp. (e.g., *Staphylococcus epidermidis*) are the microorganisms most commonly isolated from clean surgical processes that subsequently become infected. In surgical sites involving the gastrointestinal tract, the respiratory tract, or the female reproductive tract, multiple microbial species are generally involved (mixed etiology), usually including microorganisms (anaerobic and aerobic) endogenous to the system surgically involved.

In recent years, antibiotic-resistant strains of bacteria, both gram-positive and gram-negative, increasingly have been involved in surgical site infections. Because an increasing number of patients are immunocompromised, fungi, especially *Candida albicans* and *Candida tropicalis*, are more common in surgical site infections [4]. Also, incidence of surgical site infections due to unusual microorganisms is increasing, e.g., *Rhizopus rhizopodiformis* infections due to contaminated adhesive dressing tape [5]. Moreover, recent research has shown that the role of anaerobes in mixed infections has probably been greatly underestimated due primarily to insensitive techniques of sampling and culturing [6].

Microbial surgical site infections are due to endogenous microorganisms of the patient and/or exogenous sources.

II. ENDOGENOUS ETIOLOGY

The patient's own resident microflora, which are opportunistic, account for the majority of surgical site infections. Coagulase-negative staphylococci, as well as *S. aureus*, account for the majority of these infections [1]. As long as the skin integrity is not compromised, these organisms do not usually cause infections. However, because they become pathogenic when introduced into unnatural environs (so-called privileged sites), much care must be taken to use preoperative skin preparations that are effective. Better yet, for elective surgical procedures, patients should be instructed to use an effective presurgical wash at home for several days prior to surgery. Such preparations greatly reduce the microbial populations at the proposed surgical site prior to prepping [4,7].

III. EXOGENOUS ETIOLOGY

The hands of surgical staff members often have been implicated in surgical infections due to direct (hand–to–surgical site) contact. Usually, the surgical gloves have become compromised—ripped, torn, or punctured—or have preexisting punctures, allowing skin microorganisms to pass from the occluded, gloved hands to the surgically involved tissues [8]. In order to reduce the probability of this source of infection, it is important that the surgical staff perform an effective presurgical scrub using a hand-cleansing product proven to be antimicrobially effective.

Exogenous contamination also can derive from the hair and scalp of the surgical staff members [8], as well as the nares, oropharynyx, and skin of the face and neck [1,2,5]. The hospital environment also provides a potential source for surgical infections but is considered of relatively lower risk. The use of effective disinfectants for wet-mopping the floors of operating rooms between surgeries is a wise practice [2]. The air itself can also be a significant contributor to surgical infections, usually related to ''clean'' surgical procedures.

IV. IMPORTANT RISK FACTORS IN SURGICAL SITE INFECTIONS

In 1965, a study by Altemeir and Culbertson determined the risk factors associated with surgical infections: surgical infection risk (1) varies directly in proportion to the dose level of the microbial contaminants, (2) varies directly with the virulence of the microorganism(s), and (3) varies inversely with the immunological competence of the patient in controlling microbial invaders [9].

A surgical site in poor condition—one that is poorly vascularized, contains damaged tissue, is contaminated with foreign bodies, or contains necrotic tissue—is at a higher risk of infection given the same microbial contamination. Surgical-site integrity is also influenced by underlying disease in the patient and surgical trauma (skill of the surgeon) [10]. Other factors include:

1. Length of hospitalization
2. Preoperative shave
3. Length of operation
4. Surgical procedure
5. Other infection in the patient
6. Abdominal drain

A. Length of Hospitalization

Studies over the years have consistently demonstrated the negative effects of prolonged hospital stays [11]. The longer the stay, the greater the probability of a postsurgical infection. While the exact mechanism of this phenomenon is not known, several factors are involved. The longer the patient requires hospitalization in general, the more the patient has been compromised by the stress of underlying disease and trauma from the surgery and, hence the more taxed will be the patient's immune system. Generally, during the hospital stay, the number of procedural interventions allowing microbial access to areas of the body not normally disposed to microbial access is higher than for outpatients. This also is true of chemotherapeutic interventions that adversely affect the patient's immunological resistance (e.g., corticosteroid therapy) or alter the normal resident microflora (e.g., antibiotic therapy) [1]. Research has shown that patients hospitalized for cardiovascular surgery very quickly became colonized with coagulase-negative staphylococci and that these microorganisms were often responsible for surgical-site complications (e.g., prosthetic valve endocarditis) [12].

B. Preoperative Shave

While not conclusive, shaving a proposed skin site prior to surgery appears actually to increase the risk of postsurgical infections. In a study of 406 randomly selected patients, the infection rate after shaving was 5.6%, compared with 0.6% when hair was removed by a depilatory [13]. When no hair was removed the infection rate was also 0.6%.

The infection rate is also dependent upon when the shave is performed, at least when a razor is used. The infection rate of patients razor-shaved just prior to surgery was 3.1%, compared with a rate of 7.1% when shaving was performed within 24 hours of surgery. In patients razor-shaved more than 24 hours prior to surgery, the infection rate was greater than 20%.

In another study a similar pattern was observed [14]. Razor-shaving produced the highest infection rate, 2.5%. Clipping the area only resulted in an infection rate of 1.7%. Patients who were shaved with an electric razor showed an infection rate of 1.4%, and those who were not shaved or clipped at all had the lowest infection rate of 0.9%.

Hamilton and Lone applied a scanning electron microscopic procedure to understand this phenomenon [15]. Electron micrographs of razor-shaved individuals revealed frank skin cuts, the clipping of hair produced less skin trauma, and depilatory agents caused no injury to the skin. The increased infection rates observed were likely the result of compromise to the natural barrier function of the skin.

C. Length of Surgical Procedure

The length of a surgical procedure has long been known to be an important factor in postsurgical infection rates: the longer the surgery, the greater the risk of postsurgical infection. Cruse and Foord observed the relationship between infection rate in clean surgical wounds and time in surgery (Table 2) [14].

Controlled efficacy studies have shown that, after skin prepping (abdominal area and inner aspect of the thigh at the groin), significant recolonization is present several hours thereafter. However, this phenomenon is very product- and application-dependent. Currently, 4% chlorhexidine gluconate and 10% povidone

Table 2 Relationship Between Infection Rates in Clean Surgical Wounds and Time in Surgery

Length of surgery, hr	1	2	3
Incidence of postsurgical infection	1.3%	2.7%	3.6%

iodines seem to provide the greatest persistent antimicrobial effects. Yet new products worth exploring are constantly being developed [4].

An important consideration is the microbial colony counts on the gloved hands of surgical staff members. Numbers of bacteria increase as time passes after the surgical scrub procedure as a result of the occlusion by the gloves. A break or tear in the gloves several hours after the surgery begins can increase dramatically the size of an opportunistic pathogen inoculum into a surgical site [8]. Again, antimicrobial efficacy of a surgical scrub is not only product-dependent, but depends strongly, too, on quality and length of the scrub procedure. The two product formulations currently most effective are 10% povidone iodine and 4% chlorhexidine gluconate. Chlorhexidine gluconate has a theoretical advantage when surgical staff members have been scrubbing with the product for at least 2 consecutive days. The CHG binds to the stratum corneum and keeps the microbial bioburden down, even decreasing it over a period of continuous use.

Other aspects of prolonged surgeries that may be contributory to postsurgical infection are (1) increased tissue damage from drying, prolonged retraction, and surgical manipulation, (2) more suture required to close the surgical wound layers, as well as increased trauma from electrocoagulation, and (3) increased immunological trauma from both blood loss and systemic shock [10].

D. Surgical Technique

The skill of the surgeon also plays a key role in incidence of postsurgical infections [16,17]. As can be imagined, technique will relate directly to the degree of surgical wound trauma and, therefore, the susceptibility of the wound to infection. Surgeon-related iatrogenic infection is minimized by good control of bleeding, gentle retraction and handling of tissue, not inadvertently entering a viscus, and minimization of dead spaces. The number of specific procedures a surgeon has performed correlates directly with the incidence of infection.

E. Presence of Other Infection

The presence of other infections prior to surgery has been linked to an estimated threefold increase in postsurgical infections [18].

F. Abdominal Drains

While surgical drains are useful, they also present potential access for infectious organisms. Studies have shown that in a significant number of cases, *S. aureus* and *S. epidermidis* are present in samples from the interior of these drains [14].

G. The Patient Factor

Patient demographic characteristics are also important in the development of postsurgical infections. Of the many facets that may predispose to postsurgical infections, advanced age is a primary risk factor. Other facets include immunocompetence, obesity, and general health status of the patient [14].

V. PREVENTION OF SURGICAL-SITE INFECTIONS

Two areas in critical need of further study are the effectiveness of patient preoperative preparations and of surgical scrub products and procedures.

A. Patient Preoperative Preps

It is important that a preoperative skin prep be effective in greatly reducing the populations of normal skin flora, as well as numbers of contaminative microorganisms at the proposed surgical site. At dry skin sites, the microbial populations generally average about 1×10^3 CFU/cm^2. At moist skin sites, average microbial counts are about 1×10^5 CFU/cm^2. There is, however, tremendous variability between humans at these sites.

Currently, the FDA requires that preoperative skin preps reduce the microbial populations on the skin of moist sites (e.g., inguinal crease at the upper, innermost aspect of the thigh) by 3 $\log_{10}$ within 10 minutes of prepping and that populations not exceed baseline counts for 6 hours. At dry sites (e.g., abdomen), a 2 $\log_{10}$ reduction in microorganisms from the baseline number is expected within 10 minutes of prepping, and counts cannot exceed baseline within 6 hours.

Hence, if a patient had 1×10^6 microorganisms per cm^2 at the proposed operative site, as long as the population did not exceed 10^6 after 6 hours, the preparation could be marketed. Clearly, an increased risk of infection can be inferred, the longer the surgery and the greater the population of microorganisms a patient generally harbors at the operative site.

The most commonly used antimicrobial compound that displays strong residual antimicrobial properties is chlorhexidine gluconate (CHG) [4,7]. When products containing CHG are used on the skin repeatedly over time, the bacterial populations decline progressively to levels much lower than the preuse baseline.

Such residual activity is not observed nearly as graphically following use of povidone iodine, parachlorometaxylenol (PCMX), triclosan, or quaternary ammonium products. The duration of the skin, the pressure applied, the method (e.g., circumferentially outward or back and forth), and the amount of product used are all important variables in the skin preparation's efficacy.

B. Preoperative Shower Wash

One easy and potentially effective adjunct to a preoperative skin prep is a preoperative shower wash [7] that provides residual antimicrobial properties, i.e., upon repeated and prolonged use the product binds to the skin, thereby maintaining reduced population levels. Then, when the preoperative skin prep is applied, the baseline counts are already much reduced, and the prep nearly sterilizes the region. As mentioned, CHG-based products are at the forefront of residual activity.

C. Surgical Scrub Wash

For a product to qualify as a surgical scrub, the FDA requires following a series of 11 surgical scrubs with the product over the course of 5 consecutive days, $\log_{10}$ microbial reductions of 1 $\log_{10}$ after day 1, 2 $\log_{10}$ after day 2, and 3 $\log_{10}$ after day 5 within one minute following the first scrub on each of those days. Moreover, the microbial counts cannot exceed baseline counts in samples taken after 6 hours of glove occlusion on any of the 3 test days on which sampling occurs.

A hypothetical scenario has been raised concerning this. Given that there are 1×10^5 CFU of resident microflora per hand, if a patient has the first surgery on a Monday (first scrub of week by the surgeon) versus on a Friday (at least 11 consecutive scrubs intervening), a 10^2 difference in microbial population would be present. This, theoretically, could greatly affect the risk of a surgeon's hands contaminating a surgical site. What about a senior surgeon performing only several surgeries a week? What risk factor would be presented?

These scenarios are not written to negate the requirements of the FDA, for the efficacy requirements are truly burdensome for the agency. Instead, these should be addressed by infection-control practitioners at the hospital level. Undoubtedly, the tendency of many surgeons will be to ignore such concerns. Several manufacturers, however, have come up with novel ways to address these concerns. The first is to include an antimicrobial agent on the inside of the surgeon's gloves to help control regrowth of microbial populations after the scrub has been completed. Other manufacturers supply an antimicrobial cream for use on the hands *after* the scrub to provide increased antimicrobial action. The use of such novel approaches could control microbial counts more effectively—an important benefit, particularly for extended surgical procedures. But while these options are of interest, they are still in the research-and-development stage. One problem to be overcome, however, is the potential incompatibility of a surgical scrub compound with a product applied to the hands or impregnated into the gloves. If one product neutralizes—inactivates—the other, a serious contamination problem could occur.

Until better topical antimicrobials are produced, the field must rely on heavy antimicrobial therapy for the prevention of postsurgical infections. Marton and Burke demonstrated that about 90% of the patients they sampled had opportunistically pathogenic microbial contamination, including *S. aureus*, at the surgical site at the time of closure. This is a sobering fact. But heavy-duty antibiotic therapy can hardly be the final answer, with the increase in *Clostridium difficile* colonization of the large intestine in postsurgical patients as a result of elimination of the normal flora, and the ever-growing problem of antibiotic-resistant strains of bacteria harbored in the hospitals themselves.

REFERENCES

1. ES Wong. Surgical site infections. In: CG Mayhall, ed. Hospital Epidemiology and Infection Control. 2nd ed. Philadelphia: Lippincott, Williams & Wilkins, 1999, pp. 189–210.
2. CG Mayhall. Hospital Epidemiology and Infection Control. 2nd ed. Philadelphia: Lippincott, Williams & Wilkins, 1999.
3. R Berg. The APIC Curriculum for Infection Control Practice. Vol. III. Dubuque IA: Kendall/Hunt Publishing, 1988, pp. 1162–1390.
4. DS Paulson. Handbook of Topical Antimicrobial Testing and Evaluation. New York: Marcel Dekker, 1999.
5. LV Karanfil, RRM Gershon. Evaluating and selecting products that have infection control implication. In: CG Mayhall, ed. Hospital Epidemiology and Infection Control. 2nd ed Philadelphia: Lippincott, Williams & Wilkins, 1999, pp. 1367–1372.
6. PG Bowler, BI Duerden, DG Armstrong. Wound microbiology and associated approaches to wound management. Clin Microbiol Rev, 14(2):244–269, 2001.
7. DS Paulson. Efficacy evaluation of a 4% chlorhexidine gluconate solution as a full body shower wash. Am J Infect Control 21(4): 205–209, 1993.
8. DS Paulson. To glove or to wash: a current controversy. Food Qual, (June/July): 27–33, 1996.
9. WA Altemeir, NR Culbertson. Surgical Infection. In C Moyer et al, eds. Surgery, Principles and Practices, 3rd ed. Philadelphia: Lippincott, Williams & Wilkins, 1965, pp. 879–891.
10. J Carlet, M Garrouste-Orgeas, J Timsit, P Moine. Infective complications after trauma and surgery. In D Armstrong, J Cohen, eds. Infectious Diseases. Vol. I. St. Louis: Mosby, 1999, pp. 178–203.

11. National Academy of Sciences Research Council. Post-operative wound infections: the influence of ultraviolet irradiation of the operating room and of various other factors. Ann Surg 160 (suppl 2):1–132, 1964.
12. DH Culver, TC Horan, RP Gaynes. Surgical wound infection rates by wound class, operative procedure, and patient risk index. Am J Med 91 (suppl 3B):152–157, 1991.
13. R Serophian, BM Reynolds. Wound infections after preoperative depilatory versus razor preparatory. Am J Surg 121:251–254, 1971.
14. PJE Cruse, R Foord. The epidemiology of wound infection: a 10-year perspective study of 62,939 wounds. Surg Clin North Am 60:27–40, 1980.
15. WH Hamilton, FJ Lone. Preoperative hair removal. Can J Surg 20:269–275, 1977.
16. JG Bartlett, RE Condon, SL Gorbach, JS Clarke, RL Nichols, S Ochi. Veterans Administration cooperative study on bowel preparations for elective colorectal operations: impact of oral antibiotic regimen on colonic flora, wound irrigation cultures and bacteriology of septic complications. Am Surg 188:249–254, 1978.
17. PJ Miller, MA Sarcy, DL Kaiser, RP Wenzel. The relationship between surgeon experience and endometritis after cesarean section. Surg Gynecol Obstet 165: 535–539, 1987.
18. LD Edwards. The epidemiology of 2056 remote infections and 1966 surgical wound infections occurring in 1865 patients: a four-year study of 40,923 operations at Rush-Presbyterian-St. Luke's Hospital, Chicago. Ann Surg 184:758–766, 1976.
19. DS Paulson. Designing a handwash efficacy program. Pharm Cus Quality (April): 17–20, 1996.

11
Antimicrobial Body Washes

David W. Hobson
Chrysalis Biotechnology, Inc., San Antonio, Texas

Lawton Anthony Seal
Healthpoint, Ltd., San Antonio, Texas

I. INTRODUCTION

Surgical site infection (SSI) is among the most commonly reported forms of nosocomial infection and has been estimated to account for 14–16% of all nosocomial infections among hospital inpatients [1]. These infections have been documented to produce increased lengths of hospital stay (averaging 8.2 days in one study) as well as significant additional treatment costs estimated to be in the billions of dollars [2,3]. In addition, even though the full impact of the risk of SSIs for the increasing numbers of ambulatory surgeries has not been completely studied, at least one recent study indicates that the rate for SSIs for outpatient surgeries may be of similar magnitude to that reported for inpatient procedures [4].

Preoperative bathing to help prevent SSIs aid has been practiced since the late nineteenth century and has two identified purposes: (1) physical cleansing and the removal of dirt, debris, microorganisms, and other residues and (2) antisepsis to reduce the total amount of microbial flora in and surrounding the surgical site. This practice, using various antimicrobial formulations, is often recommended prior to particularly invasive procedures such as cardiac surgery [5,6], and manufacturers of antimicrobial cleansers formulated for topical use often include procedures on their labeling for this purpose.

Surprisingly, however, the value of preoperative bathing in reducing SSIs has not yet been completely established scientifically, even though the practice would appear to have a sound basis with respect to skin microbiology and the fact that the majority of SSIs are associated with skin flora. One recent review that examined 13 different articles on the effect of preoperative bathing regimens on SSI rates concluded that ''evidence of the value of preoperative whole body patient bathing with an agent such as chlorhexidine (CHG) is weak and this practice is likely to be cost effective only in situations of high patient compliance and high risk'' [7]. Nevertheless, some clinical studies have produced evidence that preoperative bathing is beneficial in reducing SSI rates, and the practice continues [8–11].

In recent years, povidone iodine (PVPI) and CHG have been the prevalent antimicrobial agents used for preoperative skin disinfection and are often the antimicrobials of choice for preoperative full body shower washing, but other antimicrobials such as triclosan are also used. Some studies evaluating the efficacy of CHG in reducing SSIs in very large patient populations have shown little or no significant differences in the rates of postsurgical infection between groups of patients using CHG-containing formulations and those using nonantimicrobial soaps for whole body bathing [12–14]. The infection rates reported for the treatment groups in these different studies was quite broad and ranged from 2.4 to 51.2%, and the numbers of showers or baths taken by the patients prior to surgery was three or less. In 1993, Paulson reported the results from a small study evaluating the efficacy of a 4% CHG formulation as a full-body shower wash and showed that use of this formulation over a 5-day period produced significantly reduced bacterial counts ($p < 0.05$) on both the abdominal and inguinal regions of the body [15]. In this study, five volunteers completed a 14-day microbial stabilization period, a 7-day baseline period, and a 5-day test period. Subjects followed a standard protocol, performed five shower washes, and were sampled at both the abdominal and inguinal regions immediately after the shower wash as well as at 3 and 6 hours later on days 1, 2, and 5. The results from this study showed very dramatic reductions from the baseline bacterial counts for both the inguinal and abdominal sites (Figs. 1 and 2). Paulson recognized the ambiguous nature of the results reported for CHG in the literature and concluded that a 1- or 2-day presurgical application period with CHG may be too short to establish the levels of residual antimicrobial effect necessary to be of value in reducing postsurgical infection rates. This conclusion needs to be further examined in a large-scale clinical trial, however.

At present, typical use instructions for a commercial CHG formulation are as follows:

The patient should wash the entire body, including the scalp, on two consecutive occasions immediately prior to surgery.

Each procedure should consist of two consecutive thorough applications of the CHG-containing formulation followed by thorough rinsing.

If the patient's condition allows, showering is recommended for whole-body bathing.

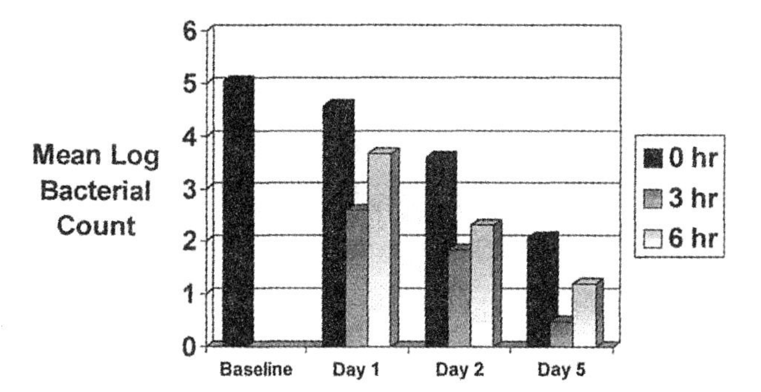

Figure 1 Effects of a 5-day antimicrobial body wash regimen from a sampling of the inguinal region using a product containing 4% chlorhexidine gluconate (CHG). Data shown are the numbers of organisms per cm^2 given as $\log_{10}$ values at different sampling times after the body wash was completed on days 1, 2, and 5 ($n = 5$). (Data from Ref. 15.)

The recommended use procedure is:

Wet the body, including hair.

Wash the hair using 25 mL of the formulation and the body with another 25 mL of the formulation.

Rinse.

Repeat.

Rinse thoroughly after second application.

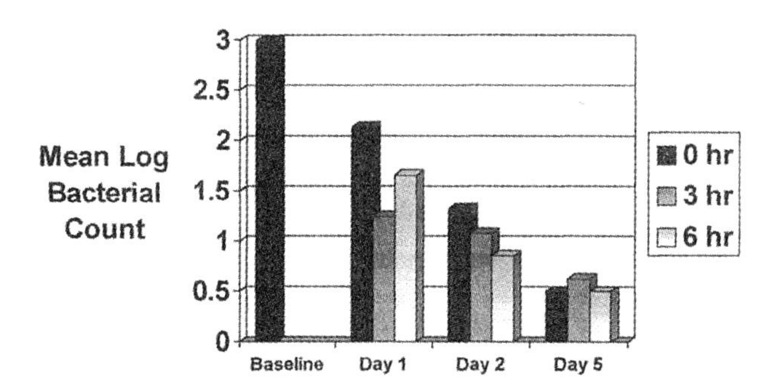

Figure 2 Effects of a 5-day antimicrobial body wash regimen from a sampling of the abdominal region using a product containing 4% chlorhexidine gluconate (CHG). Data shown are the numbers of organisms per cm^2 given as $\log_{10}$ values at different sampling times after the body wash was completed on days 1, 2, and 5. The number of subjects was five ($n = 5$). (Data from Ref. 15.)

Even though manufacturers of antimicrobial products have provided instructions for whole body use, there is currently no established consensus protocol in healthcare practice settings for preoperative whole body showering or bathing, with or without using an antimicrobial formulation. When written standards mention preoperative antiseptic showering, there is generally no procedure mentioned, rather only an abbreviated note to the fact that the procedure is advised or required. For example, the most recent guidelines for SSI prevention by the Centers for Disease Control and Prevention (CDC) contain the following statement: ''Require patients to shower or bathe with an antiseptic agent on at least the night before the operative day'' [16]. The U.S. Food and Drug Administration's (FDA) Tentative Final Monograph for Antiseptic Drug Products does not clearly address preoperative showering or bathing procedures and describes the ''patient preoperative skin preparation'' as simply a preventive preparation aimed at reducing the patient's skin colonization before the incision is made [17].

There is still a clear need to better characterize and define the area of antiseptic body washes for preoperative purposes. There are at least three areas where research effort may hold promise for future improvements in this area:

1. Improvement in the statistical design of large-scale clinical trials such that there are few confounding variables and enough power to detect a statistically significant difference between test populations.
2. Better defining the minimum required performance characteristics of preoperative antimicrobial showering or bathing regimens to include log reduction minimums and the spectrum of required antimicrobial activity.
3. Development of vastly improved antimicrobial formulations that have broad-spectrum activity, high-level disinfection capability, and persistence such that log reductions that meet or exceed performance minimums can be obtained with a very high frequency of occurrence.

Future significant advancements in whole body preoperative showering or bathing will likely include much more sophisticated study designs compared to those of the past. These designs will take into consideration issues related to the areas for improvement described above and will also very likely include analysis of the types of organisms giving rise to different types of SSIs as well as antibiotic- and/or antiseptic-resistant microbes. For this reason, the following discussion regarding the microbiology that defines the need for as well as the success and failure of antiseptic body washes is particularly germane to any discussion of this topic.

II. MICROBIOLOGY OF ANTISEPTIC BODY WASHES

Employing an antiseptic body wash affords the surgery patient an opportunity to participate in the process and to impact positively the outcome as it relates to surgical site infections. Presurgical washing should remove any dirt and organic debris, reduce the resident and transient microflora beyond that achieved by only prepping the surgical site, and therefore reduce the rate of cutaneous infection [18].

The casual observer would not necessarily recognize skin as being soiled, not realizing that a square-inch patch of skin may harbor 100,000 (10^5) or more microorganisms. Residential flora varies in number and species from site to site on the body, as well as by geographical location and time of year. Although generally commensal with the human body, residential flora may become a source of infection from immune response–related consequences of surgery. Where skin is oilier, Propionibacteria may establish populations as high as 10^7 cfu/cm^2. Corynebacteria prefer moist skin with populations between 10^5 and 10^8 cfu/cm^2, while drier surfaces support 10^3–10^4 cfu/cm^2 gram-positive cocci [19]. Microorganisms reside not only on the skin surface, but also deep within the keratinized layers of the stratum corneum and inside structures such as hair follicles, oil glands, and sweat ducts. Their presence in these locations increases the difficulty of removing microorganisms prior to surgery. A variety of transient organism species may be deposited on the skin during contact with contaminated materials or items. These organisms generally do not colonize and may include *Escherichia coli*, *Staphylococcus aureus*, *Enterococcus* spp., and *Pseudomonas* spp., among many others. The distinction between transient and residential flora blurs, as incidences of persistent colonization of transient organisms (thereby becoming residential flora) have been reported [20]. At times it becomes difficult to distinguish exogenous contamination from endogenous contamination by transient host flora. Extended-duration surgical procedures, prolonged healing, and reinoculation or colonization, or regrowth of suppressed organisms, increase the risk of infection. Furthermore, microorganisms within the skin structures are shielded from topical antimicrobials and can provide a reservoir for repopulation [21,22].

Agents that have been used frequently for surgical site preparation (see Table 1) include alcohols, iodophors, or chlorhexidine gluconate (CHG) [23]. These antimicrobial agents vary with regard to efficacy, irritation, or toxicity. Alcohols (70–90%) are the most rapid acting, with excellent coverage for gram-positive, gram-negative, and mycobacteria species. In addition, they are active against many fungi and viruses and, with the exception of some drying effect on the skin, are safe and generally nontoxic. Iodophors (0.5–10%) are nearly as efficacious as the alcohols and have a rate of kill characterized as intermediate. However, organic matter adversely affects their efficacy, and they are known to induce significant skin irritation. CHG (0.5–4%), like iodophors, is characterized

Table 1 Summary of Agents Approved for Use as the Principal Active Ingredient in Preoperative Preparation Formulas

Antimicrobial Agent	Spectrum of Activity	Rapidity	Safety	Development of Resistance	Debris Removal	Persistence
Alcohol	Very broad	Fast	Long history, very safe	Non existent	Generally poor detergency	Short, in minutes
CHG	Broad	Moderate	Requires data to demonstrate safety	Low	Good when formulated with surfactants	Relatively long, in hours
Iodophors	Broad, can be neutralized by blood or other organic materials	Fast	Generally safe; skin irritation and sensitization can occur	Low to negligible	Very good in povidone-iodine formulations	Intermediate to short, in minutes

by an acceptable antimicrobial range and rapidity of action that is intermediate. It provides excellent residual activity, but is noted for significant ototoxicity, keratitis, and eye injury and should not be instilled into the ears or eyes or employed near these areas. Occasionally, other antimicrobials such as triclosan and parachlorometaxylenol (PCMX) have been employed in this process. Neither has the exceptional antimicrobial range observed with the alcohols, and resistance to gram-negative rods has been reported for each agent (triclosan, *E. coli*; PCMX, *P. aeruginosa*) [24–26].

III. PERFORMANCE AND REGULATORY ASPECTS OF ANTIMICROBIAL BODY WASHES

Even though regulatory aspects of the required safety and efficacy for antimicrobial body wash products is currently not very well defined, it seems plausible to consider the minimum performance requirements for a patient preoperative preparation described in the FDA's tentative final monograph for topical antimicrobial products as a possible benchmark for defining the desired minimum performance requirements for a preoperative antimicrobial shower or bath product [17]. These minimums require that the product achieve an initial 3 $\log_{10}$ reduction from a baseline of at least 10^5 cfu/cm^2 of skin surface from an inguinal test site immediately prior to surgery. The performance of topical antiseptics used for preoperative showering or bathing has not yet been evaluated in relation to this regulatory-defined minimum, and the only association that any antiseptic preoperative shower formulation has sufficient efficacy to reduce SSI rates has not been defined in the literature but rather has been verbally communicated as a successful recommendation for the lowering of SSI rates in outbreak investigations.

In contrast, an example of recent data obtained from the initial testing of a novel, alcohol-based, antimicrobial formulation that meets the FDA's monograph criteria is shown in Figure 3. Products that can meet such a performance minimum may very well show promise in subsequent large-scale, well-controlled clinical trials.

Another consideration in the design of new antimicrobial product formulations, the design of large-scale clinical trials, as well as for the development of a systems approach to preoperative antimicrobial bathing and showering is the compatibility of other topical products used in conjunction with the preparative process. For example, it is well established that common ingredients in body creams and lotions may inactivate antimicrobial agents such as CHG [27]. In the very near future, healthcare professionals may inquire how important it is that the antiseptic chosen for the patient's preoperative shower and the antiseptic employed for the intraoperative skin prep be compatible? Are there other potentially deleterious chemical reactions to be concerned about? What is the current science

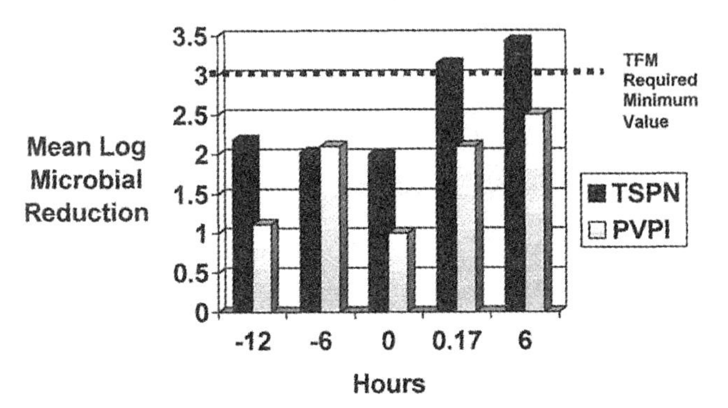

Figure 3 Test data for two different antimicrobial body wash treatments, povidone iodine (7.5%, PVPI) or a novel alcohol-based formulation (TSPN), followed with a preoperative preparation of like composition painted onto the test subjects immediately following the time = 0 hours sampling. Note that with the PVPI-based treatment regimen, the minimum 3 $\log_{10}$ reduction from baseline microbial counts at 0.17 hours (10 min) is not obtained, whereas with the TSPN system the required reduction is obtained at both the 0.17-hour sampling as well as the 6-hour (end of typical surgical procedure) sampling. The average baseline microbial count for the subjects in this study was 5.4 $\log_{10}$ organisms per cm^2 ($n = 13$).

related to compatibility concerns? Is the chosen antiseptic shower agent intended for a large-scale clinical trial compatible with other commonly used topical products, such as body lotions, creams, perfumes, colognes, and/or sunscreens? What about the presence of organic debris? How much inactivation is tolerable with currently available antiseptic agents such that negating the beneficial effect of reducing the number of microorganisms on the skin prior to surgery is avoided?

Systems approaches to antiseptic product line development and use enable the compatibility issue to be engineered out of the product during development, allowing compatible products to achieve the desired outcome by design rather than by chance. Thus, from a perioperative perspective, the preoperative antiseptic shower agent would initiate the skin flora reduction and the intraoperative skin prep antiseptic would continue to add persistence, decreasing regrowth of normal flora. A systems approach to product design and use seems a clear best practice when the selection process is objective evaluation with informed decision and intense in-service training for all users during the implementation.

In developing a standardized procedure for the patient's antiseptic preoperative shower, Bjerke et al. identified some aspects to consider [28]:

1. What antiseptic should be selected, based on what criteria/attributes? As to the choice of antiseptic, FDA under the 1994 Tentative Final

Monograph for Antiseptic Drug Products reserves the term ''antiseptic'' for a product with antimicrobial activity that has been shown to prevent skin infections in a controlled clinical trial. The antiseptic choice is based on data collected in groin and abdomen sites: in the abdominal area, a 2 $\log_{10}$ reduction of the microbial flora and suppression of bacterial growth within 10 minutes of application and no return to baseline flora count until 6 hours postapplication; in the groin area, a 3 $\log_{10}$ reduction of microbial flora from baseline and suppression of bacterial regrowth within 10 minutes of application and no return to baseline flora count until 6 hours postapplication. FDA believes that persistence of the antimicrobial effect would suppress the growth of residual skin flora not removed by preoperative prepping as well as transient microorganisms inadvertently added to the operative field during the course of surgery and reduce the risk of SSI. In this same monograph, FDA further defines ''patient preoperative skin preparations'' as a fast-acting, broad-spectrum, persistent antiseptic-containing preparation that significantly reduces the number of microorganisms on intact skin [17].

An evolution of antiseptics has influenced perioperative practices. Historically, antiseptic agents progressed from the era of alcohol and carbolic acid to hexachlorophene (HCP) and PCMX, then povidine iodine (PVP-I), followed by CHG agents [29]. Each agent possesses advantages and cautions, requiring knowledge for proper use. Now newer formulations of these antiseptics as well as advanced technologies offer an enhanced, prolonged, persistent efficacy with low toxicity to the patient when used properly. These products focus on patient safety. Of note is the reemergence of alcohol products that are waterless and water-aided and are available in rinses and gel media. Also, compatible antiseptic product lines are being identified and marketed. Thus, healthcare workers face the importance of reviewing in vitro and in vivo product data plus clinical performance of these agents. Product evaluation remains an essential process for selection of an antiseptic.

2. How and when is the antiseptic dispensed to the patient? The selected antiseptic of choice for preoperative cleansing is dispensed to the patient generally in the surgeon's office or at the preadmission clinic prior to the day of surgery in the quantity and/or media for accomplishing the showering frequency. While an impregnated sponge makes it an economical and easy dispenser of the antiseptic for patient use, 4-oz. liquid containers are given to the patient along with specific multimedia instructions for use.
3. How often should the patient use the antiseptic shower product—once or twice? Because most antiseptics only improve the more they are

used, one would conclude "the more, the better." However, patient compliance will suffer if the use instructions are not fully explained and/or demonstrated and realistic. It has been suggested that two to three applications are ideal to reduce colonization of the skin. *Note*: If the surgical prep agent is the same antimicrobial as the antiseptic shower agent, then the patient will continue the additive and cumulative effect.

4. When are the best times to accomplish preoperative antiseptic showers? At a minimum, the night before and the morning of surgery are the best times.
5. Is the whole body cleaned or just the anticipated incisional site? Protocols should follow the label claims for use. Generally, one should start at the incisional site and work outward, covering the neck down to the feet, paying particular attention to the incisional site, avoiding the head area, and cleansing the highly colonized areas last.
6. What kind of multimedia educational materials are available, or does the facility need to create their own? It is a facility choice to commercially purchase or create its own educational materials. Clear and concise instructions are imperative for patient safety in the use of these agents. Manufacturers should provide a patient-education and direction-for-use program with their products.
7. Is surgeon support key to this initiation, or does this fall into a nursing purview? As a perioperative team, all healthcare professionals should agree to and complete their specific role in this preventive initiative for quality surgical care, SSI-free outcome, and patient safety.
8. Who verifies completion of this patient responsibility, and how/where is it documented? Generally, the perioperative nurse in preadmission and circulator make the inquiries and record the findings on the surgical checklist.

Incorporating the answers to the above inquiries, the following is a *suggested* antiseptic preoperative shower/bedbath procedure:

1. A perioperative nurse orally instructs the patient and his support system in the rationale for an antiseptic preoperative shower. As a partner of the perioperative team, the patient is asked to reduce the transient and resident colonization of microbes on his skin surface with an antiseptic and, thus, lower the endogenous source for a potential surgical site infection.
2. Written directions with an illustration of a human being are given to the patient as reinforcement of the verbal information and as a reference when the patient is at home performing the shower. This illustration is marked as to what anatomical area is to be cleaned with the

antiseptic and what anatomy is cleaned normally. *Note*: Chin to toes is generally cleaned with the antiseptic. Some of the antiseptics are contraindicated in eyes, ears, and mucous membranes; read the antiseptic manufacturer's directions and follow accordingly.

Unless these are incisional sites, the following directions are appropriate: the hair is shampooed once with the patient's shampoo. The face is washed at night and in the morning with the patient's own soap. Oral hygiene is performed at night and in the morning with the patient's own toothpaste.

The antiseptic preoperative shower/bedbath is performed the night before the operation with the designated antiseptic. A clean washcloth or disposable sponge is used each time to apply the antiseptic. The intended incisional site is cleaned with extra strokes, being careful not to denude the skin resulting in case cancelation. Rinse the antiseptic off. Thoroughly towel dry before donning clean bedtime clothes.

Repeat this activity the morning of the operation, using another clean washcloth or disposable sponge to apply the designated antiseptic. Don clean street clothes if an ambulatory patient is going to the facility for same day admission or outpatient surgery. *Note*: If patient is hospitalized and bedridden, perioperative team members perform the antiseptic bedbath the night before and the morning of surgery, using the designated antiseptic and a clean washcloth or disposable sponge each time. Place a clean patient gown on the patient after each bedbath.

3. Upon admission to the operating room, the perioperative nurse inquires if the patient accomplished the antiseptic preoperative shower of the designated anatomical area the night before and the morning of surgery. The rationale is to verify that this preventive measure was accomplished as directed.
4. Tabulate the results and report them on a continuous basis, charting historical performance as well as the relationship to controlling or reducing the incidence of SSIs.

IV. CONCLUSIONS

The effectiveness of preoperative showering or bathing in reducing skin microbial colony counts is accepted to the extent that infection-control authorities support this practice [30]. However, the actual impact on the reduction of SSIs as well as the cost effectiveness of this practice remains to be proven conclusively. In order to accomplish this, improvements in study design and antiseptic formulation and the development of sound protocols need to be accomplished. It is also very important to realize that the patient also plays a very important role in this activity, as it is his or her essential responsibility to completely accomplish the regimen prior to surgery. It is likely that improvements in this area will come as incremental steps rather than large-scale change, so perseverance, research, and patience will be necessary to establish preoperative antimicrobial showering or bathing as a required practice. Careful and complete product evaluation for efficacy as well as safety will be essential to the process change desired, and industry must work very closely with clinicians and regulatory officials to provide the necessary proof that their formulations perform at least minimally as required prior to their large-scale use. The introduction of new, systems-oriented technologies allowing seamless antimicrobial protection for the patient's skin prior to surgery that combines ease of use with cumulative antimicrobial effectiveness and that results in significantly improved patient outcomes will likely be the standard for best practice in preoperative preparations in the near future.

REFERENCES

1. ETM Smyth, AM Emmerson. Surgical site infection surveillance. J Hosp Infect 45: 173–184, 2000.
2. R Coello, H Glenister, J Ferreres, C Bartlett, D Leigh, J Sedgwick, EM Cooke. The cost of infection in surgical patients: a case control study. J Hosp Infect 25:239–250, 1993.
3. WJ Martone, WR Jarvis, DH Culver, RW Haley. Incidence and nature of endemic and epidemic nosocomial infections. In: JV Bennett, PS Brachman, eds. Hospital Infections. Boston: Little, Brown and Company 1992, pp. 577–596.
4. D Villar-Compte, R Roldan, S Sandoval, R Corominas, M de la Rosa, P Gordillo, P Volkow. Surgical site infections in ambulatory surgery: a 5-year experience. Am J Infect Control 29:99–103, 2001.
5. DE Lilienfeld, D Vlahov, JH Tenney, JS McLaughlin. On antibiotic prophylaxis in cardiac surgery: a risk factor for wound infection. Ann Thorac Surg 42(6):670–674, 1986.
6. DJ Hillis, FL Rosenfeldt, WJ Spicer, GR Stirling. Antibiotic prophylaxis for coronary bypass grafting: comparison of a five-day to a two-day course. J Thorac Cardiovasc Surg 86:217–221, 1983.
7. EK Kretzer, EL Larson. Do antiseptic agents reduce surgical wound infections: preoperative bathing, operative site preparation, surgical hand scrub? In: WA Rutala, ed. Chemical Germicides in Health Care. APIC, Washington, DC and Polyscience Publications, Inc. 1994, pp. 149–161.
8. RA Garibaldi, D Skolnick, T Lerer, A Poirot, J Graham, E Krisuinas, R Lyons. The impact of preoperative skin disinfection on preventing intraoperative wound contamination. Infect Control Hosp Epidemiol 9(3):109–113, 1988.
9. AB Kaiser, DS Kernodle, NL Barg, MR Petracek. Influence of preoperative showers on staphylococcal skin colonization: a comparative trial of antiseptic skin cleansers. Ann Thorac Surg 45:35–38, 1988.

10. A Brandberg, J Holm, Hammarsten, T Schersten. Postoperative wound infections in vascular surgery: effect of preoperative whole body disinfection by shower-bath with chlorhexidine soap. In: H Maibach, R Aly, eds. Skin Microbiology Relevance to Clinical Infection. New York: Springer-Verlag, 1981, pp 98–102.
11. O Wihlborg. The effect of washing with chlorhexidine soap on wound infection rate in general surgery: a controlled clinical study. Ann Chirurg Gynacol 76:263–265, 1987.
12. W Lynch, PG Davey, M Malek, DJ Byrne, and A Napier. Cost-effectiveness analysis of the use of chlorhexidine detergent in preoperative whole body disinfection in wound infection prophylaxis. J Hosp Infect 21(3):179–191, 1992.
13. ML Rotter, SO Larsen, EM Cooke, J Dankert, F Daschner, D Greco, P Gronroos, OB Jepsen, A Lystad, B Nystrom. A comparison of the effects of preoperative whole-body bathing with detergent alone and with detergent containing chlorhexidine gluconate on the frequency of wound infections after clean surgery. J Hosp Infect 11(4):310–320, 1988.
14. GAJ Ayliffe, MF Noy, JR Babb, JG Davies, J Jackson. A comparison of pre-operative bathing with chlorhexidine-detergent and non-medicated soap in the prevention of wound infection. J Hosp Infect 4:237–244, 1983.
15. DS Paulson. Efficacy evaluation of a 4% chlorhexidine gluconate as a full-body shower wash. Am J Infect Control 21:205–209, 1993.
16. AJ Mangram, et al. ACDC HICPAC Guideline for Prevention of Surgical Site Infection, 1999. AJIC 27(2):97–133, 1999.
17. Food and Drug Administration. Topical antimicrobial drug products for over-the-counter human use: tentative final monograph for healthcare antiseptic drug products. Fed Reg 59:31432, 1994.
18. E Larson. Hygiene of the skin: When is clean too clean? Emerg Infect Dis (7)2, 2001.
19. DE Fry. Surgical Infections. Boston: Little, Brown & Company, 1995.
20. AJ Morrison Jr, J Gratz, I Cabezudo, RP Wenzel. The efficacy of several new hand-washing agents for the removal of non-transient bacterial flora from hands. Infect Control 7(5):268–272, 1986.
21. A Mangram. Guideline for prevention of surgical site infection, 1999. Infect Control Hosp Epidemiol April:250–278, 1999.
22. AN Weinberg, MN Swartz. General consideration of bacterial diseases. In: TB Fitzpatrick, et al., eds. Dermatology in General Medicine. 3rd ed. New York: McGraw-Hill, 1987, pp. 2089–2100.
23. E Larson. APIC guidelines for use of topical antimicrobial agents. AJIC 16(6):253–266, 1988.
24. WA Rutala, DJ Weber. Overview of the use of chemical germicides in healthcare. In: WA Rutala, ed. Disinfection, Sterilization and Antisepsis: Principles and practices in Healthcare Facilities. Washington, DC: Association for Professionals in Infection Control and Epidemiology, Inc., 2000, pp. 1–15.
25. LM McMurry, M Oethinger, SB Levy. Triclosan targets lipid synthesis. Nature 394: 531–532, 1988.
26. G McDonnell, AD Russell. Antiseptics and disinfectants: activity, action and resistance. Clin Micro Rev 12:147–179, 1999.

27. SW Frantz, KA Haines, CG Azar, JI Ward, SM Homan, RB Roberts. Chlorhexidine gluconate (CHG) activity against clinical isolates of vancomycin-resistant *Enterococcus faecium* (VREF) and the effects of moisturizing agents on CHG residue accumulation on the skin. J Hosp Infect 37(2):157–164, 1997.
28. NB Bjerke, DW Hobson, and LA Seal. Preopertative skin preparation: system approach, submitted for publication.
29. DW Hobson, W Woller, L Anderson, E Guthery. Development and evaluation of a new alcohol-based surgical hand scrub formulation with persistent antimicrobial characteristics and brushless application. AJIC 26:507–512, 1998.
30. AJ Mangram, et al. CDC HICPAC guideline for prevention of surgical site infection, 1999. AJIC 27(2):97–133, 1999.

12

Some Aspects of Human Disease and *Staphylococcus* Species

Daryl S. Paulson
BioScience Laboratories, Inc., Bozeman, Montana

I. INTRODUCTION

Staphylococcus species are prominent pathogenic microorganisms encountered in postsurgical infections of normal individuals, in immunocompromised patients, and in elderly hospitalized medical patients [1]. *Staphylococcus* species were among the earliest pathogenic bacteria characterized, and, aside from hospital-acquired infections, they are the most common cause of localized suppurative infections in humans.

Staphylococci are gram-positive, spherical cells ranging in size from about 0.5 to 1.5 μm in diameter. They are often characterized microscopically by their grape cluster–like morphological appearance. Staphylococcal growth is optimal under aerobic conditions, but they are facultative anaerobes. Their optimum growth temperature ranges between 30 and 37°C. *Staphylococcus* species are nonmotile and non–spore forming [2,3]. Table 1 displays distinctive biochemical characteristics of six species of medical importance.

Staphylococcus species are relatively more resistant to heat, as well as various topical disinfectants and topical antimicrobials, than are most other pathogenic bacteria [4,5]. For example, many species of bacteria are killed when exposed to 60°C temperature for 30 minutes. However, comparable microbial reductions in *Staphylococcus* spp. may require exposure to a temperature of 80°C for 1 hour. And unlike many other pathogenic microorganisms, *Staphylococcus* spp. will grow at 45°C, are relatively resistant to drying, and can remain viable for extended periods of time.

Table 1 The Medically Important Species of *Staphylococcus*[a]

	S. aureus	*S. epidermidis*	*S. saprophyticus*	*S. simulans*	*S. hominis*	*S. haemolyticus*
Morphology						
Cell size, mm	0.8–1.0	05.–1.5	0.8–1.2	0.8–1.5	1.0–1.5	0.8–1.3
Colonies						
Elevation[b]	R	R	LC	R	C	LC
Light transmission[c]	T	T	O	T	O	O
Diameter (mm)	6–8	2–4	5–8	5–7	3–4	4–8
Pigment[d]	(+)	(−)	V	−	V	−
Cell wall teichoic acid[e]	R	G	R, G	G	G	G
Physiology						
Anaerobic growth	+	+	+	+	+	+
Growth in 10% NaCl	+	W	+	+	W	+
Growth at 45°C	+	+	V	+	+	+
Biochemical reactions						
Coagulase	+	−	−	−	−	−
Hemolysis	(+)	W; −	−	W; −	(W; −)	(+)
Acetylmethyl carbinol	+	+	(+)	W; −	V	+; W
Nitrate reduction	+	(+)	−	+	+	(+)
Phosphatase	+	+	(−)	(W)	(−)	(−)
Deoxyribonuclease	+	(W)	−	(W)	W; −	W; −

Acid from carbohydrate[f]						
Sucrose	+	+	+	+	+	+
Trehalose	+	−	+	(+)	+	+
Turanose	(+)	V	+	−	(+)	V
Mannitol	+	−	+	(+)	−	V
Xylitol	−	−	(+; W)	−	−	−
Antibiotic sensitive[g]						
Novobiocin	+	+	−	+	+	+
Penicillin G	V	V	+	V	V	V
Tetracycline	+	(+)	+	V	V	+
Erythromycin	+	+	(+)	(+)	+	+

[a] Symbols alone = frequency of 90%; parentheses = 70–89% frequency; + = positive; W = weakly positive; − = negative; V = variable.
[b] R = Raised; LC = low-convex; C = convex.
[c] T = Translucent; O = opaque.
[d] Pigmented colonies range from orange to yellow-orange, to yellow, to gray or white with yellow tint.
[e] R = Ribitol type; G = glycerol type.
[f] Aerobic acid production by plate culture.
[g] These are also MRSA (methicillin-resistant *S. aureus*) with same profile except not resistant to methicillin.
Source: Data from WE Kloos and KH Schleiffer, Internat J Syst Bacteriol 25:62–79, 1975.

The majority of *Staphylococcus* spp. will grow in the presence of 10% NaCl, and some will grow in as high as 15% NaCl. This has particular relevance in the food industry, where salting is a way of preserving food.

Staphylococcus spp., particularly *S. aureus*, are prone to develop antibiotic resistance [1]. The staphylococci are, by nature, not highly fastidious in growth requirements, and they grow readily on general medium. Their growth is most profuse on blood agar, commonly used for isolation and differentiation. They will readily grow aerobically on synthetic mediums containing amino acids and vitamins, but anaerobic growth requires the addition of uracil and a fermentable carbon source, such as pyruvate. A variety of carbohydrates are used aerobically, with production of acid (see Table 1). Glucose is fermented anaerobically, with lactic acid being the primary end product.

The production of coagulase is diagnostically important for *S. aureus*. Hence, *S. aureus* strains are coagulase-positive, and all the other strains are referred to as coagulase-negative [6].

II. *STAPHYLOCOCCUS AUREUS* AND NOSOCOMIAL INFECTION

Over the years, incidence of hospital-acquired *S. aureus* infections has increased. It is the causative agent in about 50% of the catheter-related bacteremias, the majority of hospital-acquired infections from the insertion of prosthetic devices, and the majority of cases of septic arthritis and osteomyelitis [7].

III. COLONIZATION

The major reservoir of *S. aureus* in humans is apparently the anterior nares. Carriage there influences carriage at other anatomical sites, including the mucous membranes, perineum, and axillae. Individuals have been shown to harbor *S. aureus* continuously or intermittently (about 90% of a sampled population will demonstrate *S. aureus* intermittently) [3]. Factors that promote colonization include respiratory infection, prolonged hospitalization, diabetes, intravenous drug use, hemodialysis, corticosteroid treatment for allergies, cold weather exposure, and dermatological conditions such as atopic dermatitis and eczema. Patients in the upper age brackets, even when hospitalized, seem to have no higher rate of colonization. However, individuals using antibiotics over time are noted to be more frequently colonized by *S. aureus*. And once a person is colonized, the strain may be passed from person to person via hand contact or even shed skin squames. Colonization by *S. aureus* is a significant risk to later being auto-infected during a surgical procedure and/or device implantation [1]. Surgical site infections are known to be more prevalent in those colonized by *S. aureus* than in those who are not, particularly when the colonization density is greater than 1×10^6 colony-forming units per anatomical site.

Once *S. aureus* comes into contact with a new human host, it must be able to attach to a tissue or organ surface, a determining step as to whether infection occurs or not. In attachment, an interaction occurs between bacterial structural proteins on the bacterial cell wall and complementary sites on a human eukaryote cell.* Attachment is mediated by surface protein groups on the bacterial cell called microbial surface components recognizing adhesive matrix molecules (MSCRAMMs). Apparently there are three main groups of MSCRAMMs, depending on whether they bind to fibronectin, fibrinogen, or collagen [5].

S. aureus can be blocked from binding to host tissues by antigen-specific antibodies. But *staph*–host protein interaction may prevent the immune system from bacterial antigen recognition. The binding of various tissue substrates to the bacterial cell may also occur due to electrostatic charges. *S. aureus* has a net negative charge due to its protein A and ribotal teichoic acid. Hence, the prevention of infection by the host is enhanced if antibodies block one or more receptors of protein. *Staphylococcus* adherence to fibrin clots is increased with the presence of fibronectin, and bonding to fibronectin is mediated by two similar fibronectin proteins, FnBPA and FnBPB, that have specific ligand-binding areas that recognize the N or C terminus of fibronectin. When fibronectin coats surfaces, the N-terminal end enhances binding with certain strains of *S. aureus*. These strains are more likely to cause surgically induced endocarditis, for example, because fibronectin is deposited on valvular endothelial cells traumatized by surgical intervention.

Similar interactions between other surfaces and *S. aureus* can facilitate infection. Fibrinogen-coated catheters enhance *S. aureus* binding and, hence, intravenous site infections. Staphylococcal clumping factor—a surface protein—binds to fibrinogen. *S. aureus* bind to leperin and other glycosaminoglycans via two basic bacterial cell wall proteins. Glycosaminoglycans are linked with proteins to form proteoglycans and are found in relation to connective tissue, basement membranes, and eukaryote cell surfaces. While these will bind to leperinized catheters, the bonding appears not to be specific. *S. aureus* also produces a slime coat, but later in the growth cycle, and, hence, it is doubtful if it augments attachment, which occurs early in the growth cycle.

Currently, there are numerous known exoproteins produced by *S. aureus*, of which some are not only toxic, but considered virulence factors. These antigens cause activation of subpopulations of T cells with a subsequent production of cytokines, which may overwhelm the immune system, preventing concerted and

---

* In a surgical wound, the microorganisms can be implanted directly onto organs and tissues, usually mediated by bacterial cell attachment to plasma proteins within the wound itself.

coordinated antigen-antibody processing. This may elicit enterotoxins (A, B, $C_{1,2,3}$, D, E), as well as toxin-1, associated with toxic shock syndrome. The net effect is tissue damage due to cytokine activity [8].

Often, less intense reactions are encountered. The epidermolytic, exfoliative toxins A and B attack the epidermidis causing epidermal necrosis (e.g., *Staphylococcus*-scalded skin syndrome) [9]. Membrane-damaging toxins at infection sites (e.g., α-toxin, α-hemolysin) are a major factor in tissue damage after bacterial adherence has occurred. Other exoproteins, such as proteases, collagenase, hyaluronidase, and lipase, act as virulence enhancers but do not actively destroy host tissues.

IV. DRUG RESISTANCE

Penicillin, discovered in 1929 by Fleming, was first used therapeutically in 1941. The penicillin family of antibiotics, known as β-lactams, have been used extensively and successfully to treat bacterial infections, including *Staphylococcus* spp. The first reported bacterial resistance to β-lactams occurred in the 1940s, when extracts of bacteria were shown to neutralize the antimicrobial properties of penicillin. Soon after, strains of *S. aureus* were reported resistant to penicillin, in that they produced an enzyme capable of neutralizing penicillin called penicillinase (now termed β-lactamase). Currently, this form of resistance is common in as many as 93% of all *S. aureus* clinical isolates. Many coagulase-negative *Staphylococcus* spp. (CoNS) also produce β-lactamase and so are resistant to the penicillins as well [10].

The primary target of β-lactam antibiotics is the penicillin-binding proteins (PBPs), bacterial enzymes anchored in the cytoplasmic membrane that are involved in the final stages of peptidoglycan synthesis. They are responsible for the polymerization of peptide moieties of the peptidoglycan chains cross-linked in *S. aureus* species. Penicillin reduces the peptidoglycan cross-linkage, inhibiting new septum initiation. The actual lethal target of β-lactams has never been identified.

Staphylococcus spp. have various means of becoming resistant to β-lactam antibiotics. They can do so by mutation or, more efficiently and clinically relevant, by acquiring a foreign DNA element coding for methicillin resistance [5]. The *MEC* determinant confers to the staphylococci an intrinsic resistance to all β-lactams, including most cephalosporins and carbapenems.

The first methicillin-resistant *S. aureus* (MRSA) strain was isolated in 1960, just after the introduction of methicillin into clinical use. MRSA strains reside in environments where there is constant, strong antibiotic pressure (i.e., hospitals and clinics). A further property of MRSA is a tendency to accumulate additional unrelated resistance determinants and incorporate them into their genome. Their adaptability and ready response to antibiotic selection has led to resistance to almost all commonly used antibiotics, except vancomycin. This is a serious problem, and the emergence of multiple drug–resistant *S. aureus* represents a consequential response to the selective pressure imposed by antimicrobial chemotherapy [11].

Aminoglycoside antibiotics, such as gentamicin, kanamycin, streptomycin, and neomycin, inhibit protein synthesis by binding to the 30S ribosomal subunit. Aminoglycosides have been used widely to treat staphylococcal infections and are still often useful, in combination with other antistaphylococcal agents. However, resistance to aminoglycosides among staphylococci is well documented.

Chloramphenicol is a bacteriostatic agent that binds to the 50S ribosomal subunit and inhibits the transpeptidation in protein synthesis. While this agent is not widely used to treat staphylococcal infection, resistance to chloramphenicol is due to inactivation of the antibiotic by chloramphenicol acetyltransferase enzyme (CA7). Macrolides, such as erythromycin and oleandomycin; lincosamides, such as lincomycin and clindamycin; and streptogramin antibiotics also have a bacteriostatic effect on *Staphylococcus* spp. by binding to their 50S ribosomal subunit, arresting protein synthesis, but resistance to these antibiotics is also prevalent. Rifampin has also been used to treat staphylococcal infections, but when used alone, resistant strains quickly arise.

In conclusion, multidrug-resistant staphylococci are recognized worldwide as major nosocomial pathogens. Until a better molecular genetic understanding of the mechanisms of resistance is achieved, novel chemotherapeutic strategies and prudent treatment regimens must be sought to reduce this organism's impact.

REFERENCES

1. JF John, NL Bang. *Staphylococcus aureus.* In: CG Mayhall, ed. Hospital Epidemiology and Infection Control. 2nd ed. Philadelphia: Lippincott, Williams and Wilkins, pp. 325–345, 1999.
2. RF Boyd. Basic Medical Microbiology. 5th ed. Boston: Little, Brown and Company, pp. 246–262, 1995.
3. HI Maiback, R Aly. Skin Microbiology. New York: Springer-Verlag, 1981.
4. BD Davis, R Dulbecco, HN Eisen, HS Ginsberg. Microbiology. 3rd ed. Philadelphia: Harper & Row, 1980.
5. CA Mims. The Pathogenesis of Infectious Disease. 3rd ed. New York: Academic Press, 1987.
6. WK Joklik, HP Willet, DB Amos, CM Wilfert. Zinsser Microbiology. 20th ed. Norwalk, CT: Appleton and Lange, 1992.

7. DS Paulson. Topical Antimicrobial Testing and Evaluation. New York: Marcel Dekker, Inc., 1999.
8. J Kuba. Immunology. 2nd ed. WH Freeman: San Francisco, 1991.
9. M Schaecter, G Medoff, BI Eisenstein. Mechanisms of Microbial Disease. 2nd ed. Baltimore: Williams and Williams, 1993.
10. D Armstrong, J Cohen, eds. Infectious Diseases, vol. 1. London: Mosby, 1999.
11. AJ Hartstein, ME Mulligan. Methicillin-resistant *Staphylococcus aureus*. In: CE Marshall, ed. Hospital Epidemiology and Infection Control. 2nd ed. Philadelphia: Lippincott, Williams and Wilkins, pp. 347–364, 1999.
12. BA Freeman. Burrows textbook of microbiology, 22nd ed. Philadelphia: W. B. Saunders Co., 1985.

13

Full-Body Shower Wash: Efficacy Evaluation of a 4% Chlorhexidine Gluconate

Daryl S. Paulson
BioScience Laboratories, Inc., Bozeman, Montana

Before surgical procedures are performed, it is standard that the proposed operative site be prepared with an effective antimicrobial to reduce the microbial populations residing on the skin and, thereby, the potential for surgery-associated infection. Povidone iodine and chlorhexidine gluconate (CHG) have been the two most prevalent antimicrobial choices for preoperative patient skin preparation over the years. In efficacy trials using human volunteers whose baseline counts exceed 10^5 microorganisms/cm^2 on moist skin sites, both antimicrobial products commonly demonstrate at least a 3 $\log_{10}$ reduction in resident skin flora within 10 minutes of skin preparation [1]. When a formulation contains at least 60% alcohol, these reductions are observed within seconds.

Because normal resident microbial populations of skin in the abdominal and thoracic regions average $>10^3$ organisms/cm^2, the vast majority of these flora are removed during the antimicrobial skin-preparation process. At such anatomical sites as the inguinal region, however, where microbial counts average approximately 10^5 organisms/cm^2, the preoperative patient preparation alone may not be adequate to reduce significantly the possibility of postsurgical infection. In these anatomical areas with particularly high microbial counts, significant numbers of microorganisms may remain on the prepared skin site—organisms that can potentially cause postoperative infections, especially in immunocompromised patients.

Use of antimicrobials, particularly CHG, in a whole-body disinfection regimen is not new. The results and effectiveness of the procedure, however, are unclear. Seeburg and coworkers investigated the merits of CHG as a shower bath and reported that the procedure significantly reduced *S. aureus* colonization in postsurgical patients [2]. Brandberg and Andersson reported that a shower wash with nonmedicated soap increased the aerobic bacterial populations, but shower baths with CHG produced significant reductions in all microorganisms [3]. Hayek and Emerson also reported favorable results with CHG in reducing infection rates in postsurgical patients [4]. Rotter and associates reported that no significant reductions in postsurgical infection rates were observed when CHG was used as a preoperative whole-body shower wash [5]; these findings were similar to the findings of Ayliffe and coworkers [6].

In the study reported here, the objective was to evaluate the antimicrobial effectiveness of a 4% CHG used as a full-body shower wash by charting its effect on skin flora of two anatomical sites over the course of a 5-day application period. A test design was developed that would measure the immediate, persistent, and residual effects of the CHG product used in this study. The immediate antimicrobial effects are due to both the mechanical removal of washing and the immediate inactivation by the antimicrobial of microorganisms residing on the skin surface. The persistent antimicrobial effectiveness is a measure of the antimicrobial product's ability to prevent microbial recolonization on the skin surfaces, either by inhibition or by lethality. Finally, the residual antimicrobial efficacy is the measurement of the CHG product's cumulative antimicrobial properties after it has been used repeatedly [7].

I. METHODS

The study design for evaluating the CHG product applied during the shower wash combined aspects of both the surgical scrub evaluation and the preoperative skin preparation evaluation used for U.S. Food and Drug Administration product efficacy evaluations [8]. Five healthy human subjects at least 18 years of age but not older than 70 years were recruited. Insofar as possible, subjects were of mixed age, sex, and ethnic background.

The rationale for using a relatively small number of subjects was that it is more difficult to detect a significant antimicrobial effect with small sample sizes than with large sample sizes. That is because the statistical power is less than with small samples, and differences must be large to be recognized as such. Hence, small sample tests often err by concluding that no difference exists between groups when one actually does. If a significant difference was noted, it would be even more significant when a large study was conducted.

All subjects were free of clinically evident dermatitis and of injuries to the skin areas being sampled. No subject admitted to the study was using topical or systemic antimicrobial agents or any other medication known to affect the normal microbial populations of the skin.

Once the five subjects were admitted to the study, a 14-day pretrial period was observed. During this time, subjects avoided the use of medicated soaps, lotions, shampoos, deodorants, chlorinated water baths, and ultraviolet light tanning beds, as well as skin contact with solvents, acids, and bases. This regimen permitted stabilization of the normal microbial flora populations on the skin.

A 7-day baseline period followed the pretrial period. Baseline skin sampling was conducted on days 1, 3, and 5 at both the abdominal and inguinal regions.

A 5-day test period followed. Subjects used two separate 4% CHG-impregnated sponges per shower. Each subject applied water to the sponge, ad libitum, working up a lather by squeezing the sponge repeatedly. Standing away from the direct shower flow, the subjects washed their entire bodies with the CHG-lathered sponge for approximately 60 seconds. No lather was allowed to come into contact with anatomical areas above the base of the neck. This was to avoid possible toxic effects to the eardrums or eyes, should contact occur. Each subject wore a shower cap, which covered the ears completely. The CHG later was rinsed from the subject by means of the shower flow; the shower wash procedure was then repeated with a second CHG-impregnated sponge. Immediately after the shower procedure, each subject dried his or her body with a supplied sterile, soft, absorbent terry towel. Each subject was then sampled at both test sites at time 0 (within 10 minutes of showering) and again at 3 and 6 hours after the shower wash. Identical samplings were conducted on test days 1, 2, and 5. The abdominal sample was taken from the region extending approximately 1 inch to the left and right of the umbilicus. The inguinal region was sampled at the upper, most inner aspect of the thigh. A sterile gauze material, secured with adhesive tape, was used to protect the sampling sites from transient microorganism contamination. The gauze bandage was removed for the 3- and 6-hour skin samplings and then reapplied. No additional showers or baths were performed during the 5-day test period by any of the subjects.

At the sampling times, a sterile stainless steel cylinder (inside area = 3.56 cm^2) was held firmly to the skin area to be sampled. Five milliliters of sterile stripping fluid, consisting of 0.1% Triton X-100 in 0.1 mol/L phosphate-buffered saline solution (pH 7.8), was added to the cylinder, and the skin area inside the cylinder was massaged in a circumferential manner with a sterile rubber policeman for 2 minutes. One-milliliter aliquots of this solution (10-fold dilution) were removed and plated—with 10^0, 10^1, 10^2, 10^3, and 10^4 dilutions, as appropriate—in triplicate on trypticase soy agar containing 1.0% Tween 80 and 0.3% lecithin as neutralizers.

The agar plates were incubated at 30°–35°C for 48–72 hours. Only those plates yielding between 30 and 300 colonies were used in this evaluation.

II. RESULTS

All five subjects completed this study with no adverse skin irritations (edema, erythema, or rash) noted. Three of the five subjects were women; none of these reported any vaginal irritation during the course of the 5-day CHG shower procedure. (see Figs. 1 and 2 and Tables 1 and 2 for summary statistics).

The Student's *t*-test was used in this evaluation to determine the level of treatment observation significance, compared with the baseline measurement. Because multiple 95% *t*-tests were used in this evaluation, the *t*-test table values were adjusted for the multiple estimates by means of modified α values, α^*, as described by Dixon and Massey [9]: $\alpha^* = 1 - (1 - \alpha)^k$, where k is the number of confidence levels performed in the study.

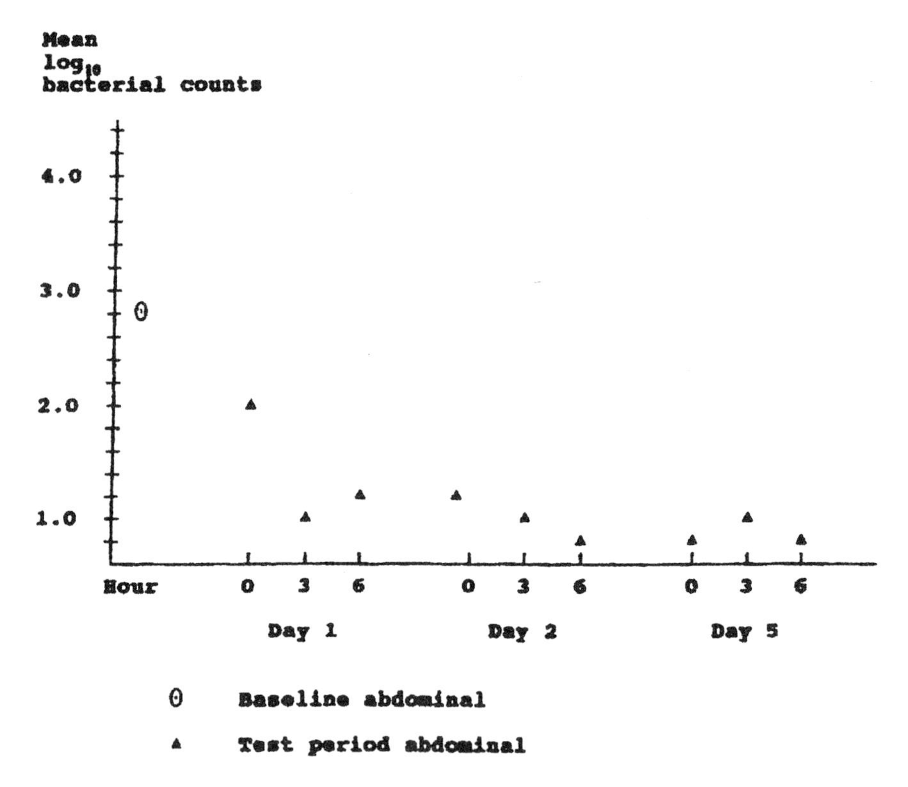

Figure 1 Antimicrobial effects of CHG, abdominal region.

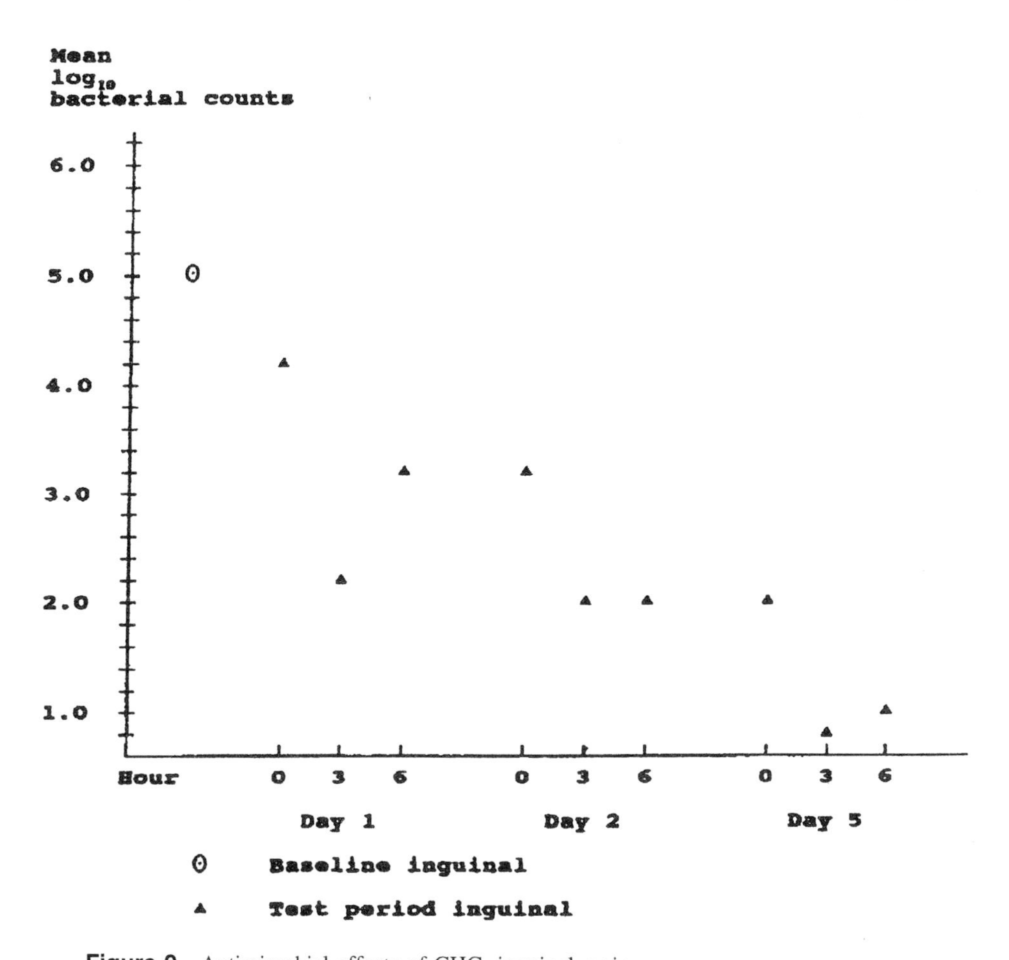

Figure 2 Antimicrobial effects of CHG, inguinal region.

Table 1 Antimicrobial Effects of CHG, Abdominal Region

Time	Day 1	Day 2	Day 5
0 hr	2.13	1.32	0.5
3 hr	1.24	1.08	0.62
6 hr	1.66	0.85	0.5

Table 2 Antimicrobial Effects of CHG, Inguinal Region

Time	Day 1	Day 2	Day 3
0 hr	4.57	3.6	2.08
3 hr	2.6	1.84	0.49
6 hr	3.66	2.32	1.21

A. Abdominal Region

As can be seen in Figure 1 and Table 1, the CHG shower wash provided significant bacterial reduction during the course of the 5-day study. A significant immediate antimicrobial effect was noted within 10 minutes of the shower wash on each of the three sampling test days, relative to the baseline measurement ($p < 0.05$). Additionally, statistically significant persistent antimicrobial effects remained through the 3- and 6-hour postshower periods (decreases from the baseline measurement) for each of the three sampling days ($p < 0.05$). A statistically significant residual effect was also noted ($p < 0.05$). As the study progressed, greater microbial reductions from the baseline were achieved.

B. Inguinal Region

From the data in Figure 2 and Table 2, it can be seen that the CHG shower wash used during the course of the 5-day test period also demonstrated statistically significant immediate, persistent, and residual antimicrobial properties ($p < 0.05$).

III. DISCUSSION

This study suggests that a CHG shower wash used over the course of a 5-day period significantly reduces the normal skin flora colonizing both the abdominal and inguinal areas. Similar and compatible findings were noted by Brandberg and Andersson [3] in both male and female subjects, as well as by Seeburg and associates [2], who were interested in the antistaphylococcal effectiveness of a shower bath with CHG. Hayek [4] reported similar positive results. Rotter and colleagues [5] and Ayliffe and associates [6], on the other hand, reported no significant benefits from preoperative full-body washes in reducing postsurgical infection rates. Perhaps the ambiguity concerning the antimicrobial efficacy of the full-body shower wash can be partially explained by this study. As Figures 1 and 2 and Tables 1 and 2 display, the full-body, CHG shower wash probably should be performed for 3–5 consecutive days, if it is to be effective on the basis of residual properties that retard recolonization of the skin microflora to normal levels. The effectiveness of the CHG full-body shower wash depends mainly on the residual antimicrobial effect, which increases the more consecutive days the antimicrobial is used. On the basis of this study, at least three, but preferably five consecutive daily washes are required to keep the normal skin flora populations significantly lower than the baseline levels through a 24-hour period, as is necessary to provide value in reducing postsurgical infection rates.

The reason for the ambiguity of results of assessment of the CHG full-body shower wash reported in the literature may be attributed to statistical variability. A 1- or 2-day presurgical application period is simply too brief to establish the necessary levels of residual antimicrobial activity to be of value in reducing postsurgical infection rates. It will be interesting to see what data other investigations produce, when the full-body shower wash procedure is employed over a period of 3–5 consecutive days.

A more effective way of delivering the antimicrobial may not be via a shower, but direct application to the proposed surgical site via a "sponge bath" type of application. This would provide at least two advantages. First, because the topical antimicrobial would not have to be diluted as in a shower, the concentration could be less—say 2% or 1%—and still be effective. Additionally, the drug would be less likely to come into contact with eyes or eardrums, eliminating a safety concern.

In conclusion, a regimen using the topical antimicrobial application may be useful to extend the application after the surgical procedure has been completed. Research is needed in this area to discover its potential merits.

REFERENCES

1. DS Paulson. A broad-based approach to evaluating topical antimicrobial products. In: JM Ascenzi, ed. Handbook of Disinfectants and Antiseptics. New York: Marcel Dekker, 1992.
2. S Seeburg, A Lindberg, BR Bergman. Preoperative shower bath with 4% chlorhexidine detergent solution: reduction of *Staphylococcus aureus* in skin carriers and practical applications. In: H Maibach, R Aly, eds. Skin Microbiology: Relevance to Clinical Infection. New York: Springer-Verlag, 1981, pp. 86–91.
3. A Brandberg, I Andersson. Preoperative whole body disinfection by shower bath with chlorhexidine soap: effect on transmission of bacteria from skin flora. In: H Maibach, R Aly, eds. Skin Microbiology: Relevance to Clinical Infection. New York: Springer-Verlag, 1981, pp. 92–97.
4. LJ Hayek, JM Emerson. Preoperative whole body disinfection: a controlled clinical study. J Hosp Infect 11(suppl B):15–19, 1988.
5. ML Rotter, S Larsen, EM Cooke, et al. A Comparison of the effects of preoperative

whole body bathing with detergent alone and with detergent containing chlorhexidine gluconate on the frequency of wound infections after clean surgery. J Hosp Infect 11:310–320, 1988.
6. GAJ Ayliffe, MF Noy, JR Babb, JG Davies, J Jackson. A comparison of preoperative bathing with chlorhexidine detergent and nonmedicated soap in the prevention of wound infection. J Hosp Infect 4:237–244, 1983.
7. DS Paulson. The anti-infective market: Where is it going? Soaps/Cosmetics/Chem Spec March: 34–36, 1990.
8. FDA. Tentative Final Monograph, June 17, 1994.
9. WJ Dixon, FJ Massey. Introduction to Statistical Analysis. 4th ed. New York: McGraw-Hill, 1983, pp. 131–133.

14
Surgical Preparation and Site Access: Testing's Stepchild

Mary Bruch
Micro-Reg, Inc., Hamilton, Virginia

I. INTRODUCTION

The testing of preoperative skin-preparation products has had the characteristics of a stepchild. More attention to demonstration of antimicrobial effectiveness has been focused on handwashing and surgical scrubbing. Now, with rapidly developing technology, attention has been focused on the preparation of access sites for new medical devices. Many of the effectiveness and testing considerations apply to skin preparations for use of these devices. This chapter will discuss where we are, how we got here, and where we hope to go.

The emphasis on asepsis in the operating theater began with Lister, who worried about many environmental factors causing postsurgical infections in his patients. It is not clear whether Lister knew of Semmelweis's work or not, but they certainly were moving in the same direction [1]. As we examine the issues, the effectiveness and testing of surgical scrubs can be looked at as opposite sides of a coin. Both these procedures are designed to prepare the patient for surgery and reduce the number of bacteria to which the patient is exposed. Since Lister's work to develop aseptic surgery, the superficial bacterial flora seems to be the only element in the operating room that is not sterilized and is the least susceptible to control.

In the convergence of ideas that resulted in the development of aseptic surgery, Semmelweis (1846) was inspired to show that the transmission of "childbed" fever could be stopped with handwashing. This breakthrough was made prior to Pasteur's identification, in 1861, of the role of bacteria in infection. We do not know if Lister (1865) communicated with Pasteur to learn of his reported use of 5% carbolic acid (phenol) to treat sewage. Lister realized that treatment of surfaces, including the patient, could reduce postsurgical morbidity and mortality and change the approach to surgery, which then had to be done quickly to decrease a patient's exposure to excessive exogenous microorganisms.

We must pause, also, to thank Professor Gustav Neuber of Kiel, Germany, who used great insight and originality to incorporate Lister's ideas into what was the prefiguring of today's operating room (OR) suites [2]. Finally, as the last part of the convergence, Robert Koch (1878), using his own postulates, concluded that bacteria could cause wound infection.

Beck [3] has suggested that, even with the slow progress of developing modern surgical techniques, the basic concepts of infection prevention remain the same: cautery, washing with water and wine (alcohol), the renewed use of alcohol, and the procedure of irrigation and delayed wound closure (originated by Lister and revived by Dakin) [4,5].

II. WHERE THE BACTERIA ARE

Before we discuss how to test removal and destruction of the microbial skin flora, we must determine where the bacteria reside and how bacterial populations are maintained on the skin [6,7]. Gibbs and Stuttard [8] have described the skin as uneven in topology and thickness (Table 1), ridged, with openings of pores and ducts (Fig. 1). The skin may act as a physical barrier, though not as a chemical barrier. Both apocrine and eccrine secretions are vital to the microbial population, as well as repopulation after the removal of superficial flora.

The skin's surface is constantly changing as a result of growth and shedding skin squames. The breakdown products from cell maturation and death, in addition to fluids bathing the skin (i.e., sebum and perspiration), are the nutritional

Table 1 Regional Variations in the Thickness of the Stratum Corneum

Skin region	Thickness μ(cm $\times$ 10^{-4})
Abdomen	15.0
Volar forearm	16.0
Back	10.5
Forehead	13.0
Scrotum	5.0
Back of hand	49.0
Palm	400.0
Plantar (sole)	600.0

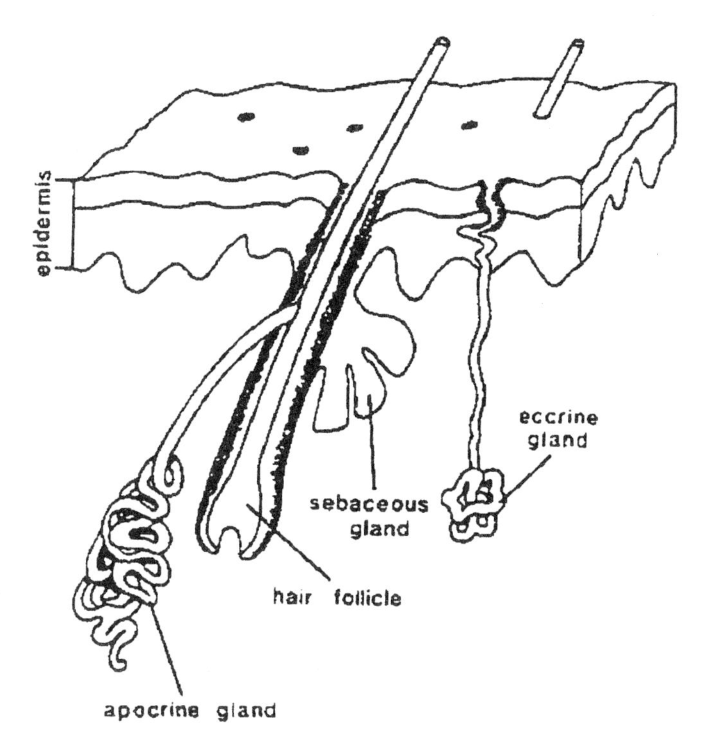

Figure 1 Diagrammatic cross section of skin, appendages, and glands.

source for the bacteria residing on the skin. The skin is normally slightly acidic with a variable pH that can help control the resident flora in conjunction with the dry skin atmosphere. The arid locales of the skin surface usually do not support gram-negative bacteria. They do exist in the axillae and groin, where the moisture level is higher. Marples and Noble have extensively described the type and location of the microbial flora normally residing on the skin [9,10]. There have been various estimates of the location and numbers of population distribution. Selwyn and Ellis have dealt with this problem by studying the work of others and drawing their own conclusions, stating that not all microbial populations are superficial and that 20–50% are deep microbial flora [11,12]. Reybrouck has described this deep fraction as the ''hidden flora'' [13].

The microbial flora of the skin is part of any risk/benefit consideration for use of topical antimicrobials. Mainly aerobic microorganisms are distributed over the surface of the body and are very numerous in particular areas of the body, including groin, axillae, and under fingernails and nail folds. The bacterial count on the skin seems to maintain an equilibrium between removal, death, and repopulation from the deep flora. Washing and scrubbing can delay the repopulation a few hours, at least, and removes large numbers of the skin bacteria, which are reestablished almost completely within 24 hours.

Dr. Phillip Price, a researcher at the University of Utah, began to investigate the microbial flora of the hands, beginning with his own hands, and the effect of handwashing on skin flora. In the late 1930s he examined methods of removing bacteria from the skin [4]. He believed that repeated handwashing removes a fractional population each time and that a straight-line reduction occurred. This did not prove exactly true, but he discovered that a fractional reduction does take place with repeated washing in a series of basins [13]. As a result of his findings, the terms ''transient'' and ''resident'' flora have become parts of our vocabulary. He is credited with defining transient flora as that array or organisms picked up from contact with the environment or contact with humans. Because it does not normally reside on the skin, it can be easily removed. Resident flora he defined as those microorganisms comprising the normal flora firmly attached to the skin and difficult, if not impossible, to remove. Much attention has been paid to transient flora by infection-control practitioners because these are the microorganisms most often involved in infection. Price also defined a third category, the deep flora, heretofore generally ignored, but recently becoming an increasingly important group of the skin's microflora, because it is the deep flora that repopulates the surface of the skin.

In considering skin or site preparation for an operative technique, the goal is to eliminate as many of both transient and resident organisms as possible, usually accomplished by extended exposure times and vigorous washing applications. However, it is now clear that rigorous agitation causes the deep resident flora to be distributed onto the skin surface. Understanding the location and relocation of the flora and its function in repopulation of the skin is necessary to achieving the effectiveness of antimicrobials.

Regardless of the fact that surgical scrubbing and skin treatment in preparation for surgery has involved the use of active antimicrobials since Semmelweis, the prevailing concept was that the effect on the microbial flora was more significant the longer and harder the scrub was done. The logic is understandable, especially up until 1937 when Price conducted his studies. Until then, the extent and location of the skin flora was largely unexamined. The idea of extraordinary cleanliness, grounded in prevalent religious attitudes of the time, fostered the early recommendations of long scrubbing and prepping times. Now we know that the excessive scrubbing does remove the transient flora, but it also has the secondary effect of causing increased shedding of superficial dead skin cells called squames. The squames carry bacteria attached to the skin cells or aggregation of cells from the stratum corneum. Until the introduction of antimicrobial/detergent (or soap), handwashing or skin preparation for surgery occurred in two steps: washing/cleansing and disinfection. Not until the 1950s was this changed with the introduction of Phisohex and Betadine (registered trademarks of Purdue Frederick, Inc. and Sterling-Winthrop), containing iodophor and hexachlorophene, respectively.

The microbial flora of the skin is part of any risk/benefit consideration for use of topical antimicrobials. Mainly, aerobic microorganisms are distributed over the surface of the body and are very numerous in particular areas, including the groin, axillae, and under fingernails and nail folds. The bacterial count on the skin seems to maintain an equilibrium between removal, death, and repopulation from the deep flora. Washing and scrubbing can delay the repopulation a few hours and removes large numbers of the skin bacteria, which are reestablished almost completely within 24 hours.

If we are to determine a more effective approach to removing microflora, yet emphasize a gentler approach to effective scrubbing and prepping skin with less irritation and abrasion, then we need to look at the object of our attack: the skin. What is this cloak of skin with which we defend our body against the outside world, but which can also react with symptoms so painful and alarming that we are immediately focused on obtaining relief? Yet, the skin does harbor a microflora that must be selectively controlled. To this end, we must discuss the structure of the skin, where bacteria are located, and the physical/chemical structure of the skin, especially the stratum corneum (the topmost skin layer).

The organizational structure of the skin layers is diagrammatically represented in Figure 2. The relative thickness of the epidermis and dermis is shown. If the skin layers are drawn in a diagrammatic way, the orderly layers of cells forming the viable (live) epidermis are seen. The dead, dry, scaling cells of the stratum corneum have a structure that is likened to "bricks and mortar."

The bricks are the stratum corneum cells, which are shown as closely placed interlocking cells. The mortar is a little more complicated, but together they form

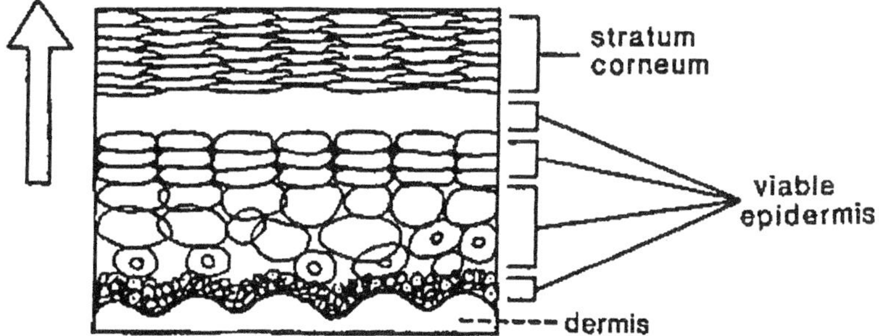

Figure 2 Layers of the epidermis.

the barrier properties of the skin. The viable epidermis is not a significant barrier to penetration. The integrity of the stratum corneum is broken with hair follicles and sweat glands. These are the sites from which the bacteria grow and repopulate the skin. As can be seen in Figure 1, the mouth of the eccrine gland and the base of the hair follicle are the inaccessible locations of the ''deep'' flora. Even if the skin surface is aggressively decontaminated with alcohol and few or no colonies are found upon sampling, the deep flora is still there. Conventional skin sampling will not detect these bacteria. Microbial colonies will soon reestablish on the skin surface.

Skin microbiology has been not only neglected, but the source of significant controversy; much of which concerns the location of the skin bacteria and the issues of skin antisepsis and disinfection. Selwyn has listed ''what and where are the target organisms'' as the primary question. He has also shown the inaccessible location of the resident flora in the opening and depths of the hair follicle.

The location and proportion of the normal, ''deep'' flora has been estimated from almost all surface to the major portion—hidden. Selwyn's own conclusion was 20–50% hidden. As discussed previously, it was Price who originally defined the ''deep'' flora [4,13,14], and Reybrouck who termed it the ''hidden'' flora [15]. Today, we realize that the boundaries of the transient and resident flora are not as clear as we might have once believed and that some transient organisms can establish themselves and be carried as part of the resident flora. We recognize that the hidden flora grows and reestablishes the population of organisms previously removed or washed off the skin [16].

The superficial flora can be removed quite efficiently and when sampled for microbial count, can even show an extremely low or zero count, but only for a short time. We cannot sterilize the skin, nor will we ever know the total microbial flora present. The measurements that we make in testing skin antimicrobials are based only on the fraction removed or released from the skin.

The accuracy with which this can be measured forms the basis for the stating of initial count in the studies reported in the literature. To make it even more frustrating, the bacteria on the skin are not evenly distributed over the skin surface, but in reality are present in microcolonies. When a sampling yields microcolonies, the bacterial count is underestimated, and when procedures for sampling dislodge and break up these small colonies, the bacterial count is exaggerated. The superficial skin organisms shed as squames reflect the remains of these colonies. Because a squame may contain a number of bacteria, the squames on surfaces and, particularly, floating in the air can initiate infection if they find a hospitable environment [17,18].

The basal layer provides a supply of cells that move up through the viable epidermis, changing form as they traverse the layers. Cells become flattened and elongated. They become progressively drier and contain protein matrices (keratin).

The function of the stratum corneum is to develop from the viable epidermis, become keratinized as part of the barrier function of this layer, and ultimately to be shed. This journey is diagrammatically presented in Figure 3. If a room is occupied by people, there are skin squames on environmental surfaces and in the air. The drying dead skin cells of the stratum corneum are shed into the air or into water with movement, rubbing, scratching, and bathing. Particle sizes range from a few cells to visible flakes (like soap flakes). Bacteria are shed and can be attached or unattached to shed squame cells.

The scrubbing process itself causes release of skin squames, most with bacteria attached. Antimicrobial products reduce the shedding squames and viable bacteria compared to soap alone. Meers and Yeo have shown that handwashing causes a shower of skin squames in the air [17].

Many chemicals are more soluble in lipids than in water. The lipid layer of the stratum corneum will dissolve and can absorb chemicals placed on the surface of the skin. In fact, this characteristic is important in transdermal delivery of drugs, for example, the skin patch for drug delivery, and in what has been termed substantivity or persistence of the applied antimicrobial.

The formulation, use, and effectiveness testing of preoperative preparations has long been a neglected area of antimicrobial use. Now, with ever-expanding numbers of devices requiring skin preparation for use, this area demands more attention to define appropriate use and testing. As we have noted, our use of skin

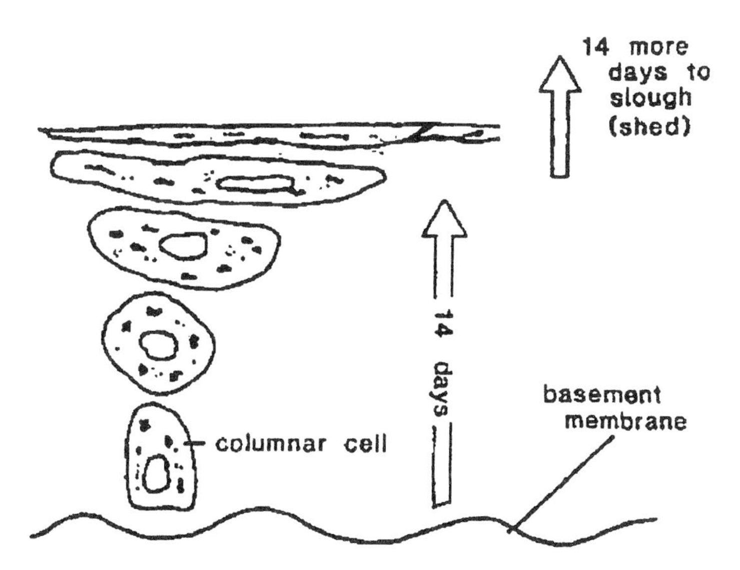

Figure 3. Diagram of progress of a basal cell through the epidermal layers to shedding as a stratum corneum cell.

preparation began with Lister, and this tradition has not changed significantly since implementation of detergents and alcohols. Newer techniques of preserving an aseptic atmosphere have been successful on one hand but fail on another. An excellent example of this dichotomy is the use of adhesive surgical drapes. Rhodeheaver has discussed the problems of adhesive tape transferring bacteria to the site and transferral of bacteria under these drapes to the wound edge [6]. Aly et al. have demonstrated that bacterial growth under wound dressings is possible [7]. The adhesive drape protects that area around the intended incision site, but their use may actually enhance the repopulation of the skin by resident bacteria after skin-site preparation by stimulating perspiration and transfer of bacteria to the wound site. As demonstrated, some positive advances have also created negative results.

As a result, we have gradually come to know that, barring "gentle flaming of the skin" (Pasteur), we cannot sterilize the skin. With this recognition, we are perpetually confronted by the residual flora that remains after antimicrobial application. When are these organisms a risk, how many remain, and how rapidly do the residual organisms repopulate the site? Although sterility of the operative or access sites would be the most desirable, testing and sampling methods can direct new work, innovation, and attention to these procedures.

The skin has a microbial flora all of the time, but it changes in variety and numbers determined by many factors such as climate, season, sex, and age. As we have discussed, part of this flora is hidden from direct access, while part is easily acquired and removed. Many imagine and picture the skin with the flora evenly distributed over the surface. This is clearly not the case; not only is a fraction of the flora hidden, but that portion distributed on the skin surface exists in microcolonies of different sizes and not in a unicellular, even fashion, nor are the microcolonies evenly distributed.

To test the effectiveness of a product, the microflora must be examined before and after the application of a test product. The critical decision is how to sample the skin so that the sample is representative.

For a number of years, the cup scrub method has been considered the "gold" standard. Some discussion of other sampling methods is warranted, especially when the relative sampling efficiency is measured by comparison to a method for recovering as much as 100% is possible (Selwyn). In this forum, standardized swabbing and the cub scrub were essentially equivalent.

A multitude of sampling techniques for the skin have been published over the last decades. We have become somewhat nearsighted in the last few years in concentrating on the cup scrub method as the only credible sampling technique. I want to look more closely at this.

Selwyn has examined our view of the skin microflora and how to sample it with some clarity [11,12]. He has asked questions about the flora and how to evaluate it. Selwyn deals with a subject that is clearly critical in infection control but is often ignored. When antiseptics are applied in testing methods, we declare that the activity is directed against either resident or transient flora, when, in fact, both are almost always present.

Selwyn and Ellis have published an important discussion of where the microflora is located and a comparison of a number of different techniques [11,12]. They used a culture of a full thickness skin biopsy to estimate their 100% count and then compared this with the results from a variety of different sampling techniques. They describe conclusions about the location of and numbers in the skin microflora. Those conclusions were that the total bacterial count per square centimeter ranged from 1,000 to 400,000, and that 20–50% percent are located deep in the crevices of the skin and hair follicles. They also stated that *Staphylococcus aureus* is never a resident on skin (nares and perineum excluded). Further, skin preparation can be achieved in 30 seconds to 2 minutes, rather than requiring the long exposures that had previously been favored.

Selwyn and Ellis emphasize, and I strongly concur, that pertinent questions include: What and where are the target microorganisms? and What are the characteristics of the available chemical agents and how do we evaluate them in vivo?

Their results further reveal the effect of agitating and rubbing the skin during sampling because the most efficient sampling techniques were the ''cylinder scrub'' (cup scrub) and standardized swabbing (rigorous). These two methods recovered 4–15% and 3–20%, respectively, compared to the numbers from skin biopsy as 100%. The techniques of tape transfer, contact plates, and velvet block gave under 1% of the 100% control. Clearly, the latter three methods sample the surface layer, presenting a picture of what flora is there and, again, as we must recognize, what the intent of the sample is in the testing.

Survival of ''resident'' bacteria occurs even when rigorous disinfection of the skin occurs prior to sampling. Also, Sidney Selwyn has described the dilemmas associated with the skin microflora, as well as published an examination of the efficiency of sampling methods. If his data are reexamined, the efficiency of sampling is poor with any method as far as sampling the complete flora of the skin is concerned.

Perhaps we should look at sampling methods that are internally consistent, rather than constantly refer comparisons to a variable pretest count. Because we can never know the exact count, or even a good estimate, of the total flora, it is difficult to draw absolute conclusions. The best hope is to prove that a selected method removes a consistent fraction at each sampling. An older method with some use history in England is the velvet block [19]. A wooden block is tightly covered with velvet and sterilized. Even contact is made with the skin to sample, and the block is then pressed onto a solid agar plate and developed colonies enumerated. The block is then repeatedly stamped onto the hard agar surface as is now done with replicate plating. Cardboard plugs have been another technique used to sample skin and operative sites and is similar to the velvet block. The end of the plug is pressed onto the skin and then repeatedly onto an agar surface. Seeberg et al. have performed studies using this technique [20].

Geelhoed [21] has published preoperative studies utilizing contact (Rodac-type) plates to sample sites before and after surgery. His results will be discussed later. This is another surface-sampling technique used extensively to sample hard surfaces where there is no underlying reservoir resupplying the surface. The contact plate will show microcolonies and give a picture of the skin surface flora. Clearly this method samples only surface microcolonies on the skin. By this method, a picture of microcolony distribution on an area of skin can be seen. These superficial colonies often may be important in situations such as preoperative preparation, when these bacteria may be organisms that migrate to an open surgical wound, especially if occlusive drapes or dressings are used.

All sampling methods have negative and positive attributes [8]. Sampling methods can be divided into techniques that actually remove some or all of the stratum corneum, such as tape-stripping, removal of layers of skin with a Castroviejo keratome, excision with a full thickness biopsy, or at autopsy. Direct sampling in vivo of the skin is desirable in that the skin is being directly examined. In comparative tests with washing procedures, the numbers of bacteria recovered are usually higher. Often the direct sample recovers microcolonies while washing techniques are assaying single bacteria removed from the skin. Clearly, if a comparison of products is made, the surface sampling methods show the same order of removal as washing methods.

Finger imprint and streaks have had a period of popularity, but in recent years, there has been a concentration on washing methods [22]. Most often it has been the cup scrub method (as published by Williamson, Kligman, and Marples) [23–25]. The focus has been on how high the numbers are and how efficiently the bacteria are removed [23]. Washing methods measure only the organisms removed and a comparison with what is assumed to be the total population. In the 1950s and 1960s, these contact methods were often used but have now been discarded. In his closer look, Selwyn has revealed that what we have adopted as our best method is only about 15% efficient, while surface techniques like contact methods in situ are about 1% or less efficient.

There has long been an approach in this type of testing that the only sampling technique is the one giving the highest recovery, usually one that involves scrubbing and washing, which we now recognize works up the resident or ''hidden'' flora to varying degrees. When this analysis is the goal of the test, these sampling procedures are appropriate. (This author's intent is to advise careful examination of the test design and sampling methods. Skin sampling of any kind is a dicey process and further has been plagued with inconsistencies and variation [26].) Some of the more static in situ surface methods show more consistent results. There is a basic inconsistency to the task of sampling in that we wish to sample the surface and the deep flora before and after application of the test prep, but techniques in situ ignore the deep flora, and more rigorous hand scrub techniques remove some variable fraction of the deeper microflora. These recovered counts are compared to pretreatment values, but the total count of the population is unknown.

Ayliffe et al. compared sampling with washing techniques (wash bowl and glass beads) and the finger streak method (five fingers drawn across an agar plate) [27]. The results were similar in the assessment of the effectiveness of reduction in bacterial count of four products and a control. These results do confirm that the finger streak method shows more variation than the washing methods. However, it is reasonable to examine the attributes of surface sampling methods, especially when they follow removal of superficial transient flora prior to sampling. In situations where there is interest in the superficial flora or the time for and ease of sampling are important, this method or other similar methods for in situ sampling such as contact plates or tape stripping should be considered. Smylie et al. have described the finger streak [28].

We must clearly define how to test and what the results mean. In the past there has been significant confusion about reduction where organisms are applied and passively dried on the skin. Results can be quite different if the artificial inoculum is rubbed into the skin [29]. Often the testing can be divided into several layers. One must test reduction in normal flora before and after treatment with cup scrub or swabbing without ignoring identification of surface, if this is indicated. In contrast, the testing can be done after artificially contaminating the skin with a cultured inoculum followed by treatment and sampling by a variety of techniques involving washing and/or surface sampling.

Marples published a method of expanding the flora on the skin using an occlusive plastic wrap, such as Saran wrap, for a number of hours [30]. Gram-negatives will grow with increased humidity. The suppressive effect on growth of the flora can be tested when applied prior to wrapping. Sampling of test areas were done using the cup scrub technique and assayed for bacterial count. Because many dressings for site access and drapes for prepping are at least semi-occlusive, the occlusion may be maintained for shorter periods. This procedure could be adapted to show the effects of dressings and drapes. For example, one might envision a probability model that could assess the contribution of the prepop prep, the use of adhesive incise drapes or impregnated ones, the use of a surgical scrub, and the use of presurgical bathing.

Presurgical body bathing, both as a bath and a shower, has enjoyed periods of vogue and disregard. From Lister to the present, the contamination of the surgical wound during and after the surgical procedure has been the subject of speculation and clinical concern.

Glove wearing, surgical draping, air filtration, presurgical hand scrubbing, and preoperative skin preparation have contributed extensively to the reduction in surgical wound infection. In highly infection-prone surgeries, such as hip joint replacement, the use of ultra-clean air (laminar flow) and space suit–type clothing have contributed further to the reduction in postsurgical infections.

But what part does the patient's own skin flora contribute to these infections? It is clear that the patient's skin flora enters the surgical wound and that subsequent infection depends upon multiple presurgical and surgical precautions and prophylactic procedures.

The presurgical bathing of the patient once or multiple times with active antimicrobial agents has been advocated and used during the last 10–20 years. Cruse's large study has fixed the value of presurgical bathing in the minds of many surgeons and infection control personnel [31]. This study utilized a hexachlorophene wash (3%) as a presurgical bath and showed a positive correlation between body bathing and postsurgical infection. However, the study has been criticized by researchers as poorly constructed for statistical analyses.

As the popularity of surgical patient bathing increased, several European and Scandinavian studies were published [20,32,33]. Some of these focused on specific areas of the body and on reduction of numbers of the skin microflora. Effects in orthopedic surgery have been the focus in some of these studies.

Seeberg et al. studied total body disinfection using a chlorhexidine solution three consecutive times for a 5-minute period (showered and dried) [20]. Subjects were exposed to two applications prior to admission and one shower immediately after admission. This work shows reduction of microbial flora, especially *S. aureus*. Lowbury and Lilly [34] and Davies et al. [35] had already shown that chlorhexidine solutions reduced bacterial flora prior to surgery.

Some restrictions on the time of showering should be imposed because of Meer's work, showing a 17-fold increase in the release of skin particles (squames) with viable bacteria after showering [17,18]. The restriction should be implemented to prevent such an increase immediately before surgery.

Brote and Nilehn [32], Jepsen [36], and Thomsen et al. [37] concluded that the post-operative contamination of the surgical wound may not be as important as previously believed. The resident flora of the patient present at the end of surgery may be a primary source of contamination. The authors stated that only a few organisms are transferred to the wound during surgery. (There is a theoretical difference of opinion here.) Dr. Rhodeheaver of the University of Virginia Medical School states that, under occlusive dressings, the surface of the wound may be bathed with bacteria-laden fluid.

Several investigators, including Brandberg and Andersson [33], have shown that showering with a nonantimicrobial soap actually increases the level of the microbial flora. Earlier studies by Cruse and Ford [31] and Dineen [38] had recognized and established the patient's own skin as an importance source of postoperative wound infection.

In their large study, Brandberg and Andersson [33] examined prepping alone (showering and washing with Hibiscrub®) and prepping with 0.5 chlorhexidine gluconate in 70% alcohol compared with this same prep plus three to eight presurgical showers with a 4% chlorhexidine gluconate product added in vascular surgery cases. There was a 17.5% infection rate reported in the nontreated group and 8% in the chlorhexidine group. However, the obvious reduction seen in the averages is not statistically significant, so that a practitioner may want to make his or her own conclusion(s) about the usefulness of presurgical bathing in this setting. With low infection rates, reliable studies require extremely large numbers of patients to make valid statistical conclusions.

Most recently, Rotter et al. [39] reported a prospective multicenter, double-blind, placebo-controlled study. Two immediately consecutive presurgical baths with a 4% w/v chlorhexidine gluconate solution were tested. Twenty-five milliliters of detergent was applied all over the body (including the face, around the eyes, and, in the second bath, a shampoo). The authors concluded that the bathing of patients twice prior to surgery with an antimicrobial detergent with chlorhexidine did not reduce the incidence of infection of clean wound procedures as shown with extensive statistical analysis.

Reports of still other recent preoperative bathing studies do show correlations with reduction in postoperative infection [40–42]. With all the statistical innuendo, methodological differences, and types of surgery, it is impossible to come to a firm conclusion regarding the effect of total body bathing with an antimicrobial soap on infection.

Controversy over alteration of the microbial flora of the skin flared in the FDA Over-the-Counter (OTC) Review of Antimicrobial Drugs, especially with reference to total body exposure to antimicrobial/deodorant soaps. The relationship of routine bathing (rinsing) with these soaps with infection was implied in several studies [43–45] and shown conclusively in the study by Taplin [46] in Costa Rica in a population with a high infection rate utilizing a 2% solution of chlorhexidine in water (therefore, 2% available chlorhexidine).

Two distinct approaches have been taken in the attempts to show a positive effect for the surgical patient: (1) the reduction of postsurgical infection and (2) the reduction of microbial flora, either on the whole body or in heavily contaminated areas such as the groin and axillae, or of specific microorganisms, such as *S. aureus* or *S. epidermidis*.

Whatever attitude practitioners may have formed concerning the value of presurgical bathing, if they choose to use an antimicrobial active ingredient in a bath or shower, a specific protocol must be developed incorporating the information reviewed above.

It does seem clear that if the antimicrobial is used, there is a reduction in the microbial skin flora.

When implementing presurgical bathing, the physician must decide (1) how many baths to use, (2) how much attention to pay specific areas, (3) when to stop the bathing prior to surgery, and (4) whether to shampoo or not. More than one presurgical bath is recommended for greater suppression of the microbial skin flora. The number of baths should be determined by the practitioners. Three to five baths would provide greater residual action and decrease the body's microbial load.

The analyses of published presurgical bathing studies with antimicrobial washing products yield variable results. It is essential to recognize whether the intent and design of the study is to show a statistical reduction in postsurgical wound infections or to determine the reduction of the microbial flora of the skin. Both types of studies have been published.

The data accumulated from studies with very large patient numbers indicate that bathing does not significantly reduce the postsurgical infection rate [39]. When other studies are factored in, the answer is that there is possibly a reduction. It is the superficial skin flora that contaminates the patient's own surgical wound. It is this population that moves across the skin under surgical dressings (often when wet). That is the population of interest to us. This is certainly not the source of all bacterial wound contamination. The surgical staff and the air in the operating theater are also common sources [47].

Ideally, clinical investigation should produce data showing that preoperative preparation of the surgical patient or treatment of access sites results in the reduction of postsurgical or posttreatment infection. Unfortunately, attempts to show the preventive effects of surgical scrubbing or prepping have not produced clear-cut results (E. Larson, personal communication). Much of the clinical investigation has relied on recovery of skin bacteria from the sites before and after applications. Infection rates have been recorded, but showing differences, especially between test products, is practically impossible. When infection rates are low, many hundreds if not thousands of subjects would be required to show a difference in the number of infections from changes in one parameter. It must be reiterated that this is a multifactorial cause.

The intent of skin preparation is to remove as much of the bacterial flora as possible prior to incision or access. As development of new and/or novel enhancements to the scrub or paint are produced, it is unavoidable that combinations are and will be used. The preparation procedure itself should be tested but often the parts are tested to see their contribution, for example, by Geelhoed [21,48,49].

Many clinical evaluations have been done, often using sampling methods in the operative setting that primarily portrayed surface contamination of the skin or wound. It is interesting to note that the results are quite similar.

FDA in its OTC drug review that began in 1972 has a suggested protocol for patient preparation for surgery in its latest publication of the Tentative Final Monograph (TFM). This is a trial using subject volunteers but is not considered a clinical trial since the target evaluation is reduction in microbial flora in a confined test situation [52,53]. [Parts of the protocol will be discussed, but see

Ref. 53 for complete details.] The American Society for Testing Materials (ASTM) is a consensus group that develops standard test methods. They have published several versions of a STM for preoperative testing, and currently the E 35.15 Subcommittee on Antimicrobial and Antiviral Agents have revised a method for both preoperative prep and site access prep combined that is in the balloting process and was published in 2001.

The cup scrub (cylinder-scrub or detergent wash) method has had several progenitors, but Williamson and Kligman [23] published the most current and used version. Changes in the method have been made by Marples, [24] and the method has even been automated by Bibel and Lovell [22].

The assumption has been made in protocols that the detergent method is best because the counts are highest compared to contact methods. Yet in a schizophrenic bent, most truly clinical examinations of effectiveness have been performed using contact methods. It must be stressed that, as with the detergent wash method, improvements can be made in contact methods. If fact, the development of an automated procedure, the Thran Gun [51], for dispensing detergent onto the skin for sampling and vacuuming up the solution after contact, has already been produced. The author's emphasis on sampling is designed to focus attention on what criteria of effectiveness should be accepted for a preoperative or site-access preparation.

As currently regulated by FDA, these products are drugs and are considered an OTC drug product or are the subject of a new drug application (NDA). In hospital usage today, the products used in skin preop prep and for access sites are iodophors, iodophors in alcohol, alcohol, and, occasionally, a chlorhexidine detergent scrub product. The individual products in this group of products are regulated differently. All of the products with the most commonly used antimicrobials were the subject of review by the FDA's OTC panels. Because of the nature of the OTC Review, the standards and test methods are ones that were in vogue when these ingredients were initially tested; although testing with many different techniques has been performed, they will not all meet NDA test requirements. However, according to the rules of OTC Review, these products are legally marketed while in different categories in the Review until a Final Monograph is published (Proposed Monograph, 1974 [52]; Tentative Final Monograph, 1994 [53].

As currently constituted, the protocol published in the TFM requires that a reference 4% chlorhexidine (CHG) detergent scrub product be included in the test and that a statistically assessed 3 log reduction is required to claim effectiveness for a preop prep. When testing is performed using this protocol, the CHG product will often meet the 3 log reduction, but not always and not on all three sampling days.

The iodophors and alcohol as well as many popular antimicrobial ingredients were included in the OTC Review, and chlorhexidine was not. If effectiveness studies with these ingredients are examined, it will be observed that Lowbury and his colleagues have performed many studies with chlorhexidine products in England, where it is commonly used for skin disinfection (European designation for antisepsis). Their use is in distinct contrast to U.S. practices for surgical preparation. The mode of application can be important. Lowbury et al. have reported [54–56] that if the antiseptic product is applied using gloved fingers to rub it into the skin, it is more effective than if it is applied with gauze (standard procedure) or sprayed on.

The activity of the ingredients may be different. When iodophors and alcohol are tested after topical application (handwashing and prep testing), the initial reduction is approximately 2 log. Depending on product, application and protocol, the results may vary 0.5 log [57–61; G. Mulberry, Hilltop Biolabs, personal communication]. Alcohol can be more complicated. Many European countries use alcohol for handwashing and prepping.

Rotter et al. [62–64] published a test method examining alcohols for effectiveness as handrubs (see Ref. 65 for a CEN Directive based on Rotter's method). Rotter used 60% isopropyl alcohol as a control and has shown that *n*-propanol formulations are the most effective. Often, a 3 log or higher reduction can be shown in the handwashing setting in which test organisms are applied to the skin but not routinely. The problem then becomes one in which the most widely used products do not meet the criteria described for preop prep in the TFM, which is also used for testing new drug entities for NDA approval.

What is at the heart of these differences? Some chemicals, such as chlorhexidine, are substantive to the skin, that is, the chemical attaches to cells of the stratum corneum. Residual activity of the chemical affects the time and extent of repopulation of the skin after treatment, but the exact mechanism is unknown. Iodophors (or other halogen products) and alcohol are not retained by the skin, but do affect the repopulation by significantly affecting the total count. Other OTC reviewed ingredients, e.g., chlorxylenol (PCMX) and triclosan, have been formulated for skin preparation and are also substantive. These chemicals are absorbed by the skin within approximately 6 hours after application but have activity when present in the skin [66]. Because absorption is a diffusion reaction, the concentration may vary, especially if a second application is made.

The majority of collected data with products containing substantive ingredients have been obtained using chlorhexidine. When ingredients such as an iodophor, iodine, CHG, or other single ingredients are combined with alcohol, the activity is enhanced and may show some persistent residual effect from the added chemical. The immediate action of substantive chemicals may not be rapid on initial contact.

Van Abbe [67] has stated that the dermatologist-coined term ''substantivity'' conveys the idea of prolonged association between a material and a substrate, an association that is greater or more prolonged than would be expected with simple mechanical deposition. The skin of the surgical patient is a source of wound contamination, either from outside sources or from the patient.

Ulrich and Beck [57] described a study with several regimens including alcohol with an iodophor-impregnated film. The traditional scrubs with iodophors were effective (2–3 log) on skin contaminated with specific organisms. The newer procedure had a significant reduction in the time required for patient prep, with the highest microbial reduction, illustrating the effects that can be achieved with a process instead of a single product prep.

Geelhoed and Sharpe [48] also performed a similar study (using contact plate sampling) with essentially the same design and results with the group treated with alcohol wipes and an impregnated film showing the highest reduction in numbers at wound closure but not statistically different from the conventional scrub groups. He does analyze the savings in time and cost when the alcohol-film procedure was used.

These incise drapes and, in fact, impregnated drapes are still used in surgery. Edlich [68] has furthered the view that drapes cause bacterial growth and believed that adding iodine to the drape did not improve a bad situation. As with many elements in surgical site preparation, there are mixed reviews. Edlich reviews many elements that contribute to surgical sepsis and, in doing so, emphasizes the multifactorial nature of infection following surgery or site-access prep. Beck et al. [69] has also weighed in on this procedure.

Ayliffe [58,60] has published summaries of results from hand sampling from testing surgical scrubs and skin treatment and reiterates the in vivo clinical use of contact plates for sampling body sites. Again, there is reinforcement of an approximate 2 log reduction with iodophors and with chlorhexidine.

Lowbury in England probably has published the most studies on chlorhexidine for topical use as a surgical scrub and as preop skin preparation [70–76]. When one examines these publications, there is great variety in methodology over the years. Detergent scrub products and tinctures containing chlorhexidine have been available and used for many years before their availability in the United States. Currently, there is no tincture product with CHG for skin preparation use in the United States. Lowbury, in considering whether topical antimicrobials are effective clinically in the prevention of infection, reported that when the Birmingham Accident Hospital had made changes in their antimicrobial use, they found a near statistically significant reduction in major sepsis.

Raahave in Scandinavia has also published studies with clinical relevance that include use of CHG and combinations, loading of bacteria in wounds, and use of velvet pads and contact plates for sampling [77–79]. He also assessed the much-discussed use of clinical outcome investigation with topical antimicrobials. Withholding antimicrobial use in a clinical trial was considered unethical. He also discussed injection-site disinfection as a marginal decision, but one that could be considered an ''insurance policy'' because he also recognized its ritualistic aspects.

Access-site preparation has been ignored as a separate effectiveness study. Instead, the results of handwashing, preop preparation, and general skin disinfection studies have been the source—often intuitive—for regimens for preparation of the patient for insertion, injection, indwelling catheters, or any new devices as they are developed. ASTM is attempting to delineate a test method that is useful for testing the effectiveness of these prep products. Some additional issues with access sites that should be considered: length of exposure, repeated application, presence of blood or other body fluids, occlusion of skin, whether required or optional, and physical limitations of the site.

The pursuit of large clinical trials in which antimicrobial agents are compared to placebo is probably not financially or ethically supportable, and it is this situation in which relatively smaller numbers of subjects can be justified. The expectation is that the antimicrobial will be effective and usually many studies have already demonstrated it. If two or more treatment regimens are compared, say for reduction in infection rates, the numbers required become very large, in the thousands, as suggested by Allen [80]. There seems to be a catch 22 to examining infection as a clinical outcome when withholding treatment cannot be justified and comparative studies engender size problems in situations where changes in already very low infection rates are nearly impossible to find reliably.

In the end, finding a reliable, reproducible skin-sampling method selected with a clear understanding of what the sampling achieves may provide the best answers to the question of effectiveness. Over the years, authors have consistently concluded that microbial reduction does give information concerning effectiveness in the clinical setting. Certainly, both FDA and ASTM are developing test methods to ensure that products are effective and yet maintain reasonable size and investment from a sponsor when compared to the outcome information.

New ideas and new products are necessarily based on the past history of development. Today there is a new challenge. Because antimicrobial chemicals are toxic to microorganisms, they have properties that may also produce toxic residues. Antiseptic disinfectant use is critically important in today's health environment. The volume of chemicals used is not economically rewarding enough to stimulate the development of new antimicrobial chemicals. Today the challenges to be met by chemical disinfectants or sterilants have been magnified by the burgeoning development of new medical devices fabricated with mixed materials requiring chemical disinfection/sterilization themselves and for use in preparation of the patient. It is unlikely that new antiseptic chemicals will be available, so the ones in current use must be used with innovation and originality in formulation or in combinations of chemicals and processes.

Modern surgical techniques have changed the environment of the surgical field and the risks the patients face from the surgical wound. Lister's original intent of trying to make the patient's skin as germ-free as the air and the various surfaces is not realistic when the effect of currently used prep regimens are examined. We do not need to persist in the old regimens and techniques for prepping simply as a ritual. The time frame for actual prepping has decreased significantly due, in large part, to increased cost pressures. The use of presurgical bathing has changed the regimen of many surgeons, especially in orthopedics. The use of adhesive drapes at the surgical site can significantly change the skin environment during surgery, where the surgical wound is made through the adhesive drape. These conditions may change the environment and bacterial population under the drape [66,68], not always to the advantage of the patient. We are entering an era where more than one product will be important in the prevention of post-surgical wound infections or at least in reducing the potential for infection of these wounds. Mary Marples [9] is notable for coining the term "ecology" of the human skin and for her early recognition of the variety and changeability of the flora that has affected our current views of its place, influence, and control. Noble has added to her early work in describing the flora of isolated body areas [10].

As we have observed, the sebaceous glands are prominent in the cross section of skin and are primary in the bacterial repopulation of the skin and as the site of the "hidden" microflora (Reybrouck). Using the term "microflora" reflects the understanding that we are interested in bacteria primarily, but that other microbes can be present.

Variation from person to person is a single note that sounds throughout the studies that have been cited. There is significant individual variation in the level of organisms on the skin, which in turn can vary depending on many environmental factors. The number and location of microcolonies vary, as does the thickness (Table 1) and level of microflora in various locations of the body.

Clearly some methods sample the superficial "transient" flora with little of the deeper resident flora. Those employing more agitation and surfactant solution sample a higher percentage of the total flora, but this sample still is only a minor fraction of what is present. The resident, deep flora remaining repopulates; the only questions are how long it takes and to what level the count is restored. Whether surface numbers or resident flora is more important depends on the particular method; whether single cells or microcolonies are sampled depends on the sampling method chosen so that microcolonies may be recognizable when agar plates are read. Much of the clinical testing from contact methods and in vivo studies have used surface sampling, often because of the exigencies of sampling the patient immediately prior to surgery and after the surgical procedure in the operating room.

FDA has suggested a protocol for study and a criteria for preoperative preparation; however, there have been no criteria for tests for preparation of access sites except for reference to a 1 log reduction for preparation of an injection site. Especially for site preparation, the level of the natural, normal flora of the area of the body under test for effectiveness must be considered. The authors mentioned above, as well as others—especially in orthopedic surgery and for presurgical bathing—have defined the numbers and types of organisms present.

An average number with statistical limits of bacteria per unit area of skin has to determine the criteria set for reduction. A similar rationale must be applied to setting the requirement for reduction in a test for preoperative preparation. We have a significant contradiction; on one side there are NDA products like chlorhexidine that have shown a 3 log reduction in a wet environment, especially after multiple uses. When many studies are examined, the results swing from approximately 2 to 3 log. On the other side are many ingredients in the OTC Review that are routinely used for prepping in hospitals and that in formulated products achieve a ≥2 log reduction, although there are not as many published tests [66]. For example, when British and European studies are examined, the results with iodophors (and some other ingredients) are for all practical purposes equivalent.

The goal of this chapter is to emphasize that tests must be carried out with an informed choice of protocol and sampling method to which criteria of effectiveness are applied and that can be met when data are statistically analyzed. The importance of the antimicrobial application to the skin as the isolated single step in prevention of postsurgical infection no longer applies. Process and regimen are the watchwords now.

In the final analysis, ease of implementation and frequency and care in execution determine the protection of the patient. Looking at all these studies, one has to feel that the more things change, the more they stay the same—that things are not so very different from Semmelweis's copper basin, Lister's and Neuber's surgical operating room, or the techniques of Dakin/Carrel or Colebrook and Maxted in the clinical prevention of infection.

REFERENCES

1. RJ Godlee. Lord Lister. Macmillan, London, 1917.
2. G Neuber. Ore aseptische Wundbehandlung Meine chirurgische Private Hospitalern. Kiel: Lipsius & Tischer, 1886.
3. WC Beck. Disinfection from antiquity to the present. Gutherie J 59:191–195.
4. B Price. The bacteriology of normal skin; a new quantitative test applied to a study of the bacterial flora and the disinfectant action of mechanical cleansing. J Infect Dis 63:301–318, 1938.
5. Subcommittee on aseptic methods in operating theaters—committee on hospital infection. Report to the Medical Research Council. EJL Lowbury, chairman. Aseptic methods in the operating suite. Lancet April 6, 13, 20:705–709, 763–768, 831–839, 1968.
6. G Rhodeheaver. Presentation to APIC Annual Meeting, Miami, FL, 1988.

7. R Aly, C Bayles, H Mailbach. Restriction of bacterial growth under commercial catheter dressings. Am J Infect Control 16(3): 95–100, 1988.
8. BM Gibbs, LW Stuttard. Evaluation of skin germicides. J Appl Bact 30(1):66–77, 1967.
9. MJ Marples. Ecology of the Human Skin. Springfield, IL: Charles C Thomas, 1965.
10. WC Noble. Microbiology of Human Skin. London: L Loyd–Luke Medical Books Ltd, London, 1981.
11. S Selwyn. Evaluating skin disinfectants in vivo by excision biopsy and other methods. J Hosp Infect 6 (suppl): 37–43, 1985.
12. S Selwyn, H Ellis. Skin bacteria and skin disinfection reconsidered. B Med J 1:136–140, 1972.
13. PB Price. Studies in surgical bacteriology and surgical technique with special reference to disinfection of the skin. JAMA 111: 1993, 1938.
14. PB Price. In: GF Reddish, ed. Antiseptics, Disinfectants, Fungicides and Chemical and Physical Sterilization. 2nd ed. Philadelphia: Lea and Febiger, 1957, p. 399.
15. G Reybrouck. Handwashing and hand disinfection. J Hosp Infect 8:5–23, 1986.
16. AM Kligman. The bacteriology of normal skin. In: HI Maibach, G Hildick-Smith, eds. Skin Bacteria and Their Role in Infection. New York: McGraw-Hill 1965, pp. 13–31.
17. PD Meers, GA Yeo. Shedding of bacteria and skin squames after handwashing. Hyg Camb 81:99–105, 1978.
18. R Speers, H Bernard, F O'Grady, RA Shooter. Increased dispersal of skin bacteria into the air after shower-baths. Lancet i: 478, 1965.
19. HR Lilly, EJL Lowbury. Disinfection of skin: an assessment of new preparations. Br Med J 3:674–676, 1971.
20. DS Seeberg, A Lindberg, BR Bergman. Preoperative shower bath with 4% chlorhexidine detergent solution: reduction of *Staphylococcus aureus* in skin carriers and practical application. In: H Maibach, R Aly, eds. Skin Microbiology. Relevance to Clinical Infection. New York: Springer Verlag, 1981.
21. GW Geelhoed. Preoperative skin preparation: Evaluation of efficacy, convenience, timing and cost. Surg Gynecol and Obstet. 157:265–268, 1983.
22. DJ Bibel, and OJ Lovell. Skin flora maps: a tool in the study of cutaneous ecology, J Invest Dermatol 66:265–269, 1976.
23. P Williamson, AM Kigman. A new method for the quantitative investigation of cutaneous bacteria. Invest Dermatol 45:498–500, 1965.
24. P Williamson. Quantitative esimation of cutaneous bacteria. In: H Maibach, G Hildick-Smith, eds. Skin Bacteria and their Role in Infection. New York, McGraw-Hill, 1965.
25. RR Marples. Newer methods of quantifying skin bacteria. In: H Maiback, R Aly, eds. Skin Microbiology: Relevance to Clinical Infection. New York: Springer-Verlag, 1981.
26. CM Shaw, JA Smith, ME McBride, WC Duncan. An evaluation of techniques for sampling skin flora. J Invest Dermatol 54:160–163, 1970.
27. GAJ Ayliffe, JR Babb, K Bridges, HA Lilly, EJL Lowbury , J Varney, NO Wilkins. Comparison of two methods for assessing the removal of total organisms and pathogens from the skin. J Hyg Camb 75:259–274, 1975.

28. HG Smylie, JRC Logie, G Smith. From Phisohex to Hibiscrub. Br Med J 4:586–589, 1973.
29. EJL Lowbury, HA Lilly, GAJ Ayliffe. Preoperative disinfection of surgeons' hands: use of alcoholic solutions and effects of gloves on skin flora. Br Med J 4:369–374, 1972.
30. RR Marples. The effect of hydration on the bacterial flora of the skin. In: HI Maibach, G Hildick-Smith, eds. Skin Bacteria and Their Role in Infection. New York: McGraw-Hill, 1965, pp. 33–41.
31. P Cruse, and R Ford. The epidemiology of wound infection–a ten year prospective study of wounds. Surg Clin North Am. 60:27–40, 1980.
32. L Brote, B Nilehn. Wound infections in general surgery with special reference to the occurrence of *Staphylococcus aureus*. Scand J Infect Dis 8:89–97, 1976.
33. A Brandberg, I Andersson. Preoperative whole body disinfection by shower bath with chlorhexidine soap: effect on transmission of bacteria from skin flora. In: H Maibach, R Aly, eds. Skin Microbiology. Relevance to Clinical Infection. New York: Springer-Verlag, New York, 1981, pp. 92–97.
34. EJL Lowbury, and HA Lilly. Use of 4% chlorhexidine solution (Hibiscrub) and other methods of skin disinfection. Br Med J 3:510–515, 1973.
35. J Davies, JR Babb, GAJ Ayliffe, S Ellis. The effect on the skin flora of bathing with antiseptic solution. J Antimicrob Chemother 3:473–481, 1977.
36. OB Jepsen. Postoperative wound sepsis in general surgery. VII. Staphylococcus wound sepsis. Acta Chir Scand 138:343–348, 1972.
37. FW Thomsen, O Larsen, OB Jepsen. Postoperative wound sepsis in general surgery. Acta Chirurg Scand 136:251–260, 1970.
38. P Dineen. Influence of operating room conduct on wound infection. Surg Clin North Am 55:1283–1287, 1975.
39. ML Rotter, SO Larsen, EM Cooke, J Dankert, F Oaschner, D Greco, RG Gronroos, OB Jepsen, A Lystad, B Nystrom. A comparison of the effects of preoperative whole body bathing with detergent alone and with detergent containing chlorhexidine gluconate on the frequency of wound infections after clean surgery. J Hosp Infect 11(4): 310–320, 1988.
40. RA Garibaldi. Prevention of intraoperative wound contamination with chlorhexidine shower and scrub. J Hosp Infect 11(B):5–9, 1988.
41. I MacKenzie. Preoperative skin preparation and surgical outcome. J Hosp Infect 11(B):27–32, 1988.
42. LJ Hayek, JM Emerson. Preoperative whole body disinfection—a controlled clinical study. J Hosp Infect 11(B):15–19, 1988.
43. WC Duncan, BG Dodge, and JM Knox. Prevention of superficial pyogenic skin infection. Arch Dermatol 99:465–468, 1969.
44. RR Leonard. Prevention of superficial cutaneous infections. Arch Dermatol 95:520–523, 1967.
45. AR MacKenzie. Effectiveness of antibacterial soaps in a healthy population. JAMA 211:973–976, 1970.
46. D Taplin. Antibacterial soaps: chlorhexidine and skin infections. In: R Aly, H Maibach, eds. Skin Microbiology, Relevance to Clinical Infection. New York: Springer-Verlag, 1981, pp. 113–124.

47. CW Howie, T Marston. A study on the sources of postoperative staphylococcal infection. Surg, Gynecol and Obstet 115:266–274, 1962.
48. G Geelhoed, K Sharpe. The rationale and ritual of preoperative skin preparation. Contemp Surg 23:31–36, 1983.
49. AM Rathburn, LA Holland, GW Geelhoed. Preoperative skin decontamination. AORN J 44:62, 1986.
50. P Williamson, AM Kligman. A new method for the quantitative investigation of cutaneous bacteria. J Invest Dermatol 45:498–503, 1965.
51. KA Gundermann, B Christiansen, C Holler. New methods for determining preoperative and postoperative skin disinfection. J Hosp Infect 6 (suppl):51–57, 1983.
52. OTC topical antimicrobial products and drug and cosmetic products. Fed Reg 39(179):33103–33133, 1974.
53. Fed Register 21 CFR Parts 333 and 369. Tenative final monograph for healthcare antiseptic drug products; proposed rule, June 17, 1994.
54. EJL Lowbury. Topical antimicrobials: perspectives and issues. In: Skin Microbiology: Relevance to Clinical Infection. New York: Springer-Verlag, 1981, pp. 158–168.
55. EJL Lowbury, HA Lilly. Gloved hand as application of antiseptic to operation sites. Lancet July 26, 1975.
56. EJL Lowbury, HA Lilly, JP Bull. Br Med J ii 1039–1044, 1960.
57. JA Ulrich, WC Beck. Surgical skin prep regimens: Comparison of antimicrobial efficacy. Infect Surg August:569–573, 1966.
58. GAJ Ayliffe. Surgical scrub and skin disinfection. Infect Control 5(1):23–27, 1984.
59. JA Ulrich. Dynamics of bacterial skin populations. In: H Maibach, G Hildick-Smith, eds. Skin Bacteria and Their Role in Infection. New York: McGraw-Hill, 1965, pp. 219–234.
60. GAJ Ayliffe, JR Babb and AH Quoraishi. A test for "hygienic" hand disinfection. J Clin Pathol 31:923–928, 1978.
61. TA Ideblick, MM Lederman, MR Jacobs, RE Marcus. Preoperative use of povidone-iodine: a prospective, randomized study. Clin Ortheped Related Res. 213:211–215, 1986.
62. ML Rotter. Hygenic hand disinfection. Infect Control 70(5):18–22, 1984.
63. M Rotter, W Koller, G Wewalka. Povidone-iodine and chlorhexidine gluconate-containing detergents for disinfection of hands. J Hosp Infect 2:149–158, 1980.
64. ML Rotter, W Koller, G Wewalka, HP Werner, GAJ Ayliffe, JR Babb. Evaluation of procedures for hygienic hand–disinfection: controlled parallel experiments on the Vienna test model. J Hyg Camb 96:27–37, 1986.
65. CEN (Comite Europeen de Normalisation). EN 1500 Chemical disinfectants and antiseptics. Hygienic handrub—test method requirements (phase 21 step 2).
66. MK Bruch. Newer germicides: what they offer. In: H Maibach, R Aly, eds. Skin Microbiology: Relevance to Clinical Infection. New York: Springer-Verlag, 1981, pp. 98–112.
67. NJ Van Abbe. The substantivity of cosmetic ingredients to the skin, hair and teeth. J Soc Cosmet Chem 25:23–31, 1974.
68. RF Edlich. Surgical sepsis: a delicate balance. AORN J 35(4):786–790, 1982.
69. WC Beck, J P Geffert, M Hansen. The incise drape: boon or hazard. Am Surg 47: 343–346, 1981.

70. EJL Lowbury. Skin preparation for operation. Br J Hosp Med 10:627–634, 1973.
71. EJL Lowbury. Gram negative bacilli on the skin. Br J Dermatol 81(suppl):55–61, 1969.
72. EJL Lowbury. Removal of bacteria from the operation site. In: H Maibach, G Hildick-Smith, eds. Skin Bacteria and Their Role in Infection. New York: McGraw-Hill, 1965, pp. 263–275.
73. EJL Lowbury. Wits versus genes: the continuing battle against infection. J Trauma 19:33–45, 1979.
74. EJL Lowbury. Bacterial infectiion and hospital infection of patients with influenza (review). Postgrad Med J 39:582–585, 1963.
75. EJL Lowbury. Hands—a Source of Infection. The Personal Factor. Royal Soc Health, London: 1963, pp. 9–19.
76. EJL Lowbury, HA Lily. Use of 4% chlorhexidine detergent solution (Hibiscrub) and other methods of skin disinfection. Br Med J 1:518–525, 1973.
77. D Raahave. Bacterial density in operation wounds. Acta Chirurg Scand 140:585–593, 1974.
78. D Raahave. Agar contact plates in evaluation of skin–disinfection. Dan Med Bull 20:204–208, 1973.
79. D Raahave. Antisepsis of the operation site with aqueous cetrimidel chlorhexidine in alcohol. Acta Chirurg Scand 140: 595–601, 1974.
80. AM Allen. Clinical trials of topical antimicrobials. In: Skin Microbiology. Relevance to Clinical Infection. New York: Springer-Verlag, 1981, pp. 77–85.
81. CW Walter. Disinfection of hands. Am J Surg 109:691–693, 1965.
82. P Story. Testing of skin disinfectants. Br Med J II:1128–1130, 1954.
83. EA Patchman, EE Vicker, MJ Brunner. Origin of glass cup scrub. J Invest Dermatol 22:389, 1954.
84. DM Updegraff. A cultural method of quantitatively studying the micro organisms in the skin. J Invest Dermatol 43:129–137, 1964.
85. JA Ulrich. Techniques of skin sampling for microbial contaminants. Health Lab Sci 1:133, 1964.
86. EA Bannan, RS Amarino. The human corneum disc test: a method for measuring activity of skin degerming products. Health Lab Sci 6(1):22–26, 1969.
87. RJ Holt. Aerobic bacterial counts on human skin after bathing. J Med Microbiol 4: 319–327, 1971.
88. WC Noble, D Sommerville. Microbiology of Human Skin. Philadelphia: WB Saunders, 1981.
89. CA Evans, RJ Stevens. Differential quantitation of surface and subsurface bacteria of normal skin by the combined use of cotton swab and the scrub methods. J Clin Microbiol 3:576–581, 1976.

15
Brushless Surgical Scrubbing and Handwashing

David W. Hobson
Chrysalis Biotechnology, Inc., San Antonio, Texas

Lawton Anthony Seal
Healthpoint, Ltd., San Antonio, Texas

I. INTRODUCTION

The practice of surgery is an ancient medical art that derives its name from the Latin word *chirurgia*, derived from the Greek *cheiros* (hand) and *ergon* (work). Thus, due to the very nature of their work, surgical professionals realize the importance of their hands to their livelihood as well as that of their patients. Surgical team members are also generally knowledgeable and appreciate the importance of good surgical handwashing practices in caring for their patients, as well as in infection control, and ascribe to some regimen for hand antisepsis that has been proven effective by research and practical experience. In these days of growing concern over transmission of diseases caused by drug-resistant and increasingly pathogenic microorganisms, surgical personnel, as well as other healthcare practitioners, should not be too surprised that implementation of proven handwashing procedures that are in compliance with recommended standards and guidelines is considered by many infection control professionals to be a cornerstone to addressing this growing challenge. Most hospitals today have formalized infection control procedures designed to establish and monitor in-house infection control practices. Due to this increasing need for vigilance in our infection control practices, handwashing standards and guidelines, which had their basis in medicine in the nineteenth century, are still being reviewed and updated, even as we move into the twenty-first century.

Historically, handwashing for antiseptic purposes had its beginning only 150 years ago. The Hungarian obstetrician Ignaz Philipp Semmelweis (1818–1865) is credited with discovering how antiseptic handwashing can reduce patient mortality from handborne infectious agents. Despite resistance to his findings and publications by hospital and medical authorities in the mid-1800s, Semmelweis strongly enforced his hand antiseptic practices and reduced maternal mortality from puerperal fever in wards under his direction approximately 90% from 13.7 to 1.3% in a 3-year period (1846–1848) [1]. Ironically, Semmelweis died in Vienna in his mid-40s from an infection contracted possibly from an operation he had performed (in those days, surgical gloving was an unknown practice). Furthermore, his death occurred just at the time when such notables in hospital sanitation and infection control as America's Oliver Wendell Holmes, Britain's Baron Joseph Lister and Florence Nightingale, and France's Louis Pasteur were beginning to make their marks in infection control. Fortunately, the contributions of Semmelweis were not forgotten, and his work is now considered to have provided the scientific basis for modern antiseptic handwashing practice.

Surprisingly, however, theories of antisepsis, including those related to handwashing practices, were initially met with resistance by most of the medical community of the mid-1800s, and it was not until the late 1800s that the discoveries of Semmelweis, Lister and others became widely accepted. By the end of the nineteenth century, Lister had introduced the use of carbolic acid to cleanse the hands prior to surgery, and the commercial development of antimicrobial agents specially formulated for handwashing followed thereafter. Since their introduction, the use of antimicrobial handwashing agents has been shown repeatedly to be more effective than plain soap in reducing the number of microbes on the skin [2–5].

Historically, the introduction of the scrub brush into the presurgical handcleansing regimen, in combination with the use of antiseptic agents, is not well documented. One may suppose, however, that brushes have been around handwash basins for many decades for use in the removal of dirt, body fluid, and tissue contaminants that were evident on the hands and under the nails before and after surgery. All one can be certain of is that when antimicrobial handwashing agents, such as those containing povidone-iodine (PVPI) and hexachlorophene (HCP), became available commercially in the United States during the 1940s and 1950s, and those containing chlorhexidine gluconate (CHG) in the 1970s, the surgical scrub brush was there and had become part of surgical scrub regimens that arc still in use today.

An interesting article published in the *American Journal of Surgery* by Engelsher in 1966 introduced a surgical scrub brush of a flexible, tubular design made of monofilamentous polypropylene to conform anatomically to the parts of the hand that it was intended to clean [6]. In this article, Engelsher notes that the literature of that time contained many articles describing techniques for proper preparation of the hands prior to surgery, including time and motion studies. It is further stated that the actual abrasive action of the bristles was the most important factor in producing surgically clean hands and that " while soap and other antiseptic detergents supplement scrubbing, the abrasive action of the bristles in removing imbedded particles of dirt, dead epithelial tissue, and other potentially virulent debris remains the most difficult operation." This article disparages the brushes used at that time in a manner that does not seek to replace the brush, but rather to make improvements to its design, indicating that the "stock-model" brush was designed generations before and was "adopted" by but not "adapted" for surgeons.

By the late 1980s and 1990s, formalized guidelines and recommended practices for handwashing were published by professional organizations, such as the Association for Professionals in Infection Control and Epidemiology, Inc. (APIC) and the Association of Operating Room Nurses (AORN), clearly supporting the routine use of antimicrobial handwashing agents when indicated by the degree of microbial contamination [7,8]. By that time, surgical scrub procedures clearly assumed the use of the brush in recommended practice. In some regimens, a sponge was allowed if the contamination was not too great, but clearly the brush was well entrenched in the established standards of practice for surgical scrubbing at that time. Manufacturers of antiseptic agents intended for use at the scrub sink readily supported these guidelines and developed prepackaged scrub brushes and sponges that had been impregnated with their antiseptic products. Alternatively, a surgical team member could elect to place a metered dose of an antiseptic of their choice from a pump-style container and use a disposable, nonimpregnated scrub brush supplied at the scrub sink. Disposable scrub brushes were even made out of polypropylene to make the brush more flexible and conforming with the hand surface, although the tubular design proposed by Engelsher apparently never caught on.

From the discussion above, it can be seen that in the classical or traditional sense, an effective scrub includes the use of an antiseptic detergent solution stored in a wall-mounted container and dispensed onto a scrub brush or impregnating a scrub brush/sponge that is easily accessible at the scrub sink. Popular antiseptic detergents are formulations containing either 4% chlorhexidine or 7.5% povidone-iodine. The scrub area, or bay, is located remote from the operating theater, and the scrub sinks and water taps are located such that the hands and forearms can be scrubbed without touching any potentially contaminated surface, the taps being activated by the elbow, foot, or a light sensor. Although debate continues as to the most effective method of presurgical hand-scrubbing, there is agreement that this procedure should include an initial scrub of 5 minutes and subsequent "scrubs" of 2 minutes.

The excessive use of a scrub brush may encourage desquamation, and hence it is used lightly and only during the initial scrub. The fingers and hands are considered as having four planes, and each plane is scrubbed sequentially from the fingertips to just below the elbows, keeping the hands elevated to allow water to drain away from the fingers.

For example, the standardized surgical scrub procedure recommended by the AORN in 1995 should include, but not be limited to the following actions [8]:

Moisten hands and forearms and wash using an approved surgical scrub agent, then rinse before beginning the surgical scrub procedure to loosen surface debris and transient microorganisms.

Clean subungual areas under running water using a nail cleaner to remove microorganisms typically residing in these areas.

Apply an antimicrobial agent with friction to wet hands and forearms in order to mechanically remove dirt and microorganisms.

Visualize the hands and arms as having four sides and scrub each side effectively in order to ensure that all areas of the hands and arms are exposed to mechanical cleaning and chemical antisepsis.

Hold hands higher than the elbows and away from the surgical attire to prevent contamination of the surgical attire and to allow water to run from the cleanest area down the arm.

The brush or sponge used should be discarded appropriately to prevent cross-contamination of the surgical scrub area.

Care should be taken to avoid splashing water onto the surgical attire to avoid subsequent contamination of the sterile surgical gown by strike-through moisture.

Over the years, the use of the scrub brush has been challenged in the scientific literature and some alternatives have been proposed. For example, in 1969 Berman and Knight proposed that the use of the brush may act to increase the transfer of bacteria from the hands to the surgical wound by removal of the outer layer of the epidermis and subsequent exposure of the underlying bacterial flora of the deeper layers [9]. These authors did not, however, attempt to prove their hypothesis with any sort of controlled clinical or preclinical study.

In 1978, using two different iodophor formulations or a plain soap followed by an alcohol foam, Galle et al. demonstrated that one hour after scrubbing a shorter, no-brush scrub procedure with any of the antiseptics used was as effective in reducing bacterial populations on the hands as was performance of a longer, standard, two-brush method [10]. These authors went on to conclude that the reduced scrub time would allow substantial financial savings to be realized, as well as reduce the likelihood of skin trauma and dermatitis.

When a waterless, alcoholic hand gel or lotion type of product is used for presurgical antisepsis following an initial brushless handwash to remove dirt and debris, the following steps have been incorporated into the procedure.

Maintain the nails with no more than a 1 mm free edge. Clean under the nails with a nail pick
Apply alcohol gel to hands that are clean and dry (washed previously with an antimicrobial or nonantimicrobial soap, if necessary to remove dirt and debris).
Place a palmful of the alcohol-based product into the palm of a hand, dip the fingernails of the opposive hand into the product, and work it under the nails.
Spread the remaining product over the hand and up to just above the elbow.
Repeat the product application process for the other hand.
After the both hands and arms are done, apply another palmful of the product to both hands just prior to surgical gloving.

In 1997 Loeb et al. reported results from a small randomized trial comparing surgical scrubbing with a brush to the use of an antiseptic soap without a brush, using the glove juice sampling procedure to obtain quantitative hand microbial counts [11]. The antiseptic soap used in this comparative, randomized, crossover trial was a 4% chlorhexidine formulation containing isopropyl alcohol, and the scrub time was 5 minutes for both treatments. These authors concluded, similarly to Galle et al., that for a short period (45 min) immediately following scrubbing with antiseptic soap with a brush, there was no significant difference in the reduction of hand bacterial counts. This study also included a simple cost analysis showing that the cost associated with the use of brushes can be significant to a healthcare institution and that a 58% savings is possible with the use of antiseptic soap alone. They, like Galle et al., hypothesized that skin irritation may be reduced with the elimination of the brush, but they did not produce any data to support this hypothesis.

Another study reported by Decker et al. in 1978 evaluated the use of a 90-second ''jet wash'' procedure compared to the more conventional, 10-minute, two-brush scrubbing procedure in a clinical setting [12]. These authors concluded that the 90-second jet wash was more effective in degerming the hands than was the 10-minute conventional scrub with a brush and noted other potential advantages of the new method, such as the amount of time saved and reduced skin irritation.

Obviously, significant time savings from preparative tasks like surgical hand-scrubbing in high-cost and sometimes high-volume areas, such as the operating room and emergency room environments, can positively impact efficiency. Such increased efficiency has potential for significantly increasing the number of surgeries that can be performed in a week, as well as for reducing the worsening of trauma associated with even momentary delays while surgical staff is in preparation. For example, if a surgical team performs six or seven procedures in a day and adheres to a 10-minute scrub regimen, they are spending about one hour a day just in the process of hand antisepsis. Reducing this time even to half would increase their availability for at least one additional procedure over a week's time. In fact, this is precisely what surgical teams, as well as infection control practitioners, have worked to accomplish over recent years. Nowadays, it is quite unlikely that you would observe many surgical team members taking 10 minutes out of their busy schedules for surgical scrubbing, when products are available with package labeling that requires much less than 10 minutes of scrubbing to obtain optimal hand antisepsis.

Time savings, however, is just a beginning. More recently, the issue of reducing skin irritation and hand dermatitis has become a focus, and new surgical hand antisepsis products are being introduced that claim to help improve hand condition. Some researchers have developed procedures for the evaluation of hand condition before and after the implementation of an improved handwashing regimen, which might include the introduction of a new product formulation or elimination of the use of the scrub brush [13,14].

In 1998 the first commercial surgical scrub formulation developed specifically for application without a brush or sponge was introduced [15]. The labeling of this scrub formulation used the term ''brushless'' in association with a new type of surgical scrub formulation and scrub regimen, conveying the message that it was possible to formulate a ''stand-alone'' surgical scrub product for use without a brush that demonstrated hand antiseptic performance and persistence characteristics comparable, and in many ways superior to current products intended for use with a brush. A 3-minute application of the new formulation using the new brushless procedure resulted in antiseptic efficacy superior to both 4% chlorhexidine gluconate and 7.5% povidone-iodine formulations applied in accordance with the scrub regimens provided in their label directions, using 6-minute or 10-minute scrub procedures, respectively.

The brushless procedure recommended for use with a brushless scrub formulated as a ''stand-alone'' product for use with water follows.

Clean under nails with a nail pick. Nails should be maintained with a 1 mm free edge, or less.

Wet hands and forearms.

Dispense a palmful of product into one hand. Spread on both hands and forearms, paying particular attention to the nails, cuticles, and interdigital spaces, rubbing into the skin for 1 minute and 30 seconds, and then rinsing.

Repeat, as above, with one additional application of a palmful of the product.

The commercial introduction of a specially formulated, brushless surgical scrub into a market of relatively old products that were intended for use with the brush fueled increased questioning among various researchers as to the necessity of the brush in surgical scrubbing. It increased, too, the commercial interest in the development and introduction of other hand disinfection products into the operating room market that are to be used without a brush [14,16,17]. Some of these formulations are actually waterless alcohol hand rubs that sanitize the hands and also contain ingredients to help slow or prevent microbial regrowth, but these have not been formulated with surfactants to be used in combination with water to remove dirt, tissue, and other debris, while at the same time providing surgical antisepsis and cleanliness comparable to that provided by a surgical scrub with a brush. In some cases, the removal of dirt and debris prior to the use of a waterless alcohol product may be accomplished using a nonantimicrobial soap in a two-step process [17]. It is important to note, however, that the use of the term ''brushless'' in conjunction with a surgical scrub formulation was intended to apply only to product formulations that could function in a ''stand-alone'' manner as the primary surgical scrubbing agent. Therefore, a complete surgical scrub formulation containing antiseptic agents, as well as surfactants, that can be used effectively with water to remove the dirt, debris, body fluids, etc., as well as provide effective and persistent antisepsis over a 6-hour period should not be confused with waterless, alcohol-based products that claim to be reentry surgical hand disinfectants. This can be confusing, because although alcohol may be the principal ingredient in both types of formulations, one must make a distinction between a brushless surgical scrub and a reentry hand disinfectant by careful examination of the product labeling and use directions.

Although much of the current activity in the introduction of brush-free products is being driven by the need to address such long-standing concerns as improving hand condition, reducing the amount of time required for scrubbing, and increasing compliance with good infection control practice, another very important concern is rapidly emerging: the need for healthcare facilities, worldwide, to significantly reduce their generation of medical waste.

Surgical brushes represent a significant amount of nonbiodegradable medical waste that adds to the waste stream going to commercial landfills or incinerators, if these cannot be separated out for recycling. Even recycling carries with it a cost that adds to the cost of the brush, as noted by Loeb et al. [11]. Due to increasing concern about expenses associated with handling medical waste, it is quite possible that elimination of surgical scrub brushes from the medical waste stream may, alone, be enough reason for a facility to switch to stand-alone brushless surgical hand scrubs. Along this same train of thought, a relatively recent article by McCuaig discussed considerations and tests conducted in the development of surgical preparative procedures for use by space station astronauts [18]. In this article, McCuaig describes the use of surgical hand antiseptics that are prepackaged with a brush. Although such products might be used in a microgravity or weightless environment, problems of the weight and space they represent to a payload specialist and to the crew members who need every bit of available room in the space station are significant. Then, of course, there is the problem of medical waste disposal. Clearly, the elimination of the surgical scrub brush will have significant impact on both our present and our future endeavors.

II. MICROBIOLOGY OF HAND ASEPSIS-RELATED TO BRUSHLESS SCRUBBING

Two expected outcomes of handwashing or surgical scrubbing include the removal of soil and debris and the reduction in the microflora populations on the hands. The contaminating flora are often categorized as being either ''transient'' or '' resident,'' in accordance with the following definitions:

Transient flora—noncolonizing organisms isolated from the skin, but not consistently present. These may include gram-negative bacteria, such as *Escherichia coli*, that survive poorly on the skin.

Resident flora—organisms considered to be permanent residents on the skin including coagulase-negative staphylococci, diphtheroids, *Propionibacterium* and *Acinetobacter* species, as well as some members of the *Klebsiella-Enterobacter* group.

Like soil and debris, the transient flora can be removed by mechanical friction with plain soap-and-water washing [19]. For the general public, plain soaps are considered adequate for this purpose. However, the hands of healthcare workers are a major source of pathogens associated with nosocomial infections, and those involved with routine exposure to blood and body fluids have greater need for antimicrobial soaps. The use of these products in handwashing and the wearing of gloves have proven effective in reducing infection rates in some scenarios [20,21].

Halstead's efforts in developing surgical gloves 100 years ago were prompted by the realization that surgical scrubbing was accompanied by damage to the skin [22]. The problems associated with poor skin condition following repeated surgical scrubs or handwashes are acknowledged even now as a major reason for noncompliance with hand hygiene protocols for healthcare workers. Hand dermatitis has a reported prevalence of approximately 30% for nurses, as compared to 5.2% for males and 10.6% for females in the general population [23].

Skin damaged by handwashing or surgical scrubbing is attributable to repeated use of soaps and detergents, the traumatic and harsh mechanical action associated with scrub brushes, and seasonal or climatic changes. The repeated use of any soap or detergent product removes lipids from and disrupts the stratum corneum, resulting in a deterioration of the skin's primary barrier function. This process is exaggerated by scrubbing with a brush or other types of harsh mechanical action and can result in significant damage to the skin. Changes in skin pH have been noted with repeated handwashing; this can alter the skin's natural antimicrobial properties. When these kinds of insults are coupled with the skin-drying conditions of a winter climate, it is understandable that poor skin condition becomes a deterrent to hand hygiene compliance. Recently, a major approach to reducing skin damage is to reduce surgical scrub times [24].

In a healthcare environment, dermatological problems have been associated with alterations in hand flora, resulting in higher rates of colonization rates by *Staphylococcus aureus*, often multiple drug–resistant strains. In addition, hand dermatitis has been associated with nosocomial transmission of these organisms [25,26]. Larson et al. investigated the relationship between damaged skin and changes in hand flora [27]. This prospective study of 40 nurses (20 with and 20 without hand irritation) revealed that those with damaged hands were significantly more likely to have *S. aureus*, gram-negative bacteria, enterococci, and yeast isolated from their hands. While not typically part of the normal hand flora, *S. aureus* was found to colonize 20% of the hands of those with damaged skin. Larson concluded that changes in the hand flora of caregivers are undesirable, given the demonstrable shift toward organisms that pose a risk for nosocomial infections.

Approximately 10^7 skin cells are shed into the air daily by each individual, with 10% of these bearing viable bacteria. Therefore, roughly 10^6 populations of microbes leave the skin each day and are available for colonization on inanimate objects or the skin of others [28]. With damaged skin, there is a potential shift toward populations of opportunistic or, frankly, pathogenic hand flora, and consequently damaged skin poses a significant risk to those with whom these healthcare workers come into contact.

Lilly et al. reported that, even with the use of antiseptic preparation, reductions in flora counts beyond an equilibrium level are not attainable [29]. More recently, others have shown that with the appropriate approach that provides both persistent and residual effects, hand flora can be reduced to near the detection limits of the assays without significant damage to the skin [15]. Such an outcome would appear to be in the best interest of both the healthcare provider and the patient, as it maintains the healthy skin environment that prevents potentially hazardous shifts in microbial flora, while significantly reducing the numbers and types of organisms present and, therefore, available for transmission.

It is well documented that proper handwashing can result in statistically significant reductions in patient morbidity and mortality from nosocomial infection and that handwashing with antimicrobial agents significantly reduces the number of potential pathogens carried on the hands [7]. Despite this evidence, recent studies indicate that handwashing practices are less than optimal in some healthcare facilities and recommend stronger mandates and monitoring of handwashing practice [30,31]. Although reports that knowledgeable healthcare practitioners disregard handwashing guidance in favor of more casual handwashing practices may be surprising to some, it must be recognized that antiseptic handwashing, including the surgical scrub, is a very repetitive, tedious, mundane and sometimes painful task that can easily be overlooked for other, more challenging and appealing tasks. Furthermore, in these times of increasing managed healthcare requirements and budget-cutting, many healthcare practitioners are experiencing increased workloads—increased patient-to-provider ratios and increased focus on documentation—that reduce the opportunity and time available for mundane tasks like thorough handwashing. In such circumstances, it is really not that surprising that proper handwashing may take a back seat to other activities in a practitioner's daily routine.

III. CURRENT PROGRESS TOWARDS BRUSHLESS TECHNOLOGIES

Interestingly, since the introduction of CHG-containing products in the early 1970s, major advances in surgical scrub formulation in the United States have been essentially nonexistent. Similarly, in Europe, various ethyl, isopropyl, and *n*-propyl alcohols have been used for surgical hand antisepsis for many years with excellent results and without the introduction of many significant new formulations [6].

In the United States, the surgical scrub procedure, as discussed earlier, is a highly regimented practice that has as its primary objectives: (1) removal of dirt, debris and body fluid contaminants, (2) achievement of the highest possible level of microbial decontamination, and (3) maintenance of this level of decontamination to the greatest possible degree for the duration of surgical procedures that may last several hours. Surgical scrub formulations designed to accomplish these objectives in a brushless fashion require that much consideration be given to both the selection of the antimicrobial product to be employed, as well as establishing the optimum technique for its application. And considerable effort should be focused on obtaining the shortest scrubbing time possible, per the considerations described previously, including staff time constraints and the desire to reduce skin trauma or exposure to potential allergens, while at the same time achieving an adequate level of surgical antisepsis [32].

In the United States, some operating rooms currently stock more than one antimicrobial product for surgical scrubbing, e.g., products containing povidone iodine (PVPI, 7.5%), chlorhexidine gluconate (CHG, usually 4%), or parachlorometaxylenol (PCMX, typically 3.75%). Other products, such as those containing hexachlorophene (HCP, 3%) and triclosan (up to 2%), can be found in use as surgical scrubs in some operating rooms. However, they tend to be available for personnel having skin sensitivity to one or more of the other, more widely used

Table 1 Some Examples of Commercial Formulations Marketed for Brushless Surgical Hand Antisepsis

Trade name	Active ingredients	Application type
Triseptin®	Ethyl alcohol—70%	Water aided
Avagard®	Ethyl alcohol—61% w/w Chlorhexidine gluconate—1% w/w	Waterless
Techni-Care®	Para-chloro-meta-xylenol—3%	Water aided
Alcare®	Ethyl alcohol—62%	Waterless

products. These products are generally labeled for use with water, a nail pick, and a scrub brush and may have a wide variety of manufacturer's suggested application procedures and times.

In contrast, alcohol-based preparations used in Europe are often used alone or after washing with soap and water or simply with water. Alcohols are recognized as having greater immediate antimicrobial action than any of the other antimicrobial ingredients and are typically used for surgical preparation of the hands by rubbing on 3–5 mL of product until dry and repeating these applications for about 5 minutes [1,6]. Unfortunately, alcohols have not been widely used in the United States as surgical scrub ingredients because, despite their excellent immediate and thorough antimicrobial action, they do not have any significant detergency for the removal of organic debris or antimicrobial persistence in the stratum corneum and also tend to be very drying to the hands.

Many of today's new, brushless, surgical antisepsis formulations described earlier in this chapter seek to combine the substantial antimicrobial benefits of alchohol with some other antimicrobial ingredient or other technology that provides persistant antimicrobial action in the stratum corneum. Some of these formulations also include emollients to reduce the drying effect of alcohol, and some also contain surfactants that allow the product to be used in combination with water to remove dirt and debris, as well as to provide a source of moisture that can be trapped and retained by humectants and emollients. Table 1 provides some examples of commercial surgical hand antisepsis products that have been marketed for use without a brush.

The principal characteristics of the "ideal" surgical scrub product can be divided into four general categories: (1) antimicrobial action, (2) safety, (3) personnel acceptance, and (4) economics. Each of these characteristics is discussed below in further detail.

The ideal handwashing agent would have a broad spectrum of antimicrobial activity against pathogenic organisms to include gram-negative and gram-positive bacteria, fungi, viruses, yeasts, bacterial spores, etc. This agent would work very rapidly and would have a duration of action such that it could be used both as a routine handwashing agent and a surgical scrub. In situations where economics or limited availability of selection are issues, departures from the ideal situation might include acceptance of a reduced spectrum of activity, if coverage still includes those classes of organisms known to be most prevalent in the particular use scenario. Another tradeoff could be accepting a decreased rapidity of antimicrobial action if the product is to be used for surgical scrubbing, where duration of action is felt to be more important. Surfactancy for removal of blood and organic matter contaminants could also be a tradeoff when use would be in situations where these types of contaminants occur infrequently and to a minimal degree.

Based on historical concerns with other products, the ideal handwashing agent would be nonirritating, and nonsensitizing, show little or no percutaneous absorption, have no appreciable ocular or ototoxicity, be safe for use over large portions of the body and for use on infants, have excellent shelf-life stability characteristics, and not be damaging to the environment. Product safety is one area where any perceived tradeoffs must be weighed very carefully, as concern for personal safety generally outweighs concern for ideal antimicrobial action. Tradeoffs can occur where the use can be clearly defined or limited to specific applications (as with the use of hexachlorophene-containing products) and controlled or monitored in some fashion. And in some locations, such as California, environmental waste issues are an increasingly significant concern.

Probably most important to achieving personnel compliance in using a new surgical scrub product is its acceptance by the user population. A product that has ideal antimicrobial action and an excellent safety profile is of little value to good infection control if the user population fails to support its use. Although it is hard to predict all possible reasons why a product might be rejected by the healthcare user population, from past experience some factors are obvious. These include drying or irritation of the skin, poor foaming action, difficulty in rinse-off, disagreeable odor, staining or discoloring of the stratum corneum, difficulty in donning surgical gloves following use, and poor packaging. Tradeoffs, if necessary, must be carefully weighed against their perceived importance to the user community.

Although some healthcare managers might support the notion that economic considerations outweigh or drive all other aspects of a desirable product, it could be argued in today's liability-oriented healthcare environment that products with clearly demonstrated and quantifiable advantages in one or more of the three categories described above provide important benefits that outweigh strict cost/benefit comparisons, for example, the typical cost difference between PVPI and CHG products versus the fact that CHG is known to exhibit better antimicrobial persistence than does PVPI. Where human lives are concerned and a tangible benefit can be demonstrated and quantified for an antiseptic handwashing product, very few people, if any, would support an increase in mortality or morbidity for purely economic reasons. To date, however, the mortality/morbidity data for most handwashing studies, including those done in surgical wards, have not demonstrated any particular antiseptic handwashing agent to be superior enough to provide an argument on that basis. It is conceivable, however, that in the next few decades, our increasing concern regarding infection control may lead to products and practices that repeatedly demonstrate statistically significant benefit over plain soap handwash and between products.

In June 1994, the U.S. Food and Drug Administration (FDA) issued via the *Federal Register* an important new document entitled "Topical antimicrobial drug products for over-the-counter human use: tentative final monograph for health care antiseptic drug products-proposed rule" [33]. This monograph proposed for the first time specific regulations designed to establish minimum antimicrobial performance characteristics for healthcare antiseptic drug products, including surgical scrubs intended for over-the-counter (OTC) use. In preparation of this document, after consideration of the available data, the FDA found that only alcohol and PVPI had sufficient safety and efficacy data to be classified as Category I, "safe and effective for OTC use" [33]. In contrast, the FDA found that healthcare antiseptic drug products, including surgical scrubs, based on CHG would continue to be considered new drugs, marketable only following approval of New Drug Applications (NDAs). Similarly, the FDA concluded that products containing PCMX and triclosan were to be considered Category III, requiring additional data to demonstrate their safety and effectiveness for use as surgical scrubs. Other FDA findings of note included reaffirmation that HCP-containing products were to remain available by prescription only due to safety considerations (i.e., neurotoxicity findings in humans and animals following use on large areas of or on broken skin) and are to be classified as Category II.

The FDA monograph, furthermore, sets forth quantifiable in vitro and in vivo performance characteristics for surgical scrubs, including (1) demonstration of in vitro activity against a broad spectrum of microorganisms, both laboratory and clinical isolates, by means of minimum inhibitory concentration and time-kill kinetic studies; (2) demonstration that microorganisms are unlikely to develop resistance to the formulation; and (3) demonstration of minimum in vivo efficacy in terms of quantifiable performance parameters for hand baseline counts (1.5×10^5 $\log_{10}$ microbial colony-forming units [CFU] per hand), immediate $\log_{10}$ reduction from baseline microbial counts (at least one $\log_{10}$ reduction from baseline at one minute postscrubbing on the first day of use), and antimicrobial persistence (at least a 3 $\log_{10}$ reduction from baseline at 1 minute after 5 days of use). Although these minimum performance characteristics appear less stringent than one might expect, relative to the desirability of removing essentially all of the baseline population of microbes from the hand, these criteria are surprisingly difficult for most current products to meet consistently. From the literature, 7.5%

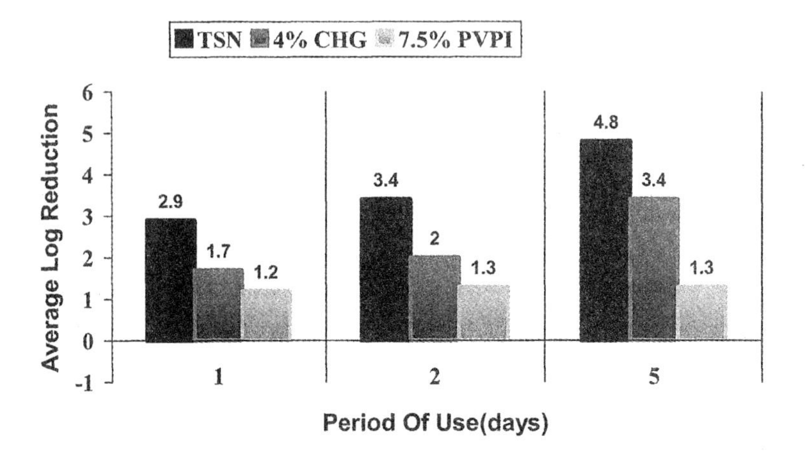

Figure 1 Example performance data for a brushless surgical hand scrub formulation containing alcohol (TSN) that meets the FDA's current tentative monograph criteria in terms of initial (Day 1) use over a six-hour period. Antimicrobial average log reductions are shown, respectively, relative to the performance of typical 7.5% povidone iodine (PVPI) and 4% chlorhexidine gluconate (CHG) formulations that were tested in accordance with their label instructions using a brush.

PVPI–containing products typically demonstrate a minimal ability to just meet these requirements, whereas products containing 4% CHG demonstrate slightly higher initial reductions, typically about 2–3 $\log_{10}$ immediately after scrubbing, and more persistent effects, about 1–2 $\log_{10}$ at 6 hours [7]. Alcohols typically show the greatest immediate log reductions (3–4 $\log_{10}$) but return to baseline levels due to recolonization of the skin prior to 6 hours postscrubbing [7].

Examples of the performance that might be expected from a brushless surgical scrub formulation that meets the FDA's tentative monograph criteria in terms of both the initial (day 1 out to 6 hours) and repetitive use (over a 5-day period) requirements are presented in Figures 1 and 2 versus the performance of typical PVPI and CHG formulations used with a brush. It is also important to note that, at present, the FDA monograph does not address the formulation or required testing of the new category of surgical reentry hand disinfectants now being introduced into the operating room. Future revision of the FDA monograph may address these products, but until then, companies wishing to introduce these types of products should consult with the FDA prior to marketing, and infection control practitioners and operating room directors should require the manufacturer to provide documentation that the formulation has been tested and meets FDA criteria for approval before adopting any new surgical hand antiseptic product.

The selection of principal antimicrobial agents for use in a surgical scrub formulation would seem to be a relatively simple process based on the years of experience with the use of a variety of different antimicrobial agents. There are, however, factors to be considered, even when the selection involves well-established products. One example might be the skin-sensitizing or irritant potential of active ingredients. PVPI is well known to produce skin reactions in some individuals [7]. Another example would be the likelihood that the active ingredients might be neutralized under actual use conditions. This is demonstrated by the fact that the antimicrobial effectiveness of iodine- and CHG-containing products may be reduced or neutralized in the presence of organic matter and hard water constituents, nonionic and anionic surfactants, and certain emollients used in soaps and hand creams [34]. Still other examples would be the generally reduced persistance of antimicrobial action of PVPI formulations on the skin, as compared to that of CHG formulations, and the greater immediate reduction in skin microflora following handwashing with alcohol-containing products relative to that of products containing only PVPI or CHG [7]. These examples are but a few of the factors that should be considered when selecting a surgical scrub product for use in a particular environment and in the development of more advanced formulations for future use. Some of the brushless hand antiseptics being introduced into the operating room are completely new, patented formulations, but many are actually alcohol hand gels that may or may not have an additional, more traditional antimicrobial such as CHG or PVPI added to enhance persistence.

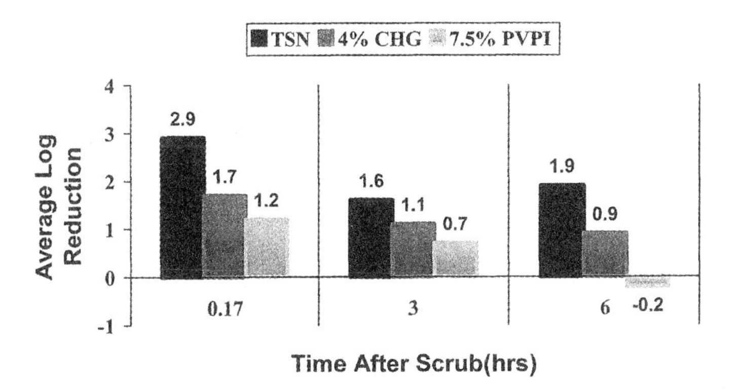

Figure 2. Example performance data for a brushless surgical hand scrub formulation containing alcohol (TSN) that meets the FDA's current tentative monograph criteria in terms of repetitive use over a five-day period. Antimicrobial average log reductions are shown, respectively, relative to the performance of typical 7.5% povidone iodine (PVPI) and 4% chlorhexidine gluconate (CHG) formulations that were tested in accordance with their label instructions using a brush.

Surgical team members beginning in this millennium now have a choice to stay with older formulations that were not specifically designed and tested for use without a brush or to adopt new "brushless" technologies. It would appear that these brushless formulations can reasonably be expected to exceed the minimum performance criteria specified in the FDA's tentative monograph and also to have many characteristics of an "ideal" antimicrobial handwashing product. By rethinking the approach to formulating a surgical hand scrub product with brushless use in mind, product development scientists have several options as to how surgical scrub formulations could be advanced. For example, new antimicrobial ingredients are being patented that may find their way into new surgical scrub formulations; two-step systems could be developed employing a brushless scrub with a gentle surfactant-containing formulation containing a substantive agent like CHG, followed by use of an alcohol-based hand product; or novel formulations based on alcohols that also have components to aid in the removal of dirt and debris, as well as impart antimicrobial persistence and skin emollience [35].

As mentioned above, hand condition is an important consideration in the formulation of new surgical scrub formulations and plainly is an important factor in personnel compliance and infection control [7]. Some recent articles have examined the importance of hand condition in depth and concluded that cleansing processes that damage the condition of the hands may actually help to colonize the hands with microorganisms and become a means of spreading disease. Responding to the need for many healthcare professionals to improve hand condition, Larson recently proposed recommendations specifying shorter, less traumatic handwashing regimens, including reduced use of brushes or other harsh mechanical action in clinical areas such as operating rooms, neonatal, and transplant units [28]. Product developers and manufacturers of surgical scrub products must now respond to ever-increasing user demands for new products that prevent hand dryness and skin eczema in addition to providing superior antimicrobial performance [36].

IV. CONCLUSIONS

It is becoming quite apparent as we move into a new millennium that the use of the brush in the surgical scrub procedure will become outmoded by the introduction of new brushless technologies. What course these technologies will take and how surgical team members respond to them remains to be seen. In addition to brushless application and improvement of hand condition, these technologies will have to address very real concerns regarding the emergence and transmission of increasingly drug-resistant and pathogenic microorganisms, as well as the long-standing compliance concerns of infection control practitioners.

In a world where healthcare systems attempt to deal with increasingly strict cost containment criteria, time-management requirements, and environmental impact concerns, the adoption and use of brushless products for routine surgical scrubbing will likely occur as an insignificant but necessary change. It is certain, however, that this is a change that will be rooted in sound science and will be supported by well-qualified experts in infection control.

From the above, it is clear that due both to more stringent regulatory and to market-driven requirements, surgical scrub products being considered for introduction into the surgical suites of the next millennium will have to be improved relative to the standards of the past. Features such as improved antimicrobial efficacy and persistence, decreased application time, elimination of the need for use of a brush, and the incorporation of more ''skin-friendly'' ingredients will likely be important considerations in the design of these new products.

REFERENCES

1. Rotter ML. Hygienic hand disinfection. Infect Control 5(1):18–22, 1984.
2. Lilly HA, Lowbury EJL. Transient skin flora: Their removal by cleansing or disinfection in relation to their model of deposition. J Clin Pathol 31:919–922, 1978.
3. Reybrouck G. Handwashing and hand disinfection. J Hosp Infect 8:5–23, 1986.
4. Larson E, Leyden JJ, McGinley KJ, Grove GL, Talbot GH. Physiologic and microbiologic changes in skin related to frequent hand washing. Infect Control 7(2):59–63, 1986.
5. Ayliffe GAJ, Babb JR, Davies JG, and Lilly HA Hand disinfection: a comparison of various agents in laboratory and ward studies. J Hosp Infect 11:226–243, 1987.
6. Engelsher HJ. A new surgical scrub brush. Am J Surg 112:964–966, 1966.
7. Larson E. APIC guideline for handwashing and hand antisepsis in health care settings. Am J Infect Control, 23:251–269, 1995.
8. Association of Operating Room Nurses, Inc. Standards and Recommended practices: Recommended practices for Surgical Hand Scrubs. 1995, pp. 185–190.
9. Berman RE, Knight RA. Evaluation of hand antisepsis. Arch Environ. Health 18: 781–783, 1969.
10. Galle PC, Homesley HD, Rhyne AL. Reassessment of the surgical scrub. Surg Gynecol Obstet 147(2):215–218, 1978.
11. Loeb MB, Wilcox L, Simaill F, Walter S, Zoubida D. A randomized trial of surgical scrubbing with a brush compared to antiseptic soap alone. Am J Infect Control 25: 11–15, 1997.
12. Decker LA, Gross A, Miller FC, Read JA, Cutright DE, Devine J. A rapid method for the presurgical cleansing of hands. Obstet Gynecol 51(1):115–117, 1978.
13. Larson E, Silberger M, Jakob K, Whittier S, Lai L, Della Latta P, Saiman L. Assessment of alternative hand hygiene regimens to improve skin health among neonatal intensive care unit nurses. Heart Lung, 29(2):136–142, 2000.
14. Larson EL, Aiello AE, Heilman JM, Lyle CT, Cronquist A, Stahl JB, Della-Latta

P. Comparison of different regimens for surgical hand preparation. AORN J 73(2): 412, 414, 417–418, 423–424, 426, 429–430, 2001.
15. Hobson DW, Woller W, Anderson L, Guthery E. Development and evaluation of a new alcohol-based surgical hand scrub formulation with persistent antimicrobial characteristics and brushless application. Am J Infect Control 26(5):507–512, 1998.
16. Jones RD, Jampani H, Mulberry G, Rizer RL, Moisturizing alcohol hand gels for surgical hand preparation. AORN J 71(3):584–587, 589–590, 592, 2000.
17. Mathias JM. Should we discard the ritual of scrubbing with the brush? OR Manager 16(9):20, 22, 2000.
18. McCuaig K. Aseptic technique in microgravity. Surg Gynecol Obstet 175(5): 466–476, 1992.
19. Larson E. Guideline for use of topical antimicrobial agents. Am J Infect Control 16(6):253–266, 1988.
20. Hospital Infection Control Practices Advisory Committee. Guidelines for isolation precaution in hospitals. Am J Infect Control 24:24–52, 1996.
21. Doebbeling BN, Stanley GL, Sheetz CT, Pfaller MA, Houston AK, Annis L, et al. Comparative efficacy of alternative handwashing agents in reducing nosocomial infections in intensive care units. N Engl J Med 327:88–93, 1992.
22. Cartwright, FF. Lister the man. Lister Centenary Conference. Br J Surg 54:405–409, 1967.
23. Springthorpe S, Sattar S. Handwashing: What can we learn from the recent research? Infect Control Today 2(4):20–26, 1988.
24. Pereire LJ, Lee GM, Wade KJ. An evaluation of five protocols for surgical handwashing in relation to skin condition and microbial counts. J Hosp Infect. 36:49–65, 1997.
25. Feingold DS. The changing spectrum of streptococcal and staphylococcal infections. In: Aly R, Beutner KR, Maibach H, eds. Cutaneous Infection and Therapy. New York: Marcel Dekker, 1997, pp. 15–25.
26. MRSA outbreak in NICU linked to nurse with hand dermatitis. Hosp Infect Control 16(1):9–10, 1989.
27. Larson E, Norton H, Carrie A, Pyrek J, Sparks S, Cagatay E, Bartkus J. Changes in bacterial flora associated with skin damage on hands of health care personnel. Am J Infect Control 26(5):512–521, 1998.
28. Larson E. Hygiene of the skin: when is clean too clean? Emerging Infectious Diseases, Centers for Disease Control (7)2, 2001 (CDC online journal).
29. Lilly HA, Lowbury EJL, Wilkins MD. Limits to progressive reduction of resident skin bacteria by disinfection. J Clin Pathol 79:107–119, 1979.
30. Bryan JL, Cohran J, Larson EL. (1995) Handwashing: a ritual revisited. In: Rutala WA, ed. *Chemical Germicides in Health Care*. Morin Heights, P.Q., Canada: Polyscience Publications, Inc., 1995.
31. Associated Press. Doctors are told: Don't forget the soap and water. Des Moines Register, 20 June, 1995.
32. Hingst V, Judizki I, Heeg P, Sonntag, HG. Evaluation of the efficacy of surgical hand disinfection following a reduced application time of 3 instead of 5 minutes. J Hosp Infect 20:79–86, 1992.
33. U.S. Food and Drug Administration. Topical antimicrobial drug products for over-

the-counter human use: tentative final monograph for health car antiseptic drug products-proposed rule. Fed Reg, 59:31441–31452, 1994.
34. Denton GW. Chlorhexidine. In: Bloch SS, ed. Disinfection, Sterilization, and Preservation. 4th ed. Philadelphia: Lea and Febiger, 1991, pp. 274–289.
35. Paulson DS, Fendler EJ, Dolan MJ, Williams RA. A close look at alcohol gel as an antimicrobial sanitizing agent. Am J Infect Control 27(4):332–338, 1999.
36. Lauharanta J, Ojajarvi J, Sarna S, Makela P. Prevention of dryness and eczema of the hands of hospital staff by emulsion cleansing instead of washing with soap. J Hosp Infect Control 17:207–215, 1991.

16
Fundamental Steps in Designing a Healthcare Personnel Handwash

Daryl S. Paulson
BioScience Laboratories, Inc., Bozeman, Montana

I. INTRODUCTION

The importance of effective antimicrobial handwashing has been known throughout the healthcare field for many years. Accurate and reliable determinations of the microbial populations residing on the hands are critical in evaluating the effectiveness of both handwash products and methods. Only when one is sure of reliable hand sampling methods can he or she attempt to assess the benefits of a personnel handwash product in terms of microbial reductions [1].

Proper research design is also crucial in gathering accurate information. There are a number of ways to perform these evaluations using quantitative research designs, and it is vital that investigators be familiar with the selection of designs at their disposal [2].

A. Types of Organisms

Microorganisms that reside on the hand surfaces are classified in two general categories. The first category consists of organisms that are accidentally "picked up" by personnel and are "transient" in that they reside on the hands only temporarily. The second consists of those microorganisms that permanently reside on the hand surfaces—the normal skin flora [1].

For healthcare personnel, transient, pathogenic microorganisms are the more significant. Contaminant microorganisms are responsible for infectious disease outbreaks passed from healthcare personnel to patients via hand contact.

For an antimicrobial product to be effective, it should afford both immediate and persistent microbial effects [1,3]. Immediate antimicrobial effects depend upon two attributes: the mechanical removal of contaminating microorganisms and the topical antimicrobial compound's ability to kill microorganisms upon contact. The persistent antimicrobial effect—the ability of the handwash to keep the microbial population at a low level after washing—is mainly dependent upon the type of antimicrobial product used.

Immediate antimicrobial effectiveness depends upon the amount of the antimicrobial hand sanitizer used, the type used, the amount of time spent washing the hands, the mechanical pressure and friction exerted in the wash, and the temperature of the water. Even when a very effective handwash regimen has been developed, healthcare personnel must comply with the wash regimen for it to be successful.

B. Antimicrobial Compounds

Several effective antimicrobial products are commonly available.

1. Iodophors

Iodines and iodophors have been used as antimicrobial agents for years and are effective against most of the microorganisms encountered [4]. They present good immediate and persistent antimicrobial properties, as well as effectiveness in removing both normal and contaminant microorganisms. However, upon frequent, repeated, and prolonged use, they tend to irritate hands.

2. Chlorhexidine Gluconate

Chlorhexidine gluconate has been used in the medical field for a number of years. It has good immediate and persistent antimicrobial effects against both normal and contaminant microorganisms. Additionally, it has the added benefit of a "residual" effect; that is, as the product is used repeatedly over time, it binds to the skin, retarding microbial growth on the hands [4].

As with iodophors, chlorhexidine gluconate may irritate hands upon repeated or prolonged use at the 4% level. At lower levels (2, 1, or 0.5%), however, irritation is generally not a problem.

3. Triclosan

Triclosan is a general purpose antimicrobial that is effective against many commonly encountered microorganisms [3]. Although it does not demonstrate the degree of antimicrobial effectiveness of the iodophors or chlorhexidine gluconates, it is effective against transient or contaminant microorganisms. Triclosan has both immediate and persistent antimicrobial effects as well as a low level of skin irritation potential.

4. Parachlorometaxylenol

Parachlorometaxylenol (PCMX), like triclosan, is a general purpose antimicrobial compound. It is effective against many microorganism types encountered in the healthcare field, but like triclosan, it does not display the high degree of antimicrobial efficacy that the iodophors and chlorhexidine gluconate do. It generally is mild to the skin, even when used repeatedly over time.

5. Alcohols

Solutions containing over 60% isopropyl or ethyl alcohol tend to provide very effective and immediate antimicrobial effects. However, they have no persistent antimicrobial properties, and without added emollients, they tend to irritate the skin upon repeated use.

C. Relationship Between Infection and Soaps

Let us discuss briefly aspects of infectious disease and the role antimicrobial hand soaps play in their prevention. In order for infectious diseases to occur, five events must take place [5]:

1. The microorganisms must come into contact with a person.
2. The microorganisms must enter a person.
3. The microorganisms must spread from the entry site.
4. The microorganisms must multiply within the person.
5. By a combination of the microbial enzymes and toxins, as well as a person's immune system, or both, tissue damage occurs.

An effective handwash disrupts the process after event 1 by removing contaminating microorganisms from the hand surfaces. While not all of the microorganisms are removed, as long as they are reduced to a number lower than the level required to cause disease in normal humans, no disease follows.

When the anti-infective properties of antimicrobial hand soaps are discussed, it is necessary to understand what types of microorganisms should be addressed. The microorganisms that normally reside on the hand surfaces pose little threat of disease transmission from person to person [7]. There are, of course, situations in which resident microorganisms cause disease. An infected cut is an example of this. However, in this case, washing serves to degerm the infected area, cleansing it of dead cells and exudate.

The principal threat of infectious disease is due to "transient," pathogenic microorganisms, temporarily colonizing hand surfaces. This occurs as a result of hand contact with contaminated substances such as mucus, blood, soil, urine, feces, and food. The contaminant microorganisms can infect the person or be passed onto others via hand contact.

II. EFFECTIVE ANTIMICROBIAL SOAP

Two areas of primary concern when evaluating antimicrobial hand soaps are the immediate degerming effectiveness and the persistent antimicrobial effectiveness [1]. Immediate antimicrobial efficacy relates to the effectiveness of the soap in both the mechanical removal of contaminating microorganisms due to the handwash procedure and the immediate inactivation of the microorganisms through contact with the active antimicrobial ingredient(s) in the soap. The persistent antimicrobial effectiveness is the product's ability to prevent, either by microbial inhibition or lethality, transient microbial recolonization of the skin surfaces after handwashing [1].

A. Evaluating Antimicrobials

To measure these areas of concern accurately and precisely is difficult [2]. Efficacy evaluations, then, must be well defined and stated clearly before conducting them.

To determine what microorganism types are susceptible to the antimicrobial soap product, as well as the rates of microbial inactivation, in vitro tests should be conducted. These include time-kill kinetic evaluations, minimum inhibition concentration evaluations, and microbial sensitivity tests. However, in order to determine the actual effectiveness, human-use studies must be conducted. There are three common ways of evaluating antimicrobial hand soaps: (1) the health care personnel handwash evaluation, (2) the modified Cade handwash procedure, and (3) the general-use handwash evaluation.

B. Health Care Personnel Handwash Evaluations

This study is utilized when evaluating antimicrobial products intended for use as healthcare personnel handwashes [1,8]. A test product, a vehicle product (test product without the active antimicrobial), and a reference product are customarily used in this evaluation.

At least a 7-day "washout" period is enforced, during which the subjects are not permitted to use antimicrobial products (including antibiotics) or expose their hands to strong acids or bases or any other compounds known to affect microbial populations. On the test day, subjects are inoculated with a marker microorganism, *Serratia marcescens*, the colonies of which appear red when plated on tryptic soy agar. This allows them to be distinguished from other microorganisms found on the skin that appear white or yellowish on the agar. The hands are then sampled using the glove juice procedure.

This first inoculation/sampling procedure constitutes the baseline measurement. The hands are again inoculated with *Serratia marcescens*, followed by a handwash using the assigned product. Following this, the glove juice sampling procedure is again employed.

The inoculation/wash procedure is repeated 10 consecutive times. Glove juice samples are taken after inoculation/wash cycles 1, 3, 7, and 10. The agar plates are then incubated at 25 ± 2°C for approximately 48 hours. This evaluation enables one to measure the microbial population counts on the hands after washes 1, 3, 7, and 10, compared to the baseline average value. To be acceptable as a healthcare personnel handwash, per FDA specifications, the product must demonstrate at least a 2 $\log_{10}$ reduction from baseline after the first wash and at least a 3 $\log_{10}$ reduction at wash 10.

C. Modified Cade Handwash Procedure

The number of subjects recruited varies, but it is common to enroll 55–65 subjects. At least a 7-day washout period is observed, as are the product-use restrictions described in the healthcare personnel handwash procedure. This is followed by a baseline measurement period. On the first day of the baseline period, subjects wash their hands five consecutive times with a nonmedicated soap.

Microbial samples are collected from washes 1 and 5, using a ''sterile basin wash'' procedure, which entails subjects washing their hands with the bland soap in a polyethylene container, such as a plastic freezer storage bag, containing 1 liter of sterile water. After washing, the wash water is well mixed and sample aliquots are plated on tryptic soy agar. The agar plates are incubated at 30–35°C for approximately 48 hours.

For days 2 and 3 of the baseline period, the subjects wash their hands, ad libitum, outside the laboratory with the bland soap and use it for bathing and showering. On day 4, those 45–50 subjects having the highest baseline hand counts from day 1 remain in the study; the others are dismissed. On day 5, subjects again wash their hands with the bland soap, as on day 1. Samples from washes 1 and 5 are again collected, as described previously, to complete the baseline period, and subjects continue using the bland soap outside the lab for all handwashing, as well as bathing and showering, on days 6 and 7. On day 1 of the test period, subjects return to the laboratory and again wash their hands five consecutive times with bland soap, with samples at washes 1 and 5. The subjects then perform supervised handwashes with their test product at the laboratory starting 30 minutes after the fifth bland wash. They wash three times with the test product, with at least 1 hour between washes. After these washes have been completed, the 50 subjects are issued test product samples, Laboratory and Home Handwash Log Forms, and use instructions.

The subjects wash three times daily at home, each wash no less than 1 hour apart, for the test period 2 and continue washing 3 times daily for the next 9 days (test days 3–11). These washes are documented on the Laboratory and Home

Wash Log Forms. Subjects also use their test product for any additional handwashes and for bathing and showering.

On day 12, subjects turn in their assigned test products and then wash five consecutive times with a bland soap. Samples are again collected using the sterile basin wash procedures for washes 1 and 5.

Ordinarily, the first wash samples collected from the baseline period are pooled, and likewise for the fifth washes. The first test wash sample is then compared to the pooled wash 1 baseline period. The fifth test wash is compared to the pooled wash 5 baseline period.

D. General Use Handwash Evaluation

Normally, 10–20 subjects are recruited for each product evaluated. A control product is sometimes used. During a washout period of at least 7 days, subjects are not permitted to use antimicrobial products or expose their hands to known compounds that have antimicrobial properties. The remaining study procedures are identical to those of the healthcare personnel handwash previously described, but the total number of washes is 5–10, with glove juice samples collected at baseline and after wash 1 and wash 5, and after wash 10 if 10 washes are used. The microbial sample counts taken after the test washes are then compared to the baseline values.

E. Optimal Evaluation Design

The optimal study design should be practical, yet provide accurate and reliable results based upon transient microorganism reductions, not resident ones. The healthcare personnel handwash is designed to evaluate the antimicrobial efficacy of products used in the healthcare field and, therefore, is excessively stringent for evaluating consumer product antimicrobial soaps.

The modified Cade handwash procedure is suboptimal, particularly in that the antimicrobial efficacy is evaluated primarily in terms of reductions in normal microbial flora, which is not appropriate for evaluating consumer antimicrobial hand soaps. Additionally, it measures the residual antimicrobial effectiveness of the test product, not the immediate effects.* Finally, a control product is rarely used, so assuring the validity of the study is not possible.

The optimum design for evaluating consumer antimicrobial hand soaps is a combination of the healthcare personnel handwash evaluation and the general

* Residual effectiveness is a measurement of the product's antimicrobial effects when utilized repeatedly over time. The antimicrobial is absorbed into the skin and, as a result, prevents recolonization of the skin by microorganisms. It is a measurement used in evaluating surgical scrub products, not consumer antimicrobial soaps.

use handwash evaluation. It is practical, accurate, and reliable, as well as based upon the removal of transient microorganisms over the course of at least five repeated inoculation/wash cycles. It is conducted exactly as the healthcare personnel handwash evaluation, except with the following changes: a test and control product are used; the control product is the test product without the antimicrobial compound; usually, instead of 10 consecutive handwashes, five are conducted; samples are collected for baseline and after product use washes 1 and 5. This design enables measurement of the immediate degerming effects (first wash), as well as the effects after five consecutive inoculation/wash cycles to assure that there is no cumulative build-up of microorganisms at wash sample 5. Additional inoculation/wash cycles and samples will add no predictive value to the evaluation.

A minimum of 15 subjects per group is recommended, as is the use of a control product. Hence, if one test product is used, a minimum of 30 subjects should be employed (15 for the test product and 15 for the control product). This is a statistically adequate number, in most cases, to detect true differences between the test and control products after washes 1 and 5, as well as the microbial reduction counts from baseline after washes 1 and 5.* A caveat is that the baseline standard deviation value must be less than 0.5 log. If the standard deviation is greater, it will be difficult to detect a true difference between the test and control products after washes 1 and 5. It will also be more difficult to detect true microbial reductions from the baseline values after wash samples 1 and 5. Hence, using a larger sample size is in the best interest of the product manufacturers.

The required log reduction values of the antimicrobial soap intended for general consumer use are hotly debated. It is this author's opinion that the reduction values should be at least 1 $\log_{10}$ from the baseline value after wash 1 and 1.5 log after wash 5.

It is necessary that antimicrobial soap product manufacturers take an active role in accurately and precisely measuring the antimicrobial effects of their antimicrobial hand soaps. It is necessary that they be evaluated under simulated wash conditions employing human subjects. Finally, it is important that other concerned parties are not only welcome to present their evaluative concerns, but that they do.

---

* The sample size determination is based on the formula:

$$n \geq S^2 \left[\frac{(Z_{\alpha/2} + Z_\beta)^2}{D^2} \right]$$

where: n = sample size; S = known standard deviation of samples taken; $Z_{\alpha/2} = 1.96$; $Z_\beta = 0.842$; and D = clinical difference of significance (20%). For an in depth discussion of sample sizes, consult the forthcoming *Applied Statistical Designs for the Researcher* (DS Paulson, Marcel Dekker, in press 2003).

III. ANTIMICROBIAL EFFICACY EVALUATION

Before choosing an antimicrobial product for routine use, it is important to evaluate its antimicrobial effects. A skin-sampling method that accurately measures the number of microorganisms on the hands is required. Because the most common function of handwashes is to remove contaminant microorganisms, a common means of measuring a product's effectiveness is to contaminate the hands artificially and then use the product to remove those microorganisms. The ''glove juice'' sampling method is the method of choice for determining the number of microorganisms present on the hands. Sterile, powder-free, latex surgical gloves are placed over the subject's hands, and about 75 mL of sterile saline, with or without neutralizers, are put into the glove. The wrist is secured, and the hand is massaged through the glove for about 60 seconds. Aliquots of the ''glove juice'' are removed, plated on agar and incubated until the colonies have grown sufficiently. To get an accurate and reliable estimate of how effective an antimicrobial product is in removing contaminant microorganisms, *S. marcescens* is used as a ''marker'' microorganism. Because *S. marcescens* colonies appear red when plated on tryptic soy agar, they can easily be distinguished from other microorganisms residing on the hands. Any nonred colonies appearing on the agar plates are not counted. The employment of *S. marcescens* prevents biasing the results by inadvertently ''mixing up'' the normal and marker microorganisms.

A. Experimental Designs

There are two types of experimental validity [2]: internal and external.

1. Internal Validity

Internal validity is experimental design validity. In particular, it deals with the way the study is carried out, how sample data are collected, and how the study is controlled, especially with respect to investigator bias. It is well known that investigators often have a ''vested interest'' in realizing that the products they employ are successful. This bias must be taken into account. Fortunately, internal validity can be controlled by using proper experimental design procedures (e.g., randomizing, blocking, and blinding the study) [9–12]. Although there are a number of aspects of internal validity, two of the most common are historical and instrumentation validity.

Historical validity assures that no event occurs between sample time measurements that biases the study results [13]. An example of negation of historical validity occurred with a group of subjects participating in preference testing of several hand-cleansing product attributes. Unknown to the investigator, the study participants continued using their personal hand soaps at home during the course of the study. It was discovered that the subjects unconsciously compared the test products to their own personal hand soaps, thus biasing the study [2].

Instrumentation validity is achieved by assuring that no biasing event occurs that could affect the ''measuring instruments'' used in the experiment. For example, a person measuring the efficacy of a hand soap preservative used agar plates (a measuring instrument) to judge the preservative's antimicrobial efficacy. Unknown to the researcher, who used different agar medium lots, the agar lots were significantly different in nutritional characteristics, affecting their abilities to support microbial growth. The differences in the bacterial growth were attributed to the antimicrobial preservatives but were really caused by the differing lots of agar medium.

2. External Validity

External validity refers to the extent the results of a specific study can be generalized to include the population at large (population validity) or general environmental conditions (environmental validity). No experimental design has built-in controls for assuring external validity [1,2].

An easy way to assure external validity of a study is to conduct the study independently at different geographic locations. If consistent results are observed and the same conclusions drawn by different investigators, the external validity of the study is probably satisfactory. Let us now turn our attention to experimental designs, both quantitative and qualitative.

B. Quantitative Research Designs

The vast majority of quantitative research designs utilize statistics [2]. Hence, it is critical to select appropriate statistical models (e.g., linear regression, analysis of variance, analysis of covariance, Student's *t*-test, or others) that complement the experimental design [9–14]. Let us now briefly address the types of statistical models available, both parametric and nonparametric.

1. Parametric Statistics

Parametric statistics, including the Student's *t*-test, linear regression, analysis of variance, and analysis of covariance, utilize parameters (e.g., the mean [average], variance, and/or standard deviation) in evaluating data (102.915, 1×10^{-5}, 7.23914, etc.). These data can be ranked, as well as subdivided into infinite intervals. To be considered ''interval'' data, they must be assessable in terms of some sort of standard physical measurement. Height, weight, blood pressure, alcohol content, and number of deaths are all interval data. Subjective perception of pregnancy, death, prestige, and social stress do not present a ''natural'' interval category, even though research designs frequently (and in error) categorize them as interval. Extreme caution must be observed in cases when quantitative designs are used to measure qualitative data [2]. Common parametric models follow.

The Student's *t*-test, probably the most common parametric statistical model, is used to compare two groups of data—for example, to compare test group data to a specific value or to compare data from two groups (e.g., a test and a control group or two test groups). It can be used as a "one-tail" test, to determine if one group of data is "better" or "worse" than another, or as a "two-tail" test, to determine simply if the data "differ."

Analysis of variance (ANOVA) is also a common parametric statistic for comparing data from more than two groups [2]. There are a number of variants of this model, depending upon the number and combination of groups, categories, and levels one desires to evaluate. Common ones include one-factor, two-factor, and three-factor designs, as well as crossover and nested designs.

Regression analysis is used to predict a response or dependent variable (y) from the value of an independent variable (x) [2]. Regression models are commonly used in product stability evaluations of attributes such as color, clarity, or fragrance.

2. Nonparametric Statistics

Nonparametric statistics do not utilize parameters (mean, variance, or standard deviation) in evaluating data [9–11]. However, they can be applied to interval data and noninterval data, both nominal and ordinal. Nominal data can be grouped, but not ranked. Data such as right/left, male/female, yes/no, and 0/1 are nominal data. Ordinal data can be both grouped and ranked. Examples include good/bad, poor/average/excellent, lower class/middle class/upper class, and low/medium/high levels of drugs.

Nonparametric statistics are often applied to interval data when sample sizes are very small. When using very small sample sizes, the variable data distribution often cannot be assured to be "normal," a requisite for using parametric statistics. A normal, "bell curve" distribution is not a requirement of nonparametric models. Hence, they are preferred in this area over parametric models. Common nonparametric models follow.

The Mann-Whitney test statistic is the nonparametric analog of the Student's *t*-test and is used to compare data from two groups [9]. Unlike the parametric Student's *t*-test which assumes a normal "bell-shaped" distribution, the Mann-Whitney statistic requires only that the sample data collected are randomly selected.

The Kruskal-Wallis model is the nonparametric analog of a one-factor ANOVA model. It is used to compare multiple groups of one factor. For example, suppose one wants to evaluate the antimicrobial effects of five different hand soaps; the Kruskal-Wallis model could be employed for this evaluation.

IV. CONCLUSION

It is important that antimicrobial products be evaluated for efficacy. There are a number of ways to perform these evaluations using quantitative research designs and statistical models. It is also vital that investigators be familiar with a selection of qualitative designs. This will prevent the researcher who has only one tool—a hammer—from viewing everything as a nail.

REFERENCES

1. DS Paulson. Topical Antimicrobials Testing and Evaluation. New York: Marcel Dekker, 1999.
2. DS Paulson. Applied Statistical Designs for the Researcher. New York: Marcel Dekker, 2002.
3. DS Paulson. Developing Effective Topical Antimicrobials. Soap/Cosmetics/Chemical Specialties, Dec. 1997, pp. 50–58.
4. DS Paulson. Section II Overview. In: DS Paulson, Ed. Handbook of Topical Antimicrobials: Industrial Applications in Consumer Products and Pharmaceuticals. New York: Marcel Dekker, 2002.
5. DS Paulson. Topical Antimicrobials and Skin Irritation: The Next Step. Soap/Cosmetics/Chemical Specialties, Nov. 1998, pp. 46–49.
6. M Schaechter, BI Einsenstein. Mechanisms of Microbial Disease. 2nd ed. Baltimore, MD: Williams and Wilkins, 1993.
7. DS Paulson. Designing a Handwash Efficacy Program. Pharmaceutical and Cosmetic Quality, Jan–Feb 1997, pp. 23–29.
8. Food and Drug Administration. Tentative Final Monograph for Health Care Antiseptic Drug Products, 21 CFR, Parts 333 and 369: 31402–31452, 1994.
9. WJ Conover. Practical Nonparametric Statistics. 3rd ed. New York: John Wiley and Sons, 1980.
10. WW Daniel. Applied Nonparametric Statistics. New York: John Wiley and Sons, 1980.
11. JD Gibbons. Nonparametric Methods for Quantitative Analysis. New York: Holt, Rinehart and Winston, 1976.
12. WJ Dixon, FJ Massey. Introduction to Statistical Analysis. 4th ed. New York: McGraw-Hill, 1983.
13. CR Hicks. Fundamental Concepts in the Design of Experiments. 4th ed. New York: Holt, Rinehart and Winston, 1993.
14. JM Neter, MH Kuter, CJ Nachtsheim, W Wasserman. Applied Linear Statistical Models. 4th ed. Chicago: Irwin, 1996.

Part IV
Food

Topical antimicrobial handwashes and hand sanitization are very important in the food service industry. Important, also, is gloving. Chapter 19 addresses gloving and handwashing, and Chapter 20 provides research data from various studies. Chapter 21 discusses quality assurance issues, and Chapter 22 offers another perspective on the wash/glove controversy.

17

Handwashing, Gloving, and Disease Transmission by the Food Preparer

Daryl S. Paulson
BioScience Laboratories, Inc., Bozeman, Montana

I. INTRODUCTION

One of the most controversial issues in the food industry is the ''bare hands'' legislation that prohibits bare-hand contact with ready-to-eat foods. Advocates of this legislation argue that because vinyl gloves provide a microbiologically impenetrable physical barrier between food workers' hands and the food they handle, and because a significant number of food workers do not wash their hands adequately to remove potentially pathogenic microorganisms, wearing gloves should be mandatory. On the other hand, opponents argue that an effective handwash is sufficient and glove-wearing is not necessary because the wash removes the disease-causing microorganisms from the hands. Additionally, they argue that relying on a glove barrier to prevent disease is unwise because tears and rips to gloves are common. The tears and rips will readily allow microorganisms to pass through the gloves and onto the food. Both views are correct, but only partially so.

A. Microorganisms of Concern

Generally, infectious diseases are spread in the food service environment in two ways: (1) hand contact with one's own infected feces and passage of microbial contaminants to prepared foods as a result of inadequate handwashing and (2) handling of microbially contaminated objects (e.g., money, counters, soiled clothing, raw meats) and subsequent contamination of foods provided to the consumer [1].

Table 1 Organisms Identified in Foodborne Outbreaks, New York State, 1975–1998

Agent	No. of outbreaks
Hepatitis A	28 (24.57%)
Norwalk-like virus	21 (25.93%)
Staphylococcus aureus	6 (7.41%)
Shigella sonnei	5 (6.17%)
Salmonella typhimurium	5 (6.17%)
Salmonella enteritidis	4 (4.94%)
Group A streptococcus	4 (4.94%)
Giardia spp.	2 (2.47%)
Salmonella paratyphi	1 (1.23%)
Salmonella javiana	1 (1.23%)
Vibrio cholerae	1 (1.23%)
Shigella flexneri	1 (1.23%)
Cryptosporidium parvum	1 (1.23%)
Yersinia enterocolitica	1 (1.23%)
Total	81 (100%)

In discussing disease transmission, it is useful to specify the types of microorganisms of concern. Table 1 presents the microorganisms most commonly involved in foodborne illnesses in New York State from 1975 to 1998 [2].

Other microorganisms, such as enterotoxigenic strains of *Escherichia coli*, have been responsible for significant foodborne disease outbreaks in other parts of the United States.

Microorganisms that normally colonize hand surfaces pose little threat of infectious disease [3]. There are situations, such as an infected cut, in which normal, resident microorganisms may cause disease; in such situations, however, washing serves to degerm the infected area, cleansing it of dead cells and exudative material [4].

B. Etiology of Infectious Disease

For infectious diseases to be spread, the following five events must take place [5]:

1. The contaminating microorganisms must be physically transmitted to others. This can occur when, for example, food workers contaminate their hands during defecation and pass the disease-causing microorganisms to consumers via hand-to-food contact.
2. The contaminating microorganisms must physically enter a person.

This can easily occur when food contaminated by enteric (intestinal) disease-causing microorganisms is ingested.

3. The contaminating microorganisms must spread from the anatomical site of entry to other areas of the body—from the mouth to the intestinal tract—via food consumption.
4. The contaminating microorganisms must be able to multiply within the person. For example, *Salmonella* spp. and related microorganisms attach to the walls of the intestinal tract and subsequently colonize it.
5. Tissue damage must occur as a result of a combination of the microbial enzymes and toxins, as well as a person's immune response.

An effective handwash or intact barrier gloves disrupt the disease process after Event 1, either by removing the contaminating microorganisms from the hand surfaces or by imposing a physical barrier that prevents microorganisms from being transmitted from hands to foods [1].

II. DISEASE TRANSMISSION TO PATRONS VIA CONTAMINATED HANDS OF FOOD WORKERS

Let us begin with the situation of infected food workers who may transmit infectious diseases by directly contacting food with contaminated hands.

A. Barrier Gloves

The main purpose of barrier gloves is to prevent pathogenic microorganisms found on the food workers' hands from being transmitted to patrons via barehand contact with food. A vinyl or latex barrier that is intact (no holes, rips, or punctures) will provide protection from microbial transmission of hand-contaminating microorganisms. However, vinyl food-grade gloves frequently have preexisting pinhole punctures that compromise the barrier protection [6]. Further, food-grade vinyl gloves can easily be ripped, torn, or punctured while personnel perform their normal duties, and in many cases such damage remains unknown to the wearer [7]. Additionally, heat has been reported to alter significantly the integrity of barrier gloves, making them brittle and, hence, more prone to breakage. In practice, then, the actual protection provided by barrier gloves is likely much less than many people assume.

In a study conducted at our laboratory facility, volunteer human subjects' hands were inoculated with a strain of *E. coli*. The subjects then donned vinyl food-server gloves having four small needle punctures in each. Within 5 minutes, sampling of the outside of the gloved hands showed that *E. coli* had been transferred from the hands onto the outer surfaces of the gloves.

Wearing gloves may actually increase the potential for disease transmission. As one wears vinyl or latex barrier gloves over a period of time, microorganisms residing on the skin are provided a physical environment more favorable to growth than ungloved hands offer. This is because the gloves occlude the hands, thereby increasing the levels of moisture, nutrients, and other factors essential to microorganism growth [5,8]. This phenomenon has long been known in the medical field, in which antimicrobial handwashes are required prior to gloving. In studies conducted in our laboratory, we have observed that the population numbers of resident bacteria increase and that transient microorganisms, although not increasing in number, could maintain viability longer on gloved hands than on bare hands. It appears that normal resident bacteria may crowd out the transient ones, over time, via more efficient growth and attachment characteristics [9]. Hence, relying solely on barrier gloves, without accompanying handwashes, to prevent disease is not prudent.

B. Handwashing

Two quantitative antimicrobial parameters are important when discussing handwashing: immediate and persistent antimicrobial effectiveness [1,8]. Immediate antimicrobial effectiveness is the effectiveness of the handwash in terms of both the mechanical removal of contaminating microorganisms and the immediate inactivation of microorganisms through contact with the antimicrobial ingredient in the soap, lotion, or gel. Persistent antimicrobial effectiveness is the antimicrobial compound's ability to prevent transient microbial recolonization of the hand surfaces after handwashing because of either microbial inhibition or the compound's lethality.

A good handwash, then, has been shown in testing in our laboratory to be very effective in removing contaminating microorganisms. Assuring that food workers perform an effective handwash is another story.

III. RECOMMENDATIONS AND COMMENTS

Toward reducing the potential for disease transmission from food workers to patrons, I will make the following four recommendations:

1. Both gloving and handwashing should be required for those performing high-risk tasks such as handling, cooking, or wrapping food. Better yet, no direct hand/glove contact with the food should occur, with sanitized serving tongs or other utensils being used where possible.
2. Mandatory and ongoing sanitation training must be provided for all food workers.

3. High levels of personal hygiene must be enforced. Employees with an infectious disease (e.g., colds, flu, etc.) should not be allowed to have direct contact with food, with or without barrier gloves.
4. Monitoring and enforcement of, and accountability for, the three above recommendations must be maintained.

Comment 1: Both glove-wearing and handwashing with an effective antimicrobial product prior to gloving should be required for those workers performing high-risk tasks involving direct hand/glove contact with food. Although neither is fail-safe, it is probable that, when used in conjunction, they will provide more protection against disease transmission than either used alone.

In a recent study at our facility, it was observed that gloving without performing a handwash supported prolonged survival of *E. coli* populations. When an effective handwash was performed prior to gloving, however, no prolonged growth promotion of the contaminative microbes was observed on the hand surfaces over the course of 3 consecutive hours of wearing.

It is recommended, therefore, that when gloves are worn, the gloving should be preceded by an effective handwash. It can be argued that, even when a handwash of only marginal quality is performed, when an effective antimicrobial soap is used, its antimicrobial properties (immediate and persistent), in combination with the glove barrier, will provide improved protection from disease transmission.

Comment 2: Sanitation training and education should be an ongoing and continuous effort, particularly with inexperienced and/or unmotivated workers. Unquestionably, in the absence of active participation of employees, achieving adequate sanitation standards will be very difficult.

Comment 3: A high degree of personal hygiene should be required of food workers. Employees should wear clean uniforms that are changed often, should bathe or shower often, and should not perform high-risk tasks when ill. High-risk tasks include any hand/glove contact with food or with materials that come into direct contact with food.

Comment 4: A quality control program supervised by qualified personnel should be initiated at each individual food service facility to monitor and enforce the handwash/glove sanitation practices.

IV. BEHAVIORAL ASPECTS

At a meeting of the National Advisory Committee on Microbiological Criteria for Foods,* Dr. Dale Morse of the New York State Department of Health pre-

* Bare-Hand Contact of Ready-to-eat Foods at Retail, Sept. 21–24, 1999. Washington Plaza Hotel, Washington, DC.

sented very sobering data concerning microbial contamination of food due to fecal contamination by food servers. His data indicated that, after New York State's prohibition of the touching of ready-to-eat food by bare hands, the incidence of foodborne disease dropped.

Does this mean that gloves are necessarily always safer than bare hands? Not if an employee performs an adequate handwash with an effective antimicrobial product, but it does suggest that such handwashes commonly did not occur. Hence, corrective measures that must be implemented by the food service industry are *behavioral*. And human behavior depends upon motivation, which depends upon values and meaning—both cultural (shared) and individual.

A. Shared Values

Shared values are ''cultural,'' or intersubjective, values [2,10]. Culture includes shared values not only of nations, but also of an industry and of the members of a company. The importance of handwashing, gloving, and use of utensils to prevent microbial transfer to patrons must be a shared value among the members of the culture [11]. It must have emotional meaning. The establishment of a set of values (e.g., to wash one's hands) can be effective coming from management in the form of policy. However, the policy must treat employees with dignity, with clear communication, with fairness, and with due process [12]. The policy will then be translated into shared values that matter.

Sociologists and social psychologists tell us that cultural values (as well as shared beliefs, goals, and views) have meaning on at least two levels [13,14]. At the surface or manifest level, shared meaning and value is concrete [14]. Regarding handwashing, it means to remove microorganisms from the hands after defecating, or else infections can be passed on to food consumers. If an employee is tired, sick, or harried, surface values are all too easily dismissed by rationalization and justification [15]. That is, a person constructs a reason for not washing the hands properly or not washing them at all, for example, ''There wasn't any soap,'' or ''I forgot,'' or ''John didn't wash his hands either.'' This is not meant to be judgmental, but to describe human behavior in terms of known psychological phenomena.

If shared meaning can be transferred to a deeper level, justification and rationalization are less likely to occur [16,17]. In such a case, one washes his or her hands because it is important to feel part of a team, larger than the individual self. At this level, meaning binds individuals to one another. With deeper, shared values, an individual is less likely to violate the rules [16].

Deeper shared values are instilled in employees when they not only understand but feel personally that handwashing is important, not just another piece of red tape or another hoop to jump through. Additionally, the reinforcement of shared values must be an ongoing process, with positive, active managerial support of the in-house hygiene program.

B. Personal Values

In addition to shared values, there are personal ones, which are likewise important. Unfortunately, in order to attain individual compliance with the handwashing codes, managers often take a ''police state'' stance toward handwashing [12]: ''If you don't wash your hands, you'll be fired.'' This, although initially effective, most probably will not be effective over the long run, particularly if nonhandwashing is used as a way to ''passively aggress'' an employer [12,16]. If an employee feels betrayed or ''screwed over'' by an employer, he may semiconsciously respond by not washing his hands, to lash out at the employer [18].

To counter and/or prevent passive/aggressive behavior, the workplace must be psychologically safe enough for employees to voice their concerns and vent their frustrations [19], and food service employees must not feel psychologically belittled or threatened in doing so [20]. Additionally, if employees are viewed merely as tools, all too often they go through the *surface* motions without really thinking or caring about their actions, because they know they are, after all, only tools [12].

C. Integration

The science we discussed initially—prevention of disease transfer from employees to patrons—must be linked to human behavior, which, as we also discussed, includes the shared group and personal values that create meaning and are the basis for intentional behavior. A break in any of these three links blocks the desired behavior from occurring, but a strengthening of any of them promotes strength in the others (Fig. 1) [26]. Clearly, changes in behavior require leadership from management, and, fortunately, the same motivating behavior required to train and manage employees successfully can be applied equally effectively to a handwashing program.

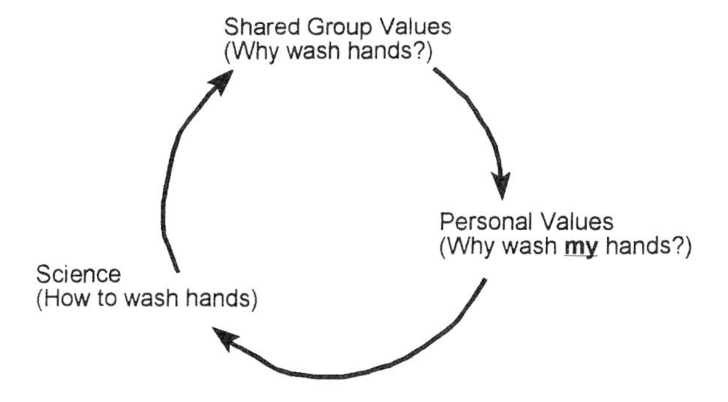

Figure 1 Behavior aspects of good handwashing practices.

V. DISEASE TRANSMISSION VIA CONTAMINATED OBJECTS

Let us now turn our attention to the situation in which disease is transmitted to patrons from food workers' hands that have become contaminated from direct contact with the work environment.

A. Etiology

The five events required for disease transmission discussed earlier (microorganisms transmitted to a person, microorganisms enter a person, microorganisms spread from entry site, microorganisms multiply within a person, tissue damage results) are likewise relevant to this situation, but because the contaminative process is one of ''picking up'' microorganisms from the work environment (money, countertops, wash sinks, clothing, nasal secreta, etc.) and passing them on via hand/glove contact with food, merely washing the hands before gloving or wearing barrier gloves will not prevent this process [22]. The ''clean'' hands/gloves become contaminated from the environment and provide a potential vehicle for disease transmission [1,6,7]. Hence, hands need to be washed and/or gloves changed frequently. Additionally, an effective program for disinfection of the work environment must be established [1,3,23].

B. Recommendations and Comments

The four recommendations that follow will contribute to reducing the potential for this type of disease transmission.

1. Handwashing and barrier glove-changing should be frequent.
2. An effective environmental disinfection/sanitation program must be established.
3. Restriction of tasks among workers must be practiced to prevent cross-contamination.
4. A program of sanitation training must be provided on a continuous basis.

Comment 1: Instead of merely requiring employees to wash and glove after using the toilet, employees must be encouraged to wash their hands and/or change their barrier gloves frequently. This is particularly true when workers move from one work task to another. For example, if an employee works at the potato fryer, then goes on a break, it should be mandatory that his hands be properly washed and a glove change performed prior to returning to the work station.

Comment 2: Countertops, refrigerator handles, floors, tables, and other equipment should be cleaned frequently with an effective, hard-surface disinfectant. This process will inactivate and remove contaminating pathogenic microorganisms.

Comment 3: Responsibility for specific tasks must be clearly identified among members of the workforce to prevent cross-contamination. Those employees working directly with food (cooking, wrapping, dispensing, etc.) should not have direct hand contact with patrons, handle money, or pass food over a common counter.

Comment 4: Environmental air, countertops and other hard surfaces, and wash sinks should be sampled on a regular basis to determine both species and population numbers of microorganisms in these areas [1,22]. Normal environmental population numbers should be established, monitored, and controlled. One of the most effective ways to monitor microbial population numbers is to utilize quality control charts for each sampling area [24]; that is, environmental samples taken at the various sites (e.g., wash sink, soap tray, wash sink water control knobs, service counter A, etc.) should be analyzed [1].

C. Control Chart Methodology

Control chart methods are very valuable in monitoring and controlling microbial contamination within a food establishment [24,25]. A control chart is a simple graph on which the microbial population numbers at a specific environmental site (sink, countertop, etc.) are recorded over time [24,25]. The control chart's vertical axis most commonly is scaled to microbial population numbers (Fig. 2). It is termed a *mean* control chart because the center horizontal line represents the average target value of the process being measured.

The upper and lower tolerance levels are customarily placed at ± 3 standard deviations from the mean, based on the data generated. The tolerance limits should be tighter than the specification limits (i.e., allowable health code microbial limits). Then, if sample results begin to exceed the tolerance limits, the problem can be corrected before specification limits are exceeded.

An example of how this works may be useful: three separate random samples (obtained by pressing Rodac plates onto the countertop randomly) are taken

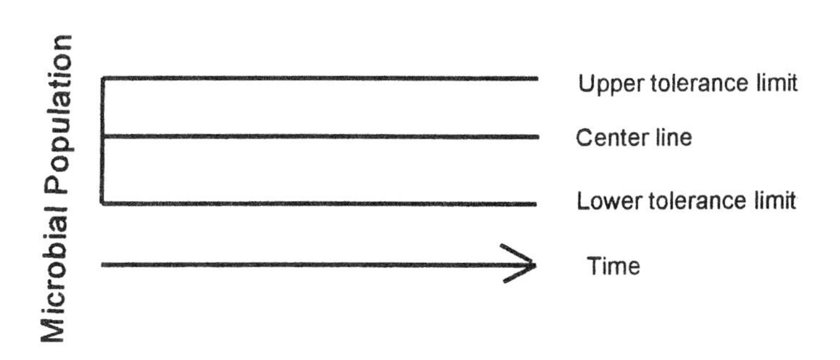

Figure 2 Mean chart of microbial counts per square of countertop after cleaning.

Table 2 Microbial Counts per Square Foot of Countertop After Cleaning

Week	1	2	3	4	5	6
	150	285	72	352	225	63
	272	195	195	495	132	395
	171	162	87	168	151	195
$\overline{X}$	198	214	118	338	169	218

The grand average (average of six weekly averages) is $\overline{\overline{X}}$ = 209, with a standard deviation of 66.76. The standard deviation in control chart methodology customarily does not divide the variability $\Sigma(\overline{X} - \overline{\overline{X}})^2$ by $n - 1$ but, instead, by n [24].

from a stainless steel countertop each week for 6 weeks (Table 2). Results obtained with the three samples per week are then averaged and plotted on a mean control chart (Fig. 3). The microbial tolerance levels are set using an internal quality control standard of, for example, 300 ± 150 microorganisms per square foot. Hence, the set upper tolerance limit is 450 microorganisms per square foot, and the lower tolerance limit is 150. The lower limit is a mathematical convention and is not really applicable to any loss of sanitation control. Additionally, swab samples of the countertop are taken and plated on selective media to screen for coliform bacteria (*E. coli*, *Shigella* spp., and *Salmonella* spp.) [10,22,26].

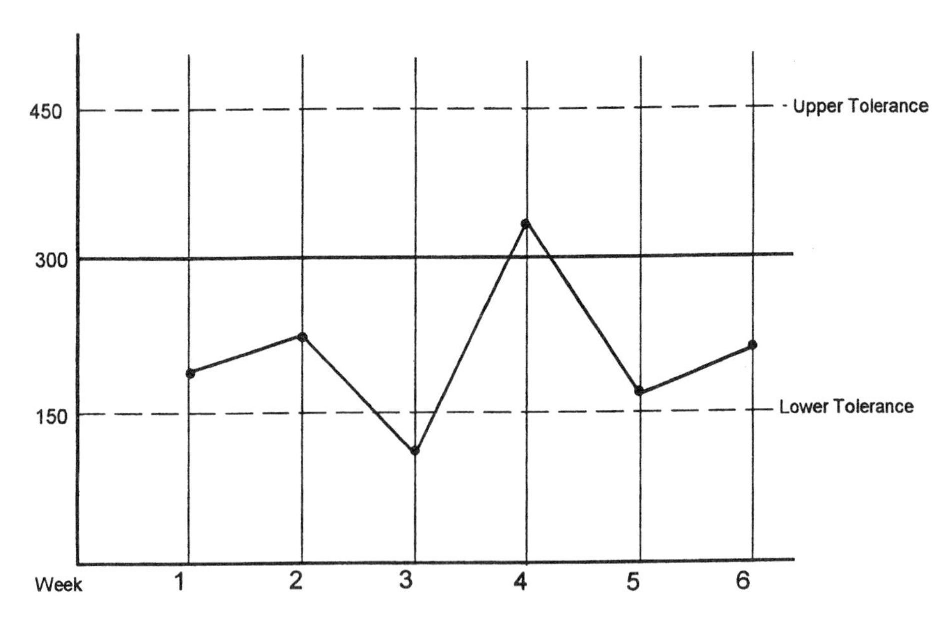

Figure 3 Mean control chart format.

Table 3 Range Values of Countertop

Week	1	2	3	4	5	6
	122[a]	123	123	327	93	332

[a] Range = highest − lowest (e.g., for week 1, 272 − 150 = 122).

The grand average (average of the six weekly averages) is $\overline{X} = 209$, with a standard deviation of 66.76. The data from Table 2, as presented in Figure 3, demonstrate that the environmental cleaning process is within set corporate limits, as well as within regulatory code. Exceeding the lower tolerance level is of no practical concern, because it more than meets the governmental standard.

D. Range Control Chart Methodology

The range chart is simply a graph on which the range (highest value minus the lowest value) of a sample set is plotted. Using the data from Table 2, we can generate the values presented in Table 3.

Corporate limits on the range are set, for example, at an average of 200 microorganisms per square foot, with 300 as the upper limit. Customarily, no requirements exist for the lower limit on the range, because it is irrelevant to the topic of excessive microbial contamination, the issue of concern. Figure 4 presents the plotted range chart.

E. Using the Control Charts

Standard quality control manuals provide detailed information on setting up and evaluating control charts. Generally, three situations can be detected and corrected using the mean and range charts in tandem:

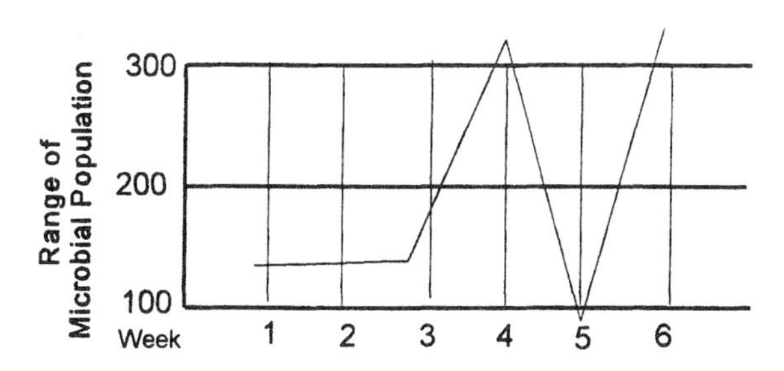

Figure 4 Range chart of microbial counts from countertop.

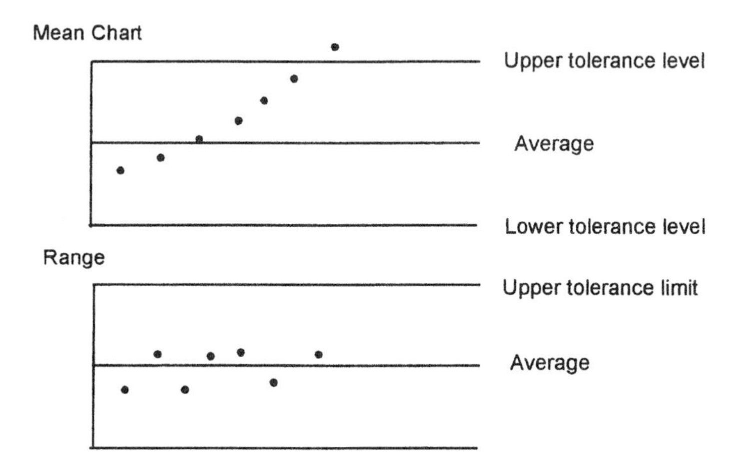

Figure 5 Mean control and range charts: Example 1.

1. Mean control chart values drift and, eventually, if corrective action is not taken, may eventually exceed tolerance limits; range chart data do not shift (no trend apparent) (Fig. 5).
2. The mean control chart does not show a shift (trend), but the range chart does (Fig. 6).

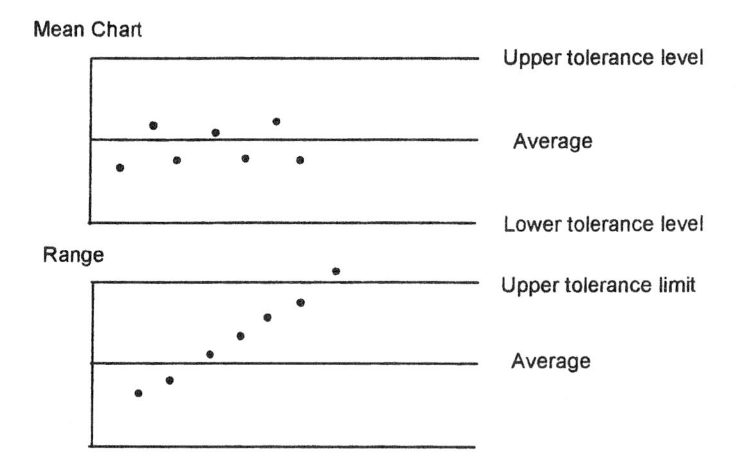

Figure 6 Mean control and range charts: Example 2.

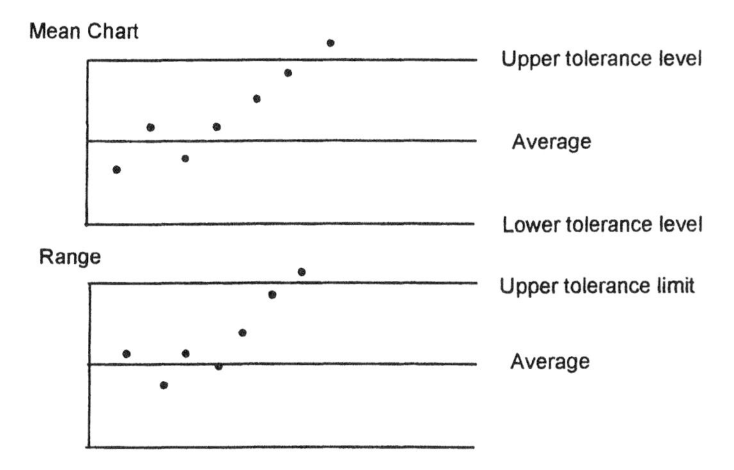

Figure 7 Mean control and range charts: Example 3.

3. Both the mean and range charts show significant shifts over time (Fig. 7).

F. Shift in Mean Chart/Range Chart Constant (Example 1, Fig. 5)

In this situation, the mean chart indicates that microbial count averages have increased over the 7 weeks plotted. Perhaps the cleaning agent is ineffective, the exposure time to the disinfectant is too short, the active ingredient in the disinfectant is degrading too quickly, or there is a seasonal increase in microorganisms. However, because the variability of the microorganisms is constant, according to the range chart, it is doubtful that variability in the cleaning process is the problem.

G. Mean Chart Constant/Range Chart Shifting (Example 2, Fig. 6)

In this situation, the Range Control Chart data suggest either a spotty, hasty cleaning process that has not covered the countertop uniformly or faulty microbial sampling procedures. Notice that as the weeks go by, the range increases, i.e., the variability in the cleaning process is increasing. Because the mean chart is constant, the environmental sanitation program is probably still adequate (e.g., disinfectant is effective, there is no seasonal increase in microorganisms, etc.).

H. Both the Mean and Range Charts Shift (Example 3, Fig. 7)

In this situation, the environmental cleaning process is out of control; the microbial population countertop levels are increasing, as is the variability between samples. This indicates (mean chart) that the disinfectant may no longer be useful in the way it is being used and/or that a seasonal increase in contaminative microorganisms is occurring. Moreover, the range chart suggests that the cleaning procedure is disinfecting some countertop areas more effectively than others—for example, the procedure may be too hasty. Further, there may be a problem with the microbial sampling procedures if cleaning thoroughness and uniformity are thought to be satisfactory.

V. CONCLUSION

What has been presented here is only a general outline for a hand and workplace sanitation program. The responsibility lies with managers, quality control personnel, and other investigators to ''flesh out'' this program with specific technical requirements that are appropriate to their particular work environments and that will assure that patrons are served food that is safe to eat.

REFERENCES

1. DS Paulson. To glove or to wash: a current controversy. Food Quality June/July: 60–63, 1996.
2. RS Guzewich, MP Ross. Evaluation of Risks Related to Microbiological Contamination of Ready-to-Eat Food by Food Preparation Workers and the Effectiveness of Interventions to Minimize Those Risks. Washington, DC: FDA Center for Food Safety and Applied Nutrition, 1999.
3. DS Paulson. A broad-based approach to evaluating topical antimicrobial products. In JM Ascenzi, ed. Handbook for Disinfectants and Antiseptics. New York: Marcel Dekker, 1996.
4. CW Van Way III, CA Buerk. Surgical Skills in Patient Care. St. Louis: Mosby, 1978.
5. M Schaechter, G Medoff, BI Eisenstein. Mechanisms of Microbial Disease. 2nd ed. Baltimore, MD: Williams & Wilkins, 1993.
6. EJ Fendler, MJ Dolan, RA Williams. Handwashing and gloving for food protection, Part I. Dairy Food Environ Sanit 18(12):814–823, 1998.
7. EJ Fendler, MJ Dolan, RA Williams, DS Paulson. Handwashing and gloving for food protection, Part II. Dairy Food Environ Sanit 18(12):824–829, 1998.
8. DS Paulson. Handbook of Topical Antimicrobial Testing and Evaluation. New York: Marcel Dekker, 1999.

9. CA Mims. The Pathogenesis of Infectious Disease. 3rd ed. New York: Academic Press, 1987.
10. K Wilber. A Brief History of Everything. Boston: Shambhala, 1996.
11. RK Merton. Social Theory and Social Construction. Glencoe, IL: Free Press, 1957.
12. C Geertz. Interpretation of Cultures. New York: Basic Books, 1973.
13. WP Anthony, PL Perrewe, KM Kacmar. Human Resource Management, 3rd ed. New York: Harcourt Brace & Company, 1999.
14. G Lenski, P Nolan, J Lenski. Human Societies. 7th ed. New York: McGraw-Hill, 1995.
15. HI Kaplan, BJ Saddock. Comprehensive Textbook of Psychiatry. 6th ed. Philadelphia, PA: Williams and Wilkins, 1995.
16. LR Wolberg. The Technique of Psychotherapy, Part I. 3rd ed. New York: Harcourt Brace and Company, 1977.
17. JV Peterson, B Nisenholz. Orientation to Counseling. 4th ed. Boston: Allyn and Bacon, 1999.
18. RS Kaplan, DP Norton. The Balanced Score Card. Boston: Harvard Business School, 1996.
19. O Fenichel. The Psychoanalytic Theory of Neurosis. New York: Norton, 1999.
20. DH Shapiro, J Astin. Control Therapy. New York: John Wiley and Sons, 1998.
21. JH Norcross, MR Goldfried. Handbook of Psychotherapy Integration. New York: Basic Books, 1992.
22. BA Forbes, DF Sahm, AS Weissfeld. Bailey and Scott's Diagnostic Microbiology. 10th ed. St. Louis: Mosby, 1998.
23. MK Bruch. Methods of testing antiseptics: Antimicrobials used topically in humans and procedures for hand scrubs. In: SS Block, ed. Disinfection, Sterilization and Preservation. 4th ed. Philadelphia: Lea & Febiger, 1991, pp. 1028–1046.
24. BL Grant, NS Leavenworth. Statistical Quality Control. 6th ed. New York: McGraw-Hill, 1988.
25. JM Juran, FM Geyna. Quality Planning and Analysis: From Product Development Through Use. 2nd ed. New York: McGraw-Hill, 1980.
26. WK Joklik, HP Willett, DB Amos, CM Wilfert. Zinsser Microbiology. 20th ed. Norwalk, CT: Appleton & Lange, 1992.
27. WC Frazier, DC Westhof. Food Microbiology. 4th ed. New York: McGraw-Hill, 1988.

18

Handwashing and Gloving for Food Protection: Examination of the Evidence

Eleanor J. Fendler, Michael J. Dolan, and Ronald A. Williams
GoJo Industries, Inc., Akron, Ohio

I. INTRODUCTION

Handwashing has been universally accepted as a means of reducing contact transmission of microorganisms for more than a century. The effectiveness of handwashing as a primary infection-control measure in healthcare has been reviewed and extensively documented [75,76]. Its effectiveness as a means of preventing the transmission of microorganisms to food via the hands is well established in the foodservice industry [85,88]. The U.S. Food and Drug Administration (FDA) Food Code 37 introduced in 1993 requires double handwash, use of a nail brush, and no-hands contact with ready-to-eat food. These requirements reflect the premise that the use of a physical barrier (gloves) on the hands of food-handling personnel prevents the transfer of pathogens to food. However, it is questionable whether there is sufficient scientific evidence to support these requirements. To answer this question, a review of the published literature related to all aspects of handwashing and gloving was undertaken.

II. METHODS

Published studies related to gloving were sought in three areas: (1) the medical literature, including healthcare, infection control, and dermatology, (2) the microbiology literature, and (3) food industry literature, including scientific and trade publications. Information sources used include the following:

1. Dialog search of technical databases
2. Dialog search of trade and industry database
3. Literature review publications and books
4. Bibliography of literature on gloves, E. D. Leach, Associated Enterprises, Inc., 1994

Articles reviewed were classified in the references into seven major headings according to their primary focus: (1) Food—articles on the general problems of food protection [1–32]; (2) Food Code/Regulatory—articles related to the Food Code and Regulatory issues [33–45]; (3) Microbiology–Skin—articles reporting studies of the microflora of the skin under various conditions [46–63]; (4) Microbiology–Efficacy—articles reporting the efficacy of handwashing and gloving in microbial control [64–93]; (5) Microbiology–Gloves—articles related to microorganisms and gloves [94–110]; (6) Gloves–Leakage—articles reporting the methods of testing, incidence, and consequences of glove leakage [111–137]; and (7) Gloves–Contact Dermatitis and Allergy—articles reporting the dermatological consequences of glove contact and skin occlusion [138–226].

III. RESULTS

A. Medical Literature

Extensive medical literature on the effectiveness of handwashing/gloving regimens exists dating from the demonstration of the importance of antisepsis by Semmelweis in 1847 and Lister in 1867. Literature on the relationship between handwashing and risk of infection from microbes has been reviewed by Larson [75,76] for the period from January 1879 to June 1993. This literature clearly demonstrates the effectiveness of handwashing in the reduction of nosocomial infections and the value of handwashing as a primary infection control measure. In 1980, the Centers for Disease Control and Prevention (CDC) began developing a series of guidelines entitled *Guidelines for the Prevention and Control of Nosocomial Infections* and in 1985 released the *Guideline for Handwashing and Environmental Control* [69]. These publications reflect the importance of compliance with handwashing/gloving regimens; however, several studies show that compliance, in general, is poor. The CDC Isolation Guidelines are a reflection of poor handwashing compliance in healthcare facilities. The effectiveness of universal precautions and body substance isolation practices tend to validate the use of gloves in conjunction with a handwashing regimen. Gloves alone have never demonstrated effective control of microbial transmission.

In addition to demonstrating the effectiveness of handwashing and gloving in preventing microbial transmission, the medical literature serves to identify and define issues related to the practice of glove use, such as compliance, importance of handwashing, single use, glove quality standards, leakage/puncture, and irritation/allergy. Recommendations and guidelines for handwashing and gloving regimens have been established and endorsed by regulatory agencies for health care settings [69,76,91]. These guidelines specify thorough handwashing and hand antisepsis with antimicrobial-containing soaps or detergents or with alcohol-based hand rubs whenever hands are soiled and before and after patient contact.

Guidelines for glove use specify the following [76]:

a. Gloves should be used as an adjunct to, not a substitute for, handwashing.
b. Gloves should be used for hand-contaminating activities. Gloves should be removed and hands washed when such activity is completed, when the integrity of the gloves is in doubt, and between patients. Gloves may need to be changed during the care of a single patient, for example, when moving from one procedure to another.
c. Disposable gloves should be used only once and should not be washed for reuse.
d. Gloves made of other materials should be made available for personnel with sensitivity to usual glove material (such as latex).

Research indicates that gloves should be changed after 3–5 minutes when used for prolonged procedures requiring high levels of stress [102,106].

Glove quality standards have been established by the FDA based on a sampling scheme and a quality assurance test known as the ''1000 mL water leak test'' described in the Code of Federal Regulations, 21 CFR 800.20 [127]. The Final Rule was published in December 12, 1990, and became effective March 12, 1991. The acceptable quality level is a maximum failure rate of 2.5% for surgeons' gloves and of 4.0% for patient examination gloves as determined in this water leak test.

Since the recent widespread increase in glove use due to the implementation of universal precautions and body substance isolation, problems associated with glove use, such as leakage and contact dermatitis, have become more evident. Considerable attention and research have been devoted to glove integrity and leakage both before and during use [111–137]. Numerous investigations have revealed a high frequency of defects (up to 60%) in unused latex and vinyl gloves as determined by air inflation–visual detection, air inflation–submersion, electrical conductivity, and fluorescein dye detection.

It is important to note that these high initial defect rates are for presumably high-quality surgical and exam gloves and that the defect rate increases sharply with use [122–126]. Penetration and leakage of gloves destroy their barrier effectiveness to prevent transmission of microorganisms [129–131,133,135]. Yangco and Yangco found that 96.4% of unused gloves allowed the passage of infected fluids [137]. Conclusions and recommendations from these studies include more stringent guidelines for manufacture with verification of compliance and more careful observation of elements of "universal precautions," such as changing gloves after each patient contact and good handwashing before and after using gloves.

The dramatic increase in the number of individuals using gloves following the adoption of the CDC "universal precautions" 16 years ago coincides with an explosion of reports in the medical literature of hypersensitivity reactions to latex products. During this same time period, the reported cases of irritant contact dermatitis, allergic reactions to plastic gloves and glove powder, and occupational asthma precipitated by glove powders and airborne latex allergens rose sharply among healthcare workers and patients.

It is now widely recognized that natural and synthetic latex, rubber additives, plastics (PVC, vinyl), organic pigments in gloves, and glove powders cause allergic contact dermatitis and contact urticaria. (Definitions of dermatological terms are given in Table 1.) Up to 30% of frequent glove wearers are believed to have some degree of acquired hypersensitivity to latex chemicals or proteins and various additives in synthetic gloves [161]. In addition to allergic contact dermatitis, protective disposable gloves can result in irritant contact dermatitis

Table 1 Definitions of Dermatological Terms

Term	Definition
Allergic contact dermatitis	Sensitization or allergic contact dermatitis is a delayed, immunologically mediated response to a chemical. Initial contact with the chemical does not appear to have any effect on the skin, but after a short delay (ca. 5 days) reexposure to the chemical causes an acute inflammatory reaction with an homogeneous "rash."
Irritant contact dermatitis	Irritant dermatitis is a nonimmunological, local inflammatory response at the site on single, repeated, or continuous contact with a chemical. It results in erythema (reddening of the skin) and edema (accumulation of fluid), which is often irregular or patchy in nature.
Contact urticaria	Irritant dermatitis is a nonimmunological, local inflammatory response at the site on single, repeated, or continuous contact with a chemical. It results in erythema (reddening of the skin) and edema (accumulation of fluid), which is often irregular or patchy in nature.

and skin barrier damage [161,208]. Gloves have been shown to result in reduced protective barrier properties of the stratum corneum due to the physical and chemical effects of skin occlusion [173,194,208]. Skin occlusion by hypoallergenic nonlatex gloves for short exposure periods (6 hours/day for 3 days) was found to have a significant negative effect on the barrier function of surfactant-compromised skin but no effect on normal skin over the same time period [201]. However, longer-term exposure (6 hours/day for 14 days) resulted in a significant negative effect on the barrier function of normal skin [202]. It was concluded that occlusion by gloves may be a substantial factor in the pathogenesis of cumulative irritant contact dermatitis [201,202]. Glove usage has also been found to result in all of the clinical types of irritant dermatitis classified by Lammintausta and Maibach [185]: (1) acute irritant dermatitis (primary irritation), (2) irritant reactions, (3) delayed acute irritant dermatitis, (4) cumulative irritant contact dermatitis, (5) traumatic irritant dermatitis, (6) pustular and acneiform dermatitis, (7) nonerythematous irritation, and (8) subjective irritation.

B. Microbiology Literature

The microbiological literature relevant to handwashing and gloving practices includes studies of the transient and resident microflora of the skin, the effects of glove occlusion on skin microflora (Table 2), hand and glove carriage, the transmission of microbes, and the antimicrobial effectiveness of handwashing agents and regimens. The microflora of normal skin including that of food handlers has been well documented [49,54,57,59,62,63]. These resident microbes present in normal skin are generally nonpathogenic and are not responsible for healthcare-related or foodborne illness. Hands and contaminated gloves, how-

Table 2 Classification of Glove Reactions

Reaction	Ref.
Barrier reduction and increased penetration of irritants/allergens by occlusion	173, 194, 201, 202, 208
Irritation form occlusion, friction, and maceration	208
Allergic reactions to glove materials (natural and synthetic latex, plastic, polymer additives, dyes, glove powder)	159
Contact dermatitis	213
Contact urticaria, angioedema, and anaphylaxis	208
Penetration of irritants through gloves	208
Others: endotoxin reactions, ethylene oxide, chemical leukoderma	208

ever, are a primary vector for transmission of **transient** microbes, both pathogenic and nonpathogenic, acquired from the environment [58,61].

Occlusion of skin by gloves affects the microbial flora on hands by greatly increasing growth rates and populations [47,48,50,59]. Price found that ''beneath rubber gloves, bacteria remaining on the skin multiply rapidly, their numbers doubling every forty (40) minutes if the hands are dry or every fifty (50) minutes if the gloves have been put on wet. If gloves are worn long enough, the cutaneous [transient] flora may increase until it exceeds by far the ordinary flora. I found that on one occasion the bacterial count of my hands and arms had increased to more than 31,000,000'' [59]. Microbiological studies have also shown that viable bacteria emerge through pinholes in surgeons' gloves [56].

An enormous number of studies have been devoted to the antimicrobial effectiveness of handwashing products and their role in preventing the transmission of pathogenic microorganisms [55,56,71,72,75,76,79]. The literature clearly demonstrates that antimicrobial handwashing agents shown can be highly effective in killing pathogens and can provide residual antimicrobial activity over a period of several hours. The importance of handwashing when using gloves is widely recognized and accepted in the healthcare field [55,76]. Antimicrobial and antiseptic products have been found to result in greater reduction in microorganisms after 3 hours of wearing gloves than immediately following the antiseptic treatment, whereas microbial counts increased when hands were washed with nonantimicrobial soap [56].

C. Food Industry Literature

Although the healthcare setting has been the primary focus of attention for research and field studies of antimicrobial efficacy, some studies have been carried out that demonstrate the effectiveness of handwashing with antimicrobial products in the food industry [65–68,69,74,77,80,82–85,87–90,92]. New York State instituted the first statewide policy of ''no bare-hand contact with ready-to-eat foods'' [41]. The state's rationale for this policy, considered to be radical by many in both government and industry, was described by Guzewich in a presentation at the 1995 annual meeting of the International Association of Milk, Food and Environmental Sanitarians. The policy is based on correcting the problem caused by food workers working when they are ill, not properly washing their hands, and preparing ready-to-eat food, thereby spreading bacterial and viral diseases [14,15]. In spite of the ''no bare-hand contact with ready-to-eat food'' policy in the Food Code, there is no direct information on the effectiveness of hand hygiene and gloving regimens in the food industry. All of the information available to date is anecdotal. Additionally, no clean epidemiology data have been found. The recent Idaho hepatitis case serves as a clear illustration. The food industry also lacks glove quality standards. Studies indicate that the gloves used in food service are generally of poor quality and have higher leakage rates than gloves used in healthcare. Although there is a keen awareness of the importance of food protection and the risks of microbial contamination and transmission, there is also a low general awareness of the importance of hand hygiene regimens by food handlers. Education and training programs and measures to promote compliance are needed in the food industry.

IV. DISCUSSION

The premise that the use of a physical barrier (gloves) on the hands of food-handling personnel prevents transfer of pathogens to food is intuitively attractive. At first glance it appears to be a simple solution, and it can be effective when practiced as part of a hand hygiene regimen, as evidenced by the healthcare experience. There are, however, numerous disadvantages and complications involved in the use of gloves for food protection from contamination by food handlers. Counterintuitive effectiveness issues arise from gloving misuse practices, such as the lack of compliance with single use and a low frequency of changing gloves. Effectiveness is also compromised by poor glove quality and the resulting high defect and leakage rates. Considering the glove to be protective can lead to low handwashing compliance and accelerated microbial growth on the occluded (gloved) hands. The glove functions as a second skin and can easily become contaminated from the activities of well or ill workers. Gloves, unlike hands washed with antimicrobial skin cleansers with persistence, lack the ability to continue killing microbes on contact. Other disadvantages of gloving, in addition to the questionable effectiveness, is the cost and the clumsiness of some manipulations when wearing gloves. An additional complication of gloving is the high potential for allergic reactions (contact dermatitis and urticaria) to latex and plastic gloves in food handlers and customers alike. Occlusion of the skin by gloves not only leads to enhanced microbial growth but also results in a decrease in skin barrier function and irritant contact dermatitis.

From this literature review, it appears that the current status of gloving is the following:

1. Gloving is a well-established infection control practice in healthcare environments.
2. Gloving is generally recognized as an adjunct to, not a replacement for, handwashing.
3. The value of gloving in food-handling settings is assumed but has not been proven.
4. Indirect data are available that indicate the potential for health hazards from gloving.

5. A total regimen for hand hygiene needs to be considered, and standards need to be established to ensure safe food handling.

V. CONCLUSION

This literature review clearly demonstrates that there is insufficient scientific evidence to support the premise that the use of a physical barrier (gloves) on the hands of food-handling personnel prevents the transfer of pathogens to food and, consequently, to support the requirement for no-hand contact with ready-to-eat food. It is our recommendation that gloving studies be performed under food service conditions to establish data to support the most effective hand hygiene regimens for food protection and minimized risk of health hazards.

REFERENCES

1. HACCP Highlights, hazards of man handling food. Food Prot Rep 6(4):1–2, 1990.
2. The safe foodhandler. In: Applied Foodservice Sanitation. 4th ed. New York: John Wiley & Sons, New York, 1992, pp. 60–76.
3. NH Bean, PM Griffin, JS Goulding, CB Ivey. Foodborne disease outbreaks, 5-year summary (1983–1987). J Food Prot 53(8):711–728, 1990.
4. FL Bryan. Risk of practices, procedures and processes that lead to outbreaks of foodborne diseases. J Food Prot 51(8):663–673, 1988.
5. FL Bryan, MP Doyle. Health risks and consequences of *Salmonella* and *Campylobacter jejuni* in raw poultry. J Food Prot 58(3):326–344, 1995.
6. JC deWit, B Broekhuizen, EH Kampelmacher. Cross-contamination during the preparation of frozen chickens in the kitchen. J Hyg (Cambridge) 83:27–32, 1979.
7. EJ Dyett. Hygiene and meat products. In: A Fox, ed.. Hygiene and Food Production. Baltimore, MD: The Williams & Wilkins Company, 1971, pp. 76–84.
8. MB Elmsley. Safety and efficiency go hand-in-glove. Baking Ind J 7(4):17, 26, 1974.
9. Foodborne illness in the United States. Fed Reg 60(23):6780–6783, 1995.
10. CW Felix. Handwash compliance. Food Prot Rep 11(7–8):1A, 1995.
11. L Gapay. A new war on tainted food, better inspections, cleanliness and thorough cooking best defense. AARP Bull 36(5):2, 8, 1995.
12. ES Geller, SL Eason, JA Phillips, MD Pierson. Interventions to improved sanitation during food preparation. J Organ Behav Manage 2(3):229–240, 1980.
13. CA Genigeorgis, D Dutulescu, JF Garayzabal. Prevalence of *Listeria* spp. in poultry meat at the supermarket and slaughterhouse level. J Food Prot 52(9):618–624, 1989.
14. JJ Guzewich. Bare hand contact with food, why isn't handwashing good enough? Abstract, IAMFES Annual Meeting, June 30–August 2, Program and Abstract Book, 1995, p. 84.

15. JJ Guzewich. The anatomy of a ''glove rule.'' Environ News Dig 61(2):4–13, 1995.
16. JM Jay. Foods with low numbers of microorganisms may not be the safest foods or, why did human listeriosis and hemorrhagic colitis become foodborne diseases? Dairy Food Environ Sanit 15(11):674–677, 1995.
17. R Martin. Food-borne disease threatens industry. Nation's Restaurant News January 8:27, 30, 1990.
18. R Martin. Food-borne illness cases underscore safety issues. Nation's Restaurant News 29(12):1, 82, 1995.
19. SA Martin, TS Wallsten, ND Beaulieu. Assessing the risk of microbial pathogens: application of a judgment-encoding methodology. J Food Prot 58(3):289–295, 1995.
20. B McCarthy. Handle with care, train staff in safe food handling to prevent food borne illness. Restaurant Inst April 22:72–74, 82, 1992.
21. G Ravenhill. Hygiene and health—the employer's responsibility. Food Flavour Ingredients Packaging Proc 1(10):38–39, 1980.
22. E Scott, SF Bloomfield. An in-use study of the relationship between bacterial contamination of food preparation surfaces and cleaning cloths. Lett Appl Microbiol 16:173–177, 1993.
23. RA Shooter, EM Cooke, MC Faiers, AL Breaden, SM O'Farrell. Isolation of *Escherichia coli*, *Pseudomonas aeruginosa*, and *Klebsiella* from food in hospitals, canteens, and schools. Lancet August 21:390–392, 1971.
24. JL Smith, PM Fratamico. Factors involved in the emergence and persistence of food-borne diseases. J Food Prot 58(6):696–708, 1995.
25. GA Steel. Operators—personal aspects of hygiene. In: R Jowitt, ed. Hygienic Design and Operation of Food Plant. Westport, CT: The AVI Publishing Company, Inc., Westport, 1980, pp. 227–234.
26. K Straus. In-house safety inspections. Restaurant Inst 103(24):73, 76, 78, 1993.
27. R Sympson. Epidemic. Restaurant Bus 94(7):84, 86, 92, 98, 1995.
28. PI Tarr. *Escherichia coli* O157:H7: overview of clinical and epidemiology issues. J Food Prot 57(7):632–636, 1994.
29. ECD Todd. Factors that contributed to foodborne disease in Canada, 1973–1977. J Food Prot 46(8):737–747, 1983.
30. ECD Todd. Economic loss from foodborne disease outbreaks associated with foodservice establishments. J Food Prot 48(2):169–180, 1985.
31. ECD Todd. Economic loss from foodborne disease and non-illness related recalls because of mishandling by food processors. J Food Prot 48(7):621–633, 1985.
32. SE Weingold, JJ Guzewich, JK Fudala. Use of foodborne disease data for HACCP risk assessment. J Food Prot 57(9):820–830, 1994.
33. Guidelines for the prevention of the transmission of viral hepatitis, Type A, in the food service area. Minneapolis: Minnesota Department of Health, 1990.
34. Where the NRA parts company with FDA on the Food Code. Food Protection Report April, 1995, pp. 6–7.
35. NRA seeks more scientific basis for hand washing rules. Food Protection Report April, 1995, pp. 6–7.
36. Sanitizing solutions. Title 21 Code of Federal Regulations part 178.1010. Washington, DC: U.S. Government Printing Office, 1994, pp. 312–319.

37. Food code. Food and Drug Administration & U.S. Public Health Service. Washington, DC: U.S. Government Printing Office, 1995.
38. Idaho unicode rejects glove rule. Food Protection Inside Report, 1995.
39. T Fahey. Good manufacturing practices (GMP) are critical for non-production employees too! Dairy Food Environ Sanit 14(1):8–10, 1994.
40. RC Hodge. Personal communication to John R. Keenan. Food and Drug Administration OTC Drug Compliance Branch, Rockville, MD, 1994.
41. P Romeo. NY lawmakers ban hand-to-food contact. Nation's Restaurant News 25(31):1, 72, 1991.
42. R Ruggless. Industry debates new FDA food code standards. Nation's Restaurant News 28(23):48, 64, 1994.
43. GJ Silverman. Establishing and maintaining microbiological standards in food service systems. In: GE Livingston, CM Chang, eds. Food-Service Systems: Analysis, Design, and Implementation. New York: Academic Press, Inc., 1979, pp. 379–404.
44. U.S. Department of Agriculture. Guidelines for obtaining authorization of compounds to be used in meat and poultry plants. Food Safety and Inspection Service Agriculture Handbook No. 562. Washington, DC: U.S. Government Printing Office.
45. U.S. Department of Agriculture. List of proprietary substances and nonfood compounds. Food Safety and Inspection Service, Miscellaneous Publication Number 1419. Washington, DC: U.S. Government Printing Office, 1993.
46. R Aly, HI Maibach, HR Shinefield, WG Strauss. Survival of pathogenic microorganisms on human skin. J Invest Dermatol 58(4):205–210, 1972.
47. R Aly, C Shirley, B Cunico, HI Maibach. Effect of prolonged occlusion on the microbial flora, pH, carbon dioxide and transepidermal water loss on human skin. J Invest Dermatol 71(6):378–381, 1978.
48. R Aly. Effect of occlusion on microbial population and physical skin conditions. Semin Dermatol 1(2):137–142, 1982.
49. R Aly. Normal flora of skin and its significance. In: DJ Dennis, RL Dobson, J McGuire, eds. Clinical Dermatology. Philadelphia: Harper and Row, 1985, pp. 1–5.
50. DJ Bibel, JR LeBrun. Changes in cutaneous flora after wet occlusion. Can J Microbiol 21:496–500, 1975.
51. A Brandberg, I Andersson. Preoperative whole body disinfection by shower bath with chlorhexidine soap: effect on transmission of bacteria from skin flora. In: HI Maibach, R Aly, eds. Skin Microbiology Relevance to Clinical Infection. New York: Springer-Verlag, pp. 92–97, 1981.
52. J Davies, JR Babb, GAJ Ayliffe. The effects on the skin flora of bathing with antiseptic solutions. J Antimicrob Chemother 3:473–481, 1977.
53. RJ Holt. Aerobic bacterial counts on human skin after bathing. J Med Microbiol 4:319–327, 1971.
54. MP Horwood, VA Minch. The numbers and types of bacteria found on the hands of food handlers. Food Res 16:133–136, 1951.
55. E Larson. Hand washing: it's essential even when you use gloves. Am J Nurs 934–939, 1989.

56. EJL Lowbury, HA Lilly, GAJ Ayliffe. Preoperative disinfection of surgeon's hands: use of alcoholic solutions and effects of gloves on skin flora. Br Med J 4: 369–372, 1974.
57. RR Marples. Antibacterial cosmetics and the microflora of human skin. In: Developments in Industrial Microbiology. Washington, DC: American Institute of Biological Sciences, 1971, pp. 178–187.
58. JVS Pether, RJ Gilbert. The survival of salmonellas on finger-tips and transfer of the organisms to foods. J Hyg (Cambridge) 69:673–681, 1971.
59. PB Price. New studies in surgical bacteriology and surgical technic. J Am Med Assoc 111(22):1993–1996, 1938.
60. G Rebell, DM Pillsbury, M Phalle, D Ginsburg. Factors affecting the rapid disappearance of bacteria placed on the normal skin. J Invest Dermatol 14:247–264, 1950.
61. E Scott, SF Bloomfield. The survival and transfer of microbial contamination via cloths, hands, and utensils. J Appl Bacteriol 68:271–278, 1990.
62. R Seligmann, S Rosenbluth. Comparison of bacterial flora on hands of personnel engaged in non-food and in food industries: a study of transient and resident bacteria. J Milk Food Technol 38(11):673–677, 1975.
63. REO Williams. Healthy carriage of *Staphylococcus aureus*: its prevalence and importance. Bacteriol Rev 27:56–71, 1963.
64. JL Bryan, J Cohran, EL Larson. Handwashing: a ritual revisited. In: WA Rutala, ed. Chemical Germicides in Health Care. Washington, DC: Association for Professionals in Infection Control and Epidemiology, Inc., 1995, pp. 163–178.
65. DO Cliver, KD Kostenbader Jr. Disinfection of virus on hands for prevention of foodborne disease. Int J Food Microbiol 1:75–87, 1984.
66. FD Crisley, MJ Foter. The use of antimicrobial soaps and detergents for hand washing in food service establishments. Milk Food Technol 28(1):278–284, 1965.
67. JC deWit. The importance of hand hygiene in contamination of food. Antonie van Leeuwenhoek 51:523–527, 1985.
68. JE Foulke. How to outsmart dangerous *E. coli* strain. FDA Consumer January/February:7–11, 1994.
69. JS Garner, MS Favero. CDC guidelines for the prevention and control of nosocomial infection: guideline for handwashing and hospital environmental control, 1985. Am J Infect Control 14(3):110–115, 126–129, 1986.
70. RRM Gershon, D Vlahov, SA Felknor, D Vesley, PC Johnson, GL Delclos, LR Murphy. Compliance with universal precautions among healthcare workers at three regional hospitals. Am J Infect Control 23(4):225–236, 1995.
71. JP Gobetti, M Cerminaro, C Shipman Jr. Hand asepsis: the efficacy of different soaps in the removal of bacteria from sterile, gloved hands. J Am Dental Assoc 113:291–292, 1986.
72. D Goldmann, E Larson. Hand-washing and nosocomial infections. N Engl J Med 327(2):120–122, 1992.
73. AC Hamilton, ed. Guidelines for Protecting the Safety and Health of Healthcare Workers. Department of Health and Human Services (NIOSH & CDC), DHHS (NIOSH) Publication #88-119, 1988.
74. TJ Humphrey, KW Martin, A Whitehead. Contamination of hands and work sur-

faces with *Salmonella enteritidis* PT4 during the preparation of egg dishes. Epidemiol Infect 113:403–409, 1994.
75. E Larson. A causal link between handwashing and risk of infection? Examination of the evidence. Infect Control Hosp Epidemiol 9(1):28–36, 1988.
76. EL Larson. APIC guideline for handwashing and hand antisepsis in health care settings. Am J Infect Control 23(4):251–269, 1995.
77. LA Lee, SM Ostroff, HB McGee, DR Johnson, FP Downes, DN Cameron, NH Bean, PM Griffin. An outbreak of *shigellosis* at an outdoor music festival. Am J Epidemiol 133(6):608–615, 1991.
78. JA Lopes. Food- and water-infective microorganisms. In: SS Block, ed. Disinfection, Sterilization, and Preservation. 4th ed. Philadelphia: Lea & Febiger, 1991, pp. 773–790.
79. EJL Lowbury, HA Lilly. Disinfection of the hands of surgeons and nurses. Br Med J 14 May:1445–1450, 1960.
80. ML Miller, LA James-Davis, LE Milanesi. A field study evaluating the effectiveness of different hand soaps and sanitizers. Dairy Food Environ Sanit 14(3):155–160, 1994.
81. SWB Newsom, C Rowland. Application of the hygienic hand-disinfection test to the gloved hand. J Hosp Infect 14:245–247, 1989.
82. L Oblinger, ed. Bacteria Associated with Foodbourne Diseases. Food Technol April:181–200, 1988.
83. DS Paulson. Evaluation of three handwash modalities commonly employed in the food processing industry. Dairy Food Environ Sanit 12(10):615–618, 1992.
84. DS Paulson. Variability evaluation of two handwash modalities employed in the food processing industry. Dairy Food Environ Sanit 13(6):332–335, 1993.
85. DS Paulson. A comparative evaluation of different hand cleansers. Dairy Food Environ Sanit 14(9):524–528, 1994.
86. DS Paulson. A statistical approach to evaluating the effectiveness of hand-cleansing products used in the food-processing industry. Dairy Food Environ Sanit 16(6): 389–392, 1996.
87. JVS Pether, RJ Gilbert. The survival of salmonellas on finger-tips and transfer of the organisms to foods. J Hyg (Cambridge) 69:673–681, 1971.
88. L Restaino, CE Wind. Antimicrobial effectiveness of handwashing for food establishments. Dairy Food Environ Sanit 10(3):136–141, 1990.
89. AZ Sheena, ME Stiles. Efficacy of germicidal handwash agents against transient bacteria inoculated onto hands. J Food Prot 46(8):722–727, 1983.
90. AZ Sheena, ME Stiles. Comparison of barrier creams and germicides for hand hygiene. J Food Prot 46(11):943–946, 1983.
91. BP Simmons. Guideline for Prevention of Nosocomial Pneumonia. Atlanta: Centers for Disease Control, 1982.
92. ME Stiles, AZ Sheena. Efficacy of germicidal handwash agents in use in a meat processing plant. J Food Prot 50(4):289–295, 1987.
93. RP Wenzel, ed. Prevention and Control of Nosocomial Infections. Baltimore: Williams & Wilkens, 1993.
94. Glove "beepers" reassure surgeons, but epidemiologists remain skeptical. Hosp Infect Control 21(6):73–76, 1994.

95. Phoenix Medical Technology, Inc.; Filing of Food Additive Petition. Federal Register No. 56, part 215. Washington, DC: U.S. Government Printing Office, 1994, p. 56656.
96. L Borgatta, M Fisher, N Robbins. Hand protection and protection from hands: handwashing, germicides, and gloves. Women Health 15(4):77–92, 1989.
97. S Brooks. To fight hepatitis, put on the gloves: in parts of Idaho and Colorado, it's mandatory. Restaurant Bus January 1:18, 1995.
98. RDA Dodds, SGE Barker, NH Morgan, DR Donaldson, MH Thomas. Self-protection in surgery: the use of double gloves. Br J Surg 77:219–220, 1990.
99. BN Doebbeling, MA Pfaller, AK Houston, RP Wenzel. Removal of nosocomial pathogens from the contaminated glove. Ann Int Med 109:394–398, 1988.
100. PM Doyle, A Alvi, R Johanson. The effectiveness of double-gloving in obstetrics and gynecology. Br J Obstet Gynecol 99:83–84, 1992.
101. C Felix. Putting on the gloves to fight foodborne disease. Nation's Restaurant News August 3:36, 1992.
102. DM Korniewicz, M Kirwin, K Cresci, C Markut, E Larson. In-use comparison of latex gloves in two high-risk units: surgical intensive care and acquired immunodeficiency syndrome. Heart Lung 21:81–84, 1992.
103. H Matta, AM Thompson, JB Rainey. Does wearing two pairs of gloves protect operating theatre staff from skin contaminations? Br Med J 297:597–598, 1988.
104. S McCue, E Berg, E Saunders. Efficacy of double-gloving as a barrier to microbial contamination during total joint arthroplasty. J Bone Joint Surg 63-A(5):811–813, 1981.
105. LK Nesse. Safeskin's strategy fits like a glove. South Fla Bus J July 15–21:1B–2B, 1994.
106. RJ Olsen, P Lynch, MB Coyle, J Cummings, T Bokete, WE Stamm. Examination gloves as barriers to hand contamination in clinical practice. J Am Med Assoc 270(3):350–353, 1993.
107. M Ritter, M French, H Eitzen. Evaluation of microbial contamination of surgical gloves during actual use. Clin Orthop Res 117:303–306, 1976.
108. R Van Warner. Taking off the gloves: sanitation starts with clean hands. Nation's Restaurant News May 4:23, 1992.
109. CW Walter, RB Kundsin. The bacteriologic study of surgical gloves from 250 operations. Surg Gynecol Obstet November: 949–952, 1969.
110. J Zimakoff, M Stormark, SO Larsen. Use of gloves and handwashing behavior among healthcare workers in intensive care units. J Hosp Infect 24:63–67, 1993.
111. M Aggarwal, TT Manson, B VanMeter, JG Thacker, RF Edlich. Biomechanics of surgical glove expansion. J Long-Term Effects Med Implants 4(2&3):133–140, 1994.
112. Study raises questions about glove permeability. CDC AIDS Weekly 11:19–20, 1988.
113. SG Arnold, JE Whitman, CH Fox, MH Cotiler-Fox. Latex gloves not enough to exclude viruses. Nature 335:19, 1988.
114. WC Beck. Barrier breach of surgical gloves. J Long-Term Effects Med Implants 4(2&3):127–132, 1994.

115. B Bennett, P Duff. The effects of double gloving on frequency of glove perforations. Obstet Gynecol 78:1019–1022, 1991.
116. SJ Brough, TM Hunt, WW Barrie. Surgical glove perforations. Br J Surg 75:317, 1988.
117. AG Dalgleish, M Malkovsky. Surgical gloves as a mechanical barrier against human immunodeficiency viruses. Br J Surg 75:171–172, 1988.
118. J DeGroot-Kosolcharoen, JM Jones. Permeability of latex and vinyl gloves to water and blood. Am J Infect Control 17:196–201, 1989.
119. J DeGroot-Kosolcharoen. Pandemonium over gloves: use and abuse. Am J Infect Control 19:225–227, 1991.
120. RDA Dodds, PJ Guy, AM Peacock, SR Duffy, SGE Barker, MH Thomas. Surgical glove perforation. Br J Surg 75:966–968, 1988.
121. LP Jordan, MF Stowers, EG Trawick, AB Theis. Glutaraldehyde permeation: choosing the proper glove. Am J Infect Control 24:67–69, 1996.
122. JN Katz, JP Gobetti, C Shipman Jr. Fluorescein dye evaluation of glove integrity. J Am Dent Assoc 118:327–331, 1989.
123. DM Korniewicz, B Laughon, A Butz, E Larson. Integrity of vinyl and latex procedure gloves. Nurs Res 38(3):144–146, 1989.
124. DM Korniewicz, M Kirwin, K Cresci, E Larson. Leakage of latex and vinyl exam gloves in high and low risk clinical settings. Am Indust Hyg Assoc 54:22–26, 1993.
125. HR Kotilainen, JP Brinker, JL Avato, NM Gantz. Latex and vinyl examination gloves: quality control procedures and implications for healthcare workers. Arch Int Med 149:2749–2753, 1989.
126. HR Kotilainen, JL Avato, NM Gantz. Latex and vinyl nonsterile examination gloves: status report on laboratory evaluation of defects by physical and biological methods. Appl Environ Microbiol 56(65):1627–1630, 1990.
127. CD Lytle, WH Cyr, RF Carey, DG Shombert, BA Herman, et al. Standard quality control testing and virus penetration. In: GA Mellsrom, JE Wahlberg, HI Maibach, eds. Protective Gloves for Occupational Use. Boca Raton, FL: CRC Press, 1994, pp. 109–127.
128. GA Mellstrom, K Wrangsjö, JE Wahlberg, B Fryklund. The value and limitations of protective gloves in medical health service: Part II. Dermatol Nurs 8(4):287–292, 1996.
129. JM Miller, CS Collier, NM Griffith. Permeability of surgical rubber gloves. Am J Surg 124:57–59, 1972.
130. M Nakazawa, M Sato, K Mizuno. Incidence of perforations in rubber gloves during ophthalmic surgery. Ophthalmol Surg 15(3):236–240, 1984.
131. LL Otis, JA Cottone. Prevalence of perforations in disposable latex glove during routine dental treatment. J Am Dent Assoc 118:321–324, 1989.
132. DW Ramsing, A Fullerton. Permeability of protective gloves to sodium lauryl sulfate. A release cambers used as an in vitro test system. Skin Res Technol 2:37–39, 1996.
133. AL Reingold, MA Kane, AW Hightower. Failure of gloves and other protective devices to prevent transmission of hepatitis B virus to oral surgeons. J Am Med Assoc 259(17):2558–2560, 1988.

134. R Russell, F Roque, F Miller. A new method for detection of the leaky glove. Arch Surg 93:245–249, 1966.
135. N Skaug. Micropunctures of rubber gloves used in oral surgery. Int J Oral Surg 5: 220–225, 1976.
136. N White, K Taylor, A Lyszkowski, C Morris. Dangers of lubricants used with condoms. Nature 335:19, 1988.
137. BG Yangco, NF Yangco. What is leaky can be risky: a study of the integrity of hospital gloves. Infect Control Hosp Epidemiol 10(12):553–556, 1989.
138. 5 Hazards of surgical glove powder. Biogel Information Pamphlet, Regent Hospital Products, London International House, U.K.
139. T Agner, J Serup. Time course of occlusive effects on skin evaluated by measurement of transepidermal water loss (TEWL). Contact Dermatitis 28:6–9, 1993.
140. S Amin, C Tanglertsampan, HI Maibach. Contact urticaria syndrome: 1997. Am J Contact Dermatitis 8(1):15–19, 1997.
141. D Assavle, C Cicioni, P Perno, P Lisi. Contact urticaria and anaphylactoid reaction from cornstarch surgical glove powder. Contact Dermatitis 19:61–78, 1988.
142. LM Barclay. Developments in low protein prevulcanized latex materials. In: Latex Protein Allergy: The Latest Position. Brickendonbury, United Kingdom: Crain Communications Ltd, Rubber Consultants, 1995, pp. 41–53.
143. RD Barlow, S Rosenbaum. Hospitals on yellow alert over allergic reactions to latex gloves. Hosp Mater Manage 16(9):10–13, 1991.
144. X Baur, D Jäger. Airborne antigens from latex gloves. Lancet 335:127–132, 1990.
145. X Baur, J Ammon, Z Chen, U Beckmann, AB Czuppon. Health risk in hospitals through airborne allergens for patients presensitized to latex. Lancet 342(6):1148–1149, 1993.
146. X Baur. Characterization and inactivation of a major latex allergen. In: Latex Protein Allergy: The Latest Position. Brickendonbury, United Kingdom: Crain Communications Ltd, Rubber Consultants, 1995, pp. 17–18.
147. DH Beezhold, WC Beck. Surgical glove powders bind latex antigens. Arch Surg 127:1354–1357, 1992.
148. DH Beezhold. Measurement of latex protein by chemical and immunological methods. In: Latex Protein Allergy: The Present Position. Brickendonbury, United Kingdom: Crain Communications Ltd, Rubber Consultants, 1993, pp. 25–31.
149. DH Beezhold, MF Fay, GL Sussman. Scientific assay methods as the basis for glove selection. J Long-Term Effects Med Implants 4(2&3):103–125, 1994.
150. DH Beezhold. Identification of latex protein allergens. In: Latex Protein Allergy: The Latest Position. Brickendonbury, United Kingdom: Crain Communications Ltd, Rubber Consultants, 1995, pp. 19–27.
151. PC Belvedere, DL Lambert. Negative effects of powdered latex gloves in clinical dentistry. J Long-Term Effects Med Implants 4(2&3):119–125, 1994.
152. I Bodycoat. Manufacture of hypoallergenic rubber products. In: Latex Protein Allergy: The Present Position. Brickendonbury, United Kingdom: Crain Communications Ltd, Rubber Consultants, 1993, pp. 43–45.
153. R Brehler, A Rütter. Food allergy in patients with IgE-mediated hypersensitivity to latex. Allergologie 9:379–382, 1995.

154. T Carrillo, M Cuevas, T Murioz, M Hinojosa, I Moneo. Contact urticaria and rhinitis from latex surgical gloves. Contact Dermatitis 15:69–72, 1986.
155. L Conde-Salazar, E del-Rio, D Guimaraens, AG Domingo. Type IV allergy to rubber additives: a 10-year study of 686 cases. J Am Acad Dermatol 29:176–180, 1993.
156. A Dean. The prevalence of latex glove allergy in healthcare workers (nurses). Abstract #123, International Conference on the Prevention of Contact Dermatitis, Zurich, October 4–7, 1995, p. 474.
157. K Ellis. Type I allergy to latex products—incidence, management and future prospects. In: Latex Protein Allergy: The Latest Position. Brickendonbury, United Kingdom: Crain Communications Ltd, Rubber Consultants, 1995, pp. 29–31.
158. T Estlander, R Jolanki, and L Kanerva. Dermatitis and urticaria from rubber and plastic gloves. Contact Dermatitis 14:20–25, 1986.
159. T Estlander, R Jolanki, L Kanerva. Allergic contact dermatitis from rubber and plastic gloves. In: GA Mellsrom, JE Wahlberg, HI Maibach, eds. Protective Gloves for Occupational Use. Boca Raton, FL: CRC Press, 1994, pp. 221–239.
160. MF Fay. Hand dermatitis—the role of gloves. Assoc Operating Room Nurs J 54(3): 451–467, 1991.
161. M Fay. Risk analysis as the basis for surgical glove selection. J Long-Term Effects Med Implants 4(2&3):141–155, 1994.
162. M Fay. Latex allergy—a recent epidemic on medical emergencies. Clin Focus 1(1): 1, 1995.
163. M Fay. Powdered gloves starch in surgical wounds. Clin Focus 1(1):2, 1995.
164. M Fay. Airborne latex allergens. Clin Focus 1(1):3, 1995.
165. EA Field, CM King. Skin problems associated with routine wearing of protective gloves in dental practice. Br Dentistry J 168:281, 1990.
166. AA Fisher. Management of allergic contact dermatitis due to rubber gloves in health and hospital personnel. CUTIS 47:301–302, 1991.
167. AA Fisher. Allergic contact reactions in health personnel. Allergy Clin Immunol 90:730–738, 1992.
168. AA Fisher. Association of latex and food allergy. CUTIS 52:70, 1993.
169. AA Fisher. The latest in latex allergy. CUTIS 53:69–75, 1994.
170. AA Fisher. Standard and special tests for the barrier integrity of medical gloves Part I: The use and abuse of vinyl gloves by healthcare workers allergic to latex. CUTIS 59(2):61–62, 1997.
171. AA Fisher. The latest in latex allergy. CUTIS 59(4):168–170, 1997.
172. T Fuchs, HJ Gonzl. Clinical manifestation and diagnosis of natural latex allergy. Allergologie 9:350–357, 1995.
173. CJ Graves, C Edwards, R Marks. The effects of protective occlusive gloves on stratum corneum barrier properties. Contact Dermatitis 33:183–187, 1995.
174. CP Hamann. Natural rubber latex protein sensitivity in review. Am J Contact Dermatitis 4(1):4–21, 1993.
175. CP Hamann, SA Kick. Update: immediate and delayed hypersensitivity to natural rubber latex. CUTIS 52(5):307–311, 1993.
176. A Heese, JV Hintzenstern, KP Peters, HU Koch, OP Hornstein. Allergic and irritant

reactions to rubber gloves in medical health services. Spectrum, diagnostic approach, and therapy. J Am Acad Dermatol 25(5):831–839, 1991.
177. A Heese, KP Peters, HU Koch, & OP Hornstein. Allergy against latex gloves. Allergologie 9:358–365, 1995.
178. JV Hintzenstern, A Heese, HU Koch, KP Peters, OP Hornstein. Frequency, spectrum, and occupational relevance of Type IV allergies to rubber chemicals. Contact Dermatitis 24:244–252, 1991.
179. W Huang, M Vistins. Alternative materials for medical goods. In: Latex Protein Allergy: The Present Position. Brickendonbury, United Kingdom: Crain Communications Ltd, Rubber Consultants, 1993, pp. 47–58.
180. L Kanerva, R Jolanki, T Estlander. Organic pigment as a cause of plastic glove dermatitis. Contact Dermatitis 13:41–43, 1985.
181. RGO Kekwick. Origin and source of latex protein allergy. In: Latex Protein Allergy: The Present Position. Brickendonbury, United Kingdom: Crain Communications Ltd, Rubber Consultants, 1993, pp. 21–24.
182. D Kleinhans. Allergy against latex: angioneurotic edema by rubber dam. Allergologie 9:383–384, 1995.
183. BB Knudsen. Biological hazards—regulatory options. In: Latex Protein Allergy: The Latest Position. Brickendonbury, United Kingdom: Crain Communications Ltd, Rubber Consultants, 1995, pp. 55–59.
184. A Lahti, K Turjanmaa. Prick and use tests with 6 glove brands in patients with immediate allergy to rubber proteins. Contact Dermatitis 26:259–262, 1992.
185. K Lammintausta, HL Maibach. Contact dermatitis due to irritation. In: RM Adams, ed. Occupational Skin Diseases. 2nd ed. Philadelphia: WB Saunders, 1990, pp. 1–25.
186. DA Levy. Diagnosis of allergy to latex proteins. In: Latex Protein Allergy: The Present Position. Brickendonbury, United Kingdom: Crain Communications Ltd, Rubber Consultants, 1993, pp. 33–39.
187. DA Levy. Measurement of hypersensitivity to latex protein. In: Latex Protein Allergy: The Latest Position. Brickendonbury, United Kingdom: Crain Communications Ltd, Rubber Consultants, 1995, pp. 1–5.
188. D Lewis. Don't let latex allergy leave you short-handed. Safety Health December: 44, 48, 1996.
189. R Leynadier, JE Autegarden, DA Levy. Management of patients with latex protein allergy. In: Latex Protein Allergy: The Present Position, Brickendonbury, United Kingdom: Crain Communications Ltd, Rubber Consultants, 1993, pp. 59–60.
190. M Littman. Latex allergy. Healthcare Purchasing News (Supplement: Infection Control) April 15:4, 13, 1996.
191. C Lovell. Relationship between protein level and allergic response. In: Latex Protein Allergy: The Present Position. Brickendonbury, United Kingdom: Crain Communications Ltd, Rubber Consultants, 1993, pp. 41–42.
192. S Makinen-Kiljunen, K Turjanmaa. Latex allergens in various types of latex gloves and sensitization of hospital personnel and patients. Allergologie 9:366–368, 1995.
193. MJ Maso, DJ Goldberg. Contact dermatoses from disposable glove use: a review. J Am Acad Dermatol 23:733–737, 1990.

194. H Matsummura, K Oka, K Umekage, H Akita, J Kawai, Y Kitazawa, S Suda, et al. Effects of occlusion on human skin. Contact Dermatitis 33:231–235, 1995.
195. S Milkovic-Kraus. Glove powder as a contact allergen. Contact Dermatitis 26:198, 1992.
196. M Morris. A comparison of natural rubber with synthetic alternatives. In: Latex Protein Allergy: The Latest Position. Brickendonbury, United Kingdom: Crain Communications Ltd, Rubber Consultants, 1995, pp. 61–66.
197. T Palosuo. Purification and molecular characterization of latex allergens. In: Latex Protein Allergy: The Latest Position. Brickendonbury, United Kingdom: Crain Communications Ltd, Rubber Consultants, 1995, pp. 11–15.
198. P Patterson. Latex allergies: allergy issues complicate buying decisions for gloves. OR Manager 11(6):1–9, 1995.
199. KP Peters, A Heese, OP Hornstein. IgE-mediated contact eczema from latex. Allergologie 9:369–373, 1995.
200. W Potter. Legislative approach to latex products. In: Latex Protein Allergy: The Present Position. Brickendonbury, United Kingdom: Crain Communications Ltd, Rubber Consultants, 1993, pp. 61–67.
201. DW Ramsing, T Agner. Effect of glove occlusion on human skin (I). Short-term experimental exposure. Contact Dermatitis 34:1–5, 1996.
202. DW Ramsing, T Agner. Effect of glove occlusion on human skin (II). Long-term experimental exposure. Contact Dermatitis 34:258–262, 1996.
203. BS Rissman, ed. Latex Allergies: Confronting the Risk to Healthcare Workers. Atlanta, GA: American Health Consultants, Inc., 1995.
204. R Russel-Fell. Introductory paper. In: Latex Protein Allergy: The Present Position. Brickendonbury, United Kingdom: Crain Communications Ltd, Rubber Consultants, 1993, pp. 3–6.
205. JE Slater. Latex allergy—the U.S. medical experience. In: Latex Protein Allergy: The Present Position. Brickendonbury, United Kingdom: Crain Communications Ltd, Rubber Consultants, 1993, pp. 7–16.
206. GL Sussman, DH Beezhold. Latex allergy—a clinical perspective. J Long-Term Effects Med Implants 4(2&3):95–101, 1994.
207. JS Taylor, J Cassettari, W Wagner, T Helm. Contact urticaria and anaphylaxis to latex. J Am Acad Dermatol 21:874–877, 1989.
208. JS Taylor. Other reactions from gloves. In: GA Mellsrom, JE Wahlberg, HI Maibach, eds. Protective Gloves for Occupational Use. Boca Raton, FL: CRC Press, 1994, pp. 255–265.
209. K Turjanmaa. Incidence of immediate allergy to latex gloves in hospital personnel. Contact Dermatitis 17:270–275, 1987.
210. K Turjanmaa. Rubber contact urticaria. Contact Dermatitis 19:362–367, 1988.
211. K Turjanmaa, T Reunala, H Alenius, J Brummer-Korvenkontio, T Palosuo. Allergens in latex surgical gloves and glove powder. Lancet 336:1588, 1990.
212. K Turjanmaa. European medical experience. In: Latex Protein Allergy: The Present Position. Brickendonbury, United Kingdom: Crain Communications Ltd, Rubber Consultants, 1993, pp. 17–19.
213. K Turjanmaa. Contact urticaria from latex gloves. In: GA Mellsrom, JE Wahlberg,

HI Maibach, eds. Protective Gloves for Occupational Use. Boca Raton, FL: CRC Press, 1994.
214. K Turjanmaa. Allergy to natural rubber latex: a growing problem. Ann Med 26: 297–300, 1994.
215. K Turjanmaa. Occupational aspects and occurrence of natural rubber latex allergy. In: Latex Protein Allergy: The Latest Position. Brickendonbury, United Kingdom: Crain Communications Ltd, Rubber Consultants, 1995, pp. 7–10.
216. HLM van der Meeren, PEJ van Erp. Life-threatening contact urticaria from glove powder. Contact Dermatitis 14:190–191, 1986.
217. R Wahl, T Fuchs. Serological diagnostic of natural latex allergy and molecular characterization of a natural latex extract. Allergologie 9:374–378, 1995.
218. J Welzel, KP Wilhelm, HH Wolff. Skin permeability barrier and occlusion: no delay of repair in irritated human skin. Contact Dermatitis 35:163–168, 1996.
219. L White. Allergy challenges latex markets. Eur Rubber J 176(1):20–23, 1994.
220. I White. Rubber glove allergy: a problem of current concern. Pamphlet, 1995, pp. 1–4.
221. K Wrangsjö, G Mellström, G Axelsson. Discomfort from rubber gloves indicating contact urticaria. Contact Dermatitis 15:79–84, 1986.
222. K Wrangsjö, JE Wahlberg, GK Axelsson. IgE-mediated allergy to natural rubber in 30 patients with contact urticaria. Contact Dermatitis 19:264–271, 1988.
223. K Wrangsjö, K Oserman, M van Hage-Hamsten. Glove-related skin symptoms among operating theatre and dental care unit personnel: (I). Interview investigation. Contact Dermatitis 30:102–107, 1994.
224. K Wrangsjö, K Oserman, M van Hage-Hamsten. Glove-related skin symptoms among operating theatre and dental care unit personnel: (II). Clinical examination, test, and laboratory findings indicating latex allergy. Contact Dermatitis 30:139–143, 1994.
225. E Yip. Residual extractable proteins and allergenicity of natural rubber products. In: Latex Protein Allergy: The Latest Position. Brickendonbury, United Kingdom: Crain Communications Ltd, Rubber Consultants, 1995, pp. 33–39.
226. JW Yunginger, RT Jones, AF Fransway, JM Kelso, MA Warner, LW Hunt. Extractable latex allergens and proteins in disposable medical gloves and other rubber products. J Allergy Clin Immunol 93:836–842, 1994.

19

Handwashing and Gloving for Food Protection: Effectiveness

Eleanor J. Fendler, Michael J. Dolan, Ronald A. Williams
GoJo Industries, Inc., Akron, Ohio

Daryl S. Paulson
BioScience Laboratories, Inc., Bozeman, Montana

I. INTRODUCTION

The potential for food workers to be a factor in transmitting foodborne disease continues to be significant; however, the most effective method to break the contamination vector between food workers and consumers is a topic of intense debate. One view maintains that food workers must eliminate bare-hand contact with ready-to-eat food (by the use of gloves, utensils, etc.) to ensure protection, while the other position holds that a well-managed handwashing and sanitizing program is sufficient to ensure protection. Previously, we explored the evidence for these widely differing opinions via a review of the published literature relating to all aspects of handwashing and gloving [2], which clearly demonstrated that there is insufficient evidence to support the premise that use of gloves on the hands of food workers prevents the transfer of microorganisms to food and consequently to support the requirement for no hand contact with ready-to-eat food. The present study was carried out to establish data under simulated food service conditions to support the most effective hand hygiene regimens for food protection and minimized risk of health hazards.

Disease transmission via the hands from food workers to consumers can involve various types of microorganisms. ''Resident'' microorganisms that normally colonize the skin pose little threat of infectious disease [3,4]. There are situations, e.g., an infected cut, in which resident microorganisms may cause disease. In such situations, however, washing serves to degerm the infected area, cleansing it of dead cells and exudate material [5]. The threat comes instead from "transient" pathogenic microorganisms that temporarily reside on the skin of the hands. Transient microbial contamination occurs when a person makes hand contact with contaminated materials such as mucus, blood, soil, urine, feces, or food. In the food industry, contamination usually occurs through contact with excretions or infected areas of one's self or others, most commonly through hand transmission. Additionally, food workers can contaminate the food they prepare or serve others through hand contact with microbially contaminated materials such as money, raw and discarded food, tableware, countertops, soiled clothing, and other items in the work environment. Both of these types of transmission are examined in this study.

First, consider the situation of infected food workers who pass their infectious diseases by directly contacting food with contaminated hands. For infectious diseases to be spread to others via a carrier, several events must take place. The first two of these events are:

1. The contaminating microorganisms must be physically transmitted to others. This can occur when food workers contaminate their hands during defecation and pass the disease-causing microorganisms to consumers via hand contact.
2. The contaminating microorganisms must physically enter a person. This is particularly easy when the food has been contaminated by enteric (intestinal) disease-causing microorganisms.

An effective handwash or intact barrier gloves disrupt the disease process after event 1 by either removing the contaminating microorganisms from the hand surfaces or using a physical barrier to prevent them from being transmitted to the prepared food.

To evaluate the effectiveness of handwashing compared to gloving, a two-phase study was designed. The first phase evaluated the ability of hand-contaminant bacteria to penetrate compromised vinyl glove barriers. The second phase evaluated the microbial contamination level picked up on the hands from handling contaminated hamburger.

II. METHODS

A. Materials

Ambidextrous disposable polyethylene gloves were used throughout the studies. Hand cleansing and sanitization were carried out using an antibacterial lotion soap and an alcohol (gel) hand sanitizer. Ground beef, buns, vegetables, and paper wrap were used in the simulated food handling.

Before the initiation of this study, the protocol study description was given to the subjects and informed consent forms were completed. The protocol, informed consent form, and any other supportive materials relevant to the safety of the subjects were reviewed and approved by an institutional review board (IRB). No subject was admitted into the study who was using topical or systemic antimicrobials or any other medication known to affect the normal microbial flora of the skin.

The 7 days prior to the test portion of the study comprised the pretest period. During this time subjects avoided the use of medicated soaps, lotions, deodorants, and shampoos, as well as avoiding skin contact with solvents, detergents, acids, and bases. They avoided contact with products on the restricted list. Subjects were provided a personal hygiene kit containing nonantimicrobial products to be exclusively used during the course of this study. Subjects also avoided using UV tanning beds and bathing in chlorinated pools and/or hot tubs. This regimen allowed for stabilization of the normal microbial populations residing on the hands.

B. Glove Juice Sampling Method

Following the prescribed procedure [1], powder-free sterile gloves were put on. At the designated sampling time, 75.0 mL of sterile stripping fluid (SSP) were instilled into the glove. The wrist was secured, and an attendant massaged the hand through the glove in a standardized manner for 60 seconds. Aliquots of the glove juice (dilution 10^0) were removed and serially diluted in Butterfield's buffered phosphate diluent (BBP).

Duplicate spread plates were prepared from each dilution using MacConkey's agar. The plates were incubated at 30–35°C for 24–48 hours. Those plates providing colony counts between 25–250 were preferentially utilized in this study. If no plates provided counts in the 25 to 250 range, the plates closest to that range were counted and used in determining the number of viable microorganisms. If 10^0 plates gave an average count of zero, the average plate count was expressed as 1.00. This was done because the $\log_{10}$ of zero is undefined, but the $\log_{10}$ of 1.00 is zero. The number of viable bacteria recovered was obtained from the formula 75 × dilution factor × mean plate count for the two plates.

A statistical analysis—two-factor analysis of variance (ANOVA)—was performed on the collected data. The significant level was set at 0.05. The optimum levels of two test configurations evaluated were determined, as well as which glove type to use in the remainder of the study (Phases 1 and 2).

1. Phase 1

Nineteen human subjects were utilized in this evaluation segment. Punctured gloves were simulated by introducing four holes into the fingertips of the glove with a 21-gauge hypodermic needle. Subject treatment was randomized. Subjects underwent a 7-day pretrial restriction period in which they avoided skin contact with products and/or processes known to affect the normal microbial populations of the skin.

On the day of the evaluation, the subjects' hands were inoculated with 5 mL of 9.50×10^8 CFU/mL *Escherichia coli* (ATCC #11229). After air-drying the inoculated hands for approximately one minute, technicians placed the assigned test glove configuration on the subject, taking precautions to avoid contaminating the outer surface of the glove. Sterile latex gloves were then placed over the test gloves, and one of the gloved hands, randomly selected, was sampled (zero time sample) using the glove juice sampling procedure. If the hand selected for time zero time sampling bore a punctured-glove configuration, the punctures were taped prior to the sampling. Subjects then proceeded to their assigned activity (or nonactivity) for one hour. Four test configurations were used, each with five subjects:

Test Configuration # 1 (Inactive/intact). The subjects assigned to this configuration remained ''inactive'' (sitting in a chair reading), wearing an ''intact'' glove for one hour.

Test Configuration #2 (inactive/punctured). The subjects remained ''inactive,'' wearing a ''punctured'' glove for one hour.

Test Configuration #3 (active/intact). The subjects were ''active'' (performed food service activities; specifically, handling buns and vegetables and folding paper wrap), wearing an ''intact'' glove for one hour. The vegetables were prescreened for *E. coli* and other coliform contamination.

Test Configuration #4 (active/ punctured). The five subjects were ''active'' (performed food service activities; specifically, handling buns and vegetables and folding paper wrap), wearing a ''punctured'' glove for one hour.

All subjects were sequestered and closely monitored in the laboratory during the duration of the test. The subjects were sampled, using the glove juice sampling method, one hour after donning the test barrier glove. After the samples were collected, the subjects were required to perform a supervised surgical scrub with a 4% chlorhexidine gluconate (CHG) solution for 4 minutes, then wash their hands with 70% isopropyl alcohol for one minute and air-dry the hands for an additional 5 minutes.

A statistical analysis was performed on the collected data. The 0.05 level of significance was utilized in a two-factor analysis, the ANOVA design of which is given in Table 1. The results from Phase 1 are shown in Table 2.

Table 1 Two-Factor Analysis-of-Variance Design

	B_1	B_2
A_1	$n = 5$	$n = 5$
A_2	$n = 5$	$n = 5$

Where: A = activity level, A_1 = inactive, A_2 = active; B = glove status, B_1 = intact, B_2 = punctured

2. Phase 2

Thirty subjects were randomly assigned to an evaluation segment (five subjects to each of the six test configurations). Subjects underwent a 7-day pretrial restriction period during which they avoided skin contact with products and/or processes known to affect the normal microbial populations of the skin.

Assay of the ground beef for *E. coli*, prior to its experimental use revealed loads of 1.30×10^3 to 2.00×10^4 CFU per ounce of ground beef. The ground beef was then inoculated with aliquots of an *E. coli* (ATCC #11229) suspension to provide a final *E. coli* concentration of 1.1×10^5 to 9.6×10^5 CFU per ounce of ground beef.

The five subjects assigned to each test configuration performed a simulated food service task (kneading ground beef) over the course of three consecutive, 1-hour periods.

Test Configuration #5 (bare hands/no washing). No handwashing was conducted by these subjects during the 3-hour course of the study. A baseline

Table 2 Effect of Glove Condition and Activity on Microbial Penetration Through Gloves

	Microbial level on test glove, mean $\log_{10}$ (S.D.)	
Test configuration	Zero hour	One hour
#1 Inactive/Intact	2.05 (1.43)	0.00 (0.00)
#2 Inactive/Punctured	0.48 (1.08)	0.87 (1.21)
#3 Active/Intact	1.87 (1.29)	0.47 (0.94)
#4 Active/Punctured	1.24 (1.21)	0.85 (1.16)

(premarker bacteria exposure) sample of the hands as well as sampling at the end of the test period was conducted.

Test Configuration #6 (gloved hands/no washing and no glove changes). No glove changes or handwashes were conducted over the 3-hour course of the study. A baseline (premarker bacteria exposure) sample of the hands and of the test glove outer surfaces as well as sampling at the end of the 3-hour test period were conducted.

Test Configuration #7 (bare hands/hourly washing). Subjects washed their hands with only the assigned antimicrobial soap product immediately before beginning the simulated food service tasks (time 0), as well as at hours 1, 2, and 3. A baseline (premarker bacteria exposure) sample of the hands as well as sampling at the end of the 3-hour test period were conducted.

Test Configuration #8 (bare hands/hourly washing and sanitizing). Subjects washed their hands with the assigned antimicrobial soap followed by a hand sanitizer application immediately before beginning the simulated food service tasks (time 0), as well as hours 1, 2, and 3. A baseline (premarker bacteria exposure) sample of the hands as well as sampling at the end of the test period were conducted.

Test Configuration #9 (gloved hands/hourly glove changes, and no handwashing). Subjects changed their gloves at hourly intervals but did not wash their hands between the glove changes. A baseline (premarker bacteria exposure) sample of the hands and outer glove surfaces was conducted. Additionally, a sample of the hands and outside glove surfaces was conducted at the end of the marker bacteria exposure period.

Test Configuration # 10 (gloved hands/hourly glove changes and handwashing between changes). Subjects changed their gloves at hourly intervals and washed their hands between glove changes with the assigned product. A baseline (premarker bacteria exposure) sampling of the outer glove surfaces was conducted. Also, a sampling of outside glove surfaces was conducted at the end of the 3-hour marker bacteria exposure period.

After the samples were collected, subjects performed a surgical scrub with a 4% CHG product for 4 minutes and then washed their hands with 70% isopropyl alcohol for one minute and air-dried them for an additional 5 minutes. *Note*: All hand sampling was conducted utilizing the glove juice sampling method.

A statistical analysis was performed on the collected data. The 0. 05 level of significance was utilized. A two-factor ANOVA statistic was used to compare bare hand and outer glove microbial count differences between the zero- and 3-hour samples times.

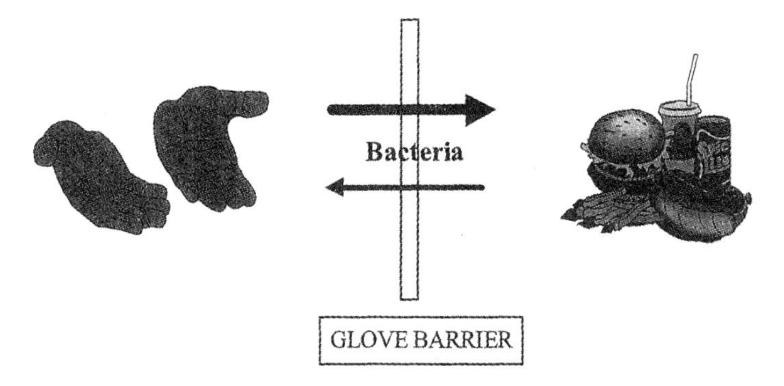

Figure 1 Glove study concept of phase 1.

III. RESULTS

A. Phase 1

The concept underlying Phase 1 of the study is illustrated in Figure 1. All values obtained for Phase 1 were obtained from the 10° dilution. If plates yielded no counts, the $\log_{10}$ value was designated 0.00 and used in the statistical analysis. The glove juice sampling procedure can detect microorganism populations only down to a $\log_{10}$ value of 1.57 because of the multiplication of the average plate count by 75 mL ($\log_{10}$ [75 × 0.5] = 1.57). As the data demonstrate, the test gloves housed within the latex gloves, regardless of test configuration, yielded contaminative bacteria at zero- and 1-hour sampling.

B. Phase 2

The concept underlying Phase 2 of the study is illustrated in Figure 2. The results from Phase 2 are presented in Table 3. A two-factor ANOVA model was used to compare times and product configurations. Figure 3 graphically displays the data obtained from the surfaces directly exposed to the inoculated ground beef for each product configuration.

IV. DISCUSSION

A. Phase 1

Although the counts were not high, clearly the *E. coli* contaminated some of the outer surface of the test gloves whether the subject was engaged in food prepara-

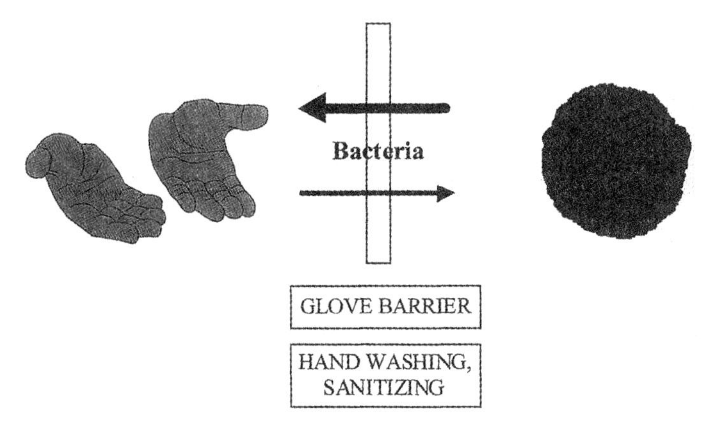

Figure 2 Glove study concept of phase 2.

Table 3 Microbial Levels of *Escherichia coli* from Contaminated Ground Beef Using Different Washing/Gloving Regimens

Configuration (regimen)			Microbial level on hands/outside of gloves, mean $\log_{10}$ (S.D.)		
			Time 0	Time 3	Time3 − Time 0
#5	Bare hands/No washing	Hands	0.53 (0.86)	6.25 (0.53)	5.72
#6	Gloved hands/No washing and no glove changes	Hands	0.21 (0.65)	2.39 (2.65)	2.18
		Gloves	0.00 (0.00)	5.70 (1.00)	5.70
#7	Bare hands/Hourly washing	Hands	0.65 (1.40)	4.16 (1.43)	3.51
#8	Bare hands/Hourly washing and sanitizing	Hands	0.00 (0.00)	0.80 (1.21)	0.80
#9	Gloved hands/ Hourly glove changes and no handwashing	Hands	0.16 (0.50)	3.04 (1.75)	2.88
		Gloves	0.00 (0.00)	6.06 (0.31)	6.06
#10	Gloved Hands/ Hourly glove changes and handwashing between changes	Hands	0.91 (1.49)	1.77 (2.37)	0.86
		Gloves	0.00 (0.00)	5.60 (0.86)	5.60

A two-factor ANOVA model was used to compare times and product configurations.

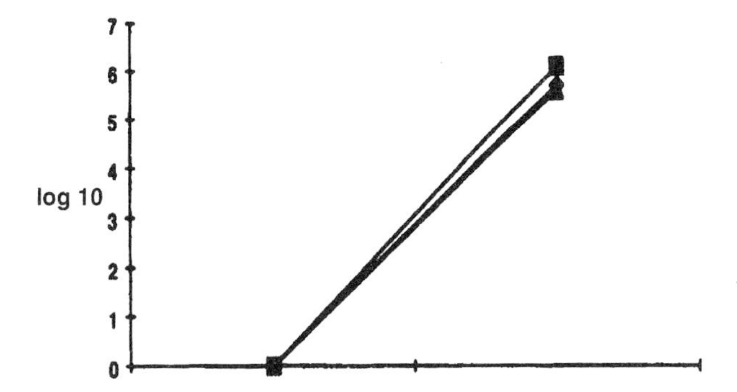

Figure 3 Data obtained from the surfaces directly exposed to the inoculated ground beef for each product configuration.

tion activities or not. Admittedly, the variables producing these results were several, but significantly, all potential sources of contamination in this test bear direct relevance to use of gloves by food handlers as barriers to transmission of infectious agents. For example, bacteria could have moved from the contaminated hand to the glove surface via breaches, whether experimentally created or because of defects in the manufacturing of the gloves. Indeed, results of preliminary testing of glove integrity, using the FDA Water Leak Test and the Glove-Check method, indicated that manufacturing defects in the vinyl gloves commonly used in food preparation were remarkably high (unpublished data, GoJo Industries, Inc.). Second, the test gloves may have been contaminated with *E. coli* during the process of their application by the technician, despite extreme measures to prevent this. The dire implications for the sterility of gloves donned by a food handler after using the toilet without a handwash are obvious. Finally, *E. coli* contaminating the food handled by subjects in the ''active'' test configurations may have penetrated the latex outer gloves through manufacturing defects, such as pinholes. Although unpublished data (GoJo Industries, Inc.) show latex gloves to be superior to vinyl gloves commonly used by food handlers, manufacturing defects are occasionally present. Hence, in all cases, the presence of bacteria on the surfaces of test gloves suggest that their value as ''barriers'' to disease is equivocal.

These results clearly have implications for gloving policies in the food industry. Use of gloves alone provides insufficient protection from transmission of pathogenic disease-causing microbes from food workers to consumers. Phase 2 of this study was carried out to determine the relative effectiveness of various handwashing and gloving regimens in preventing transmission.

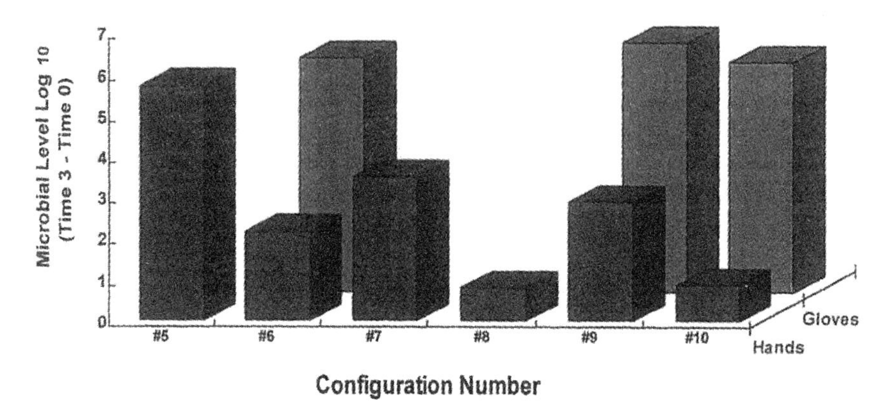

Figure 4 Microbial levels of *E. coli* on hands from contaminated ground beef using different washing/gloving regimens.

Analysis of the data for the different handwashing and gloving regimens was carried out using a two-factor ANOVA model. Both the time and product factors as well as the product versus factor interaction term were significant ($p < 0.05$). The interaction is significant because each product began at the same baseline microbial level, but the levels were different from one another at the 3-hour study completion time.

As illustrated graphically in Figure 4, bare hands/no washing (#5), gloved hands/no washing (#6), gloved hands/hourly washing (#9), and gloved hands/hourly glove changes and handwashing between changes (#10) are equivalent in microbial levels picked up from the inoculated ground beef ($p < 0.05$). Notably, results for bare hands/hourly washing (#7) and bare hands/hourly washing and sanitizing (#8) were statistically less than the results for the other four groups just mentioned ($p < 0.05$), with #8 (bare hands/hourly washing and sanitizing) being more highly statistically significant than regimen #7 (without the sanitizing).

The significantly lower microbial levels on the hands for regimens #7 and #8 as compared with that on hands with no washing (#5) and on the outer surfaces of the gloves (#6, #9, and #10) can be attributed to the residual antimicrobial activity from binding of the active in the antimicrobial handwashing product to the skin (Fig. 4).

The microbial values obtained (Table 3) from the hands for the regimens employing gloves, (#6, #9, and #10), clearly demonstrate that the polyethylene barrier gloves were unable to prevent contamination of the hands over the 3-hour course of the study. However, the microbial level found on the hands was lower when handwashing was employed between glove changes (#10) . This is also probably a consequence of the residual antimicrobial activity of the handwashing product on the hand. In conclusion, bare hands with a regimen of hourly handwashing and sanitizing provided significantly higher hand sanitization levels than any of the five other regimens, including those employing gloves.

V. CONCLUSION

The choice of and compliance with an effective regimen is essential for food protection. It is clear that a policy where gloves are employed to provide no bare-hand contact with ready-to-eat food is not a panacea and may only serve to provide a dangerous, false sense of security. Caution should be exercised in the selection of the most effective regimen for food protection. Additional studies should be conducted in food industry settings to validate the most effective regimen of hand sanitization for food protection.

REFERENCES

1. American Society for Testing and Materials. ASTM Standards on Materials and Environmental Microbiology. Philadelphia: American Society for Testing and Materials, 1993, pp. 211–213.
2. E.J. Fendler, M. J. Dolan, and R. A. Williams. Handwashing and gloving for food protection I: Examination of the evidence (this volume, Chapter 18).
3. W. C. Frazier, and D. C. Westhoff. Food Microbiology. New York: McGraw-Hill Book Co., 1988.
4. D. S. Paulson. A proposed evaluation method for antimicrobial hand soaps, Soap, Cosmet Chem Specialties June: 64–67, 1996.
5. C. W. Van Way III, and C. A. Buerk. Surgical Skills in Patient Care. St. Louis: The C. V. Mosby Co., 1978.

20

A Suggested Method for Evaluating Foodhandler/Processor Handwash Formulations

Daryl S. Paulson
BioScience Laboratories, Inc., Bozeman, Montana

I. INTRODUCTION

For years, limited effort has been directed toward assuring the antimicrobial effectiveness of handwashing products used by food-processing and food-preparation personnel [1]. However, in recent years, the high public profile accorded recurrent outbreaks of serious *Escherichia coli* O157:H7 infection, as well as notable incidents of food contamination by other microorganisms throughout the country, has been of considerable concern to food industry sanitarians, politicians, and governmental regulatory agencies.

Although there are numerous ways for consumers to become infected with disease-causing microorganisms—contaminated countertops, undercooked, microbe-laden meat, etc.—the primary focus of this chapter is microbially contaminated employees' hands [2]. A significant number of food-associated disease outbreaks are due to microorganisms picked up on the employees' hands and then passed to consumers via hand/glove contact with food. Perhaps the most common occurrence of this phenomenon is when food handlers are contaminated by enteric microorganisms from contact with their own feces or the feces of others (usually via hand-to-hand or fomes-to-hand transmission) and fail to remove these microorganisms via an effective handwash [3]. The contaminant microorganisms are then passed to the food they are preparing and, in turn, passed to consumers through its consumption. On the other hand, microorganisms that reside permanently on the hand surfaces, the normal skin flora, rarely pose any threat of infectious disease to oneself or others [4,5]. These microorganisms are more important in food spoilage, particularly in partially prepared foods such as precooked chicken and fish.

The topical antimicrobial handwash products manufactured for removal of contaminant microorganisms are generally both chemically and antimicrobially very much like those used by healthcare personnel to wash between patient examinations. Food service handwash products, however, should also effectively remove the ''organic load'' of food ingredients and fat. This is a critical point, one that can limit and even prevent products very efficacious as healthcare personnel handwashes from also being used as food handler/processor handwashes [6].

We at BioScience Laboratories, Inc. have developed an approach to testing the antimicrobial effectiveness of food handler handwash products in a worst-case situation, which we believe will provide accurate, precise, and reliable data. It is based on the current Healthcare Personnel Handwash Evaluation published in the FDA's Tentative Final Monograph (TFM) with two exceptions [7]. First, *Escherichia coli* (ATCC #11229) is substituted for *Serratia marcescens* (ATCC #14756) as the hand-contaminating microorganism, and second, the hands are inoculated not by pipette transfer, but by hand-kneading *E. coli*–contaminated hamburger. This provides a worst-case simulation of the food industry's hand-cleansing requirements.

II. MATERIALS AND METHODS

Test solutions and media include the following:

- Sterile stripping suspending fluid (SSF) [8].
- Product neutralizing fluid with 0.1% Triton X-100 and other appropriate product neutralizers to inactivate the antimicrobial action of the product collected from the hands during the glove juice sampling procedure. Otherwise, the antimicrobial compound is incubated with the microorganisms, giving the product many hours to be in contact with the microorganism. This would tend to portray the product as being more effective than it really is.
- Butterfield's phosphate buffer solution (BPB) for use as the diluent in the serial dilution schema [8].
- MacConkey agar containing appropriate test product neutralizers for use in selectively culturing the *E. coli.*
- Tryptic soy broth (TSB) is suggested for use in neutralization assay and for preparing the *E. coli* inocula to be distributed into the ground hamburger.
- High-fat hamburger (20–25% fat) is used to provide an organic load, making it more difficult for the product to remove the marker contaminative microorganisms.

III. TEST METHODS

A. Subjects

A sufficient number of overtly healthy subjects over the age of 18, but under the age of 70, should be recruited into the study to ensure that at least 18 subjects per product evaluated complete the study. A reference product should be included in the study design to assure the internal validity of the study—i.e., that the reference product provides the same efficacy in this study as it has demonstrated in the past [9]. Insofar as possible, to ensure an unbiased sampling, subject groups should be of mixed sex, age, and race. All subjects' hands must be free of clinically evident dermatoses, injuries to the hands or forearms, open wounds, hangnails, and/or any other disorders that may pose a health threat to the subject. Standard institutional review board (IRB) procedures and protocols should be in place and used throughout this evaluation.*

B. Product Neutralization

Prior to performing this evaluation, antimicrobial product neutralizers (inactivators) should be evaluated and confirmed effective for inactivating the antimicrobial compounds but not, themselves, inhibiting microbial growth. The American Society for Testing and Materials (ASTM) document entitled "Standard Practices for Evaluating Inactivators of Antimicrobial Agents Used in Disinfectant, Sanitizer, Antiseptic, or Preserved Products (ASTM E 1054-91)" provides the methodology for this test. A standard, one-way (factor) analysis of variance (ANOVA) model using a 95% confidence interval ($\alpha = 0.05$) or a series of Student's *t*-tests corrected for repeated use can be employed to assure statistically significant results from the assay.†

C. Pretest Period

A 7-day pretest period is adequate to assure elimination of any antimicrobial action residual from use of medicated personal hygiene products. During this period, subjects should be instructed to avoid using medicated hand soaps, hand wipes, hand gels, lotions, deodorants, and shampoos, as well as skin contact with solvents, detergents, acids, and bases or any other products known to affect the normal microbial populations of the skin. Each subject participant should be sup-

* IRB oversight and approval, an FDA requirement, assures the safety of the human subjects employed in a test protocol.

† When multiple *t*-tests are performed, the estimated *t*-table values need to be modified at the α term. The formula for this is $\alpha^* = 1 - (1 - \alpha)^k$, where k = number of *t*-tests performed, α = standard alpha value, and α^* = adjusted alpha value [10].

plied a personal hygiene kit containing nonmedicated soap, shampoo, deodorant, hand/skin lotion, and rubber gloves. The rubber gloves should be worn when contact with antimicrobials, solvents, detergents, acids, or bases cannot be avoided by the participant. Subjects should use the items in this kit for all relevant personal hygiene needs throughout their participation in the study. Finally, participants should avoid using ultraviolet (UV) tanning beds and swimming or bathing in biocide-treated pools or hot tubs.

IV. *ESCHERICHIA COLI* CONTAMINATION

A. Inoculum Preparation

To prepare the *E. coli* (ATCC #11229) inoculum, a 10 mL tube of tryptic soy broth should be inoculated with a loopful of a stock culture and incubated at 30 ± 2°C for 24 ± 2 hours. After the incubation period, 1.0 mL of the 10.0 mL broth culture should be aseptically transferred to a 2-L flask containing 1 L of sterile TSB, which is then incubated for 20 ± 2 hours at 30 ± 2°C and checked for purity. The resulting culture is used to inoculate each 4 oz. (113 g) raw hamburger patty to achieve a contaminant level of approximately 5.0×10^8 CFU/patty. The inoculated hamburger is then kneaded for 2 minutes by a gloved technician to distribute the *E. coli* uniformly throughout the patty. The hamburger should be quantitatively assayed for recoverable, viable *E. coli* counts at the beginning and end of the use period. That the raw hamburger often will have a bioburden prior to its inoculation is accommodated by this step.

B. Subject Safety

For their safety, the human subjects should not be permitted to leave the laboratory test area for any reason, once testing begins, since their hands will be contaminated with *E. coli*. Additionally, subjects should be required to wear protective laboratory aprons and be instructed not to touch their garments, faces, or any other body parts with their contaminated hands during the testing period.

C. Test Period

Each subject will be employed for 4–5 hours on the test day. Each subject should be required to clip his or her fingernails to a free edge of ≤1 mm. All jewelry will be removed from the hands and arms prior to beginning the test period.

D. Practice Wash

A practice wash should be performed using a nonmedicated, bland soap employing the wash procedure to be used in testing. The practice wash will ensure that each subject understands and is capable of repeatedly performing the wash procedure. The temperature of the water used for this and all subsequent wash cycles should be controlled at 40 ± 2°C.

E. Baseline Bacterial Count

On the test day, each human subject will handle and knead a hamburger patty contaminated with *E. coli* for 2 minutes. This constitutes the bacterial inoculation of the hands. This first inoculation cycle is used to provide ''baseline'' inoculation recovery levels using the glove juice sampling procedure. It should be followed with a 30-second handwash using nonmedicated soap. The subject will repeat this procedure two additional times for a total of three baseline measurements, which are then averaged.

F. Inoculation/Wash Procedures

After completion of the baseline sampling, each subject will manipulate an inoculated hamburger patty and then wash with the assigned test antimicrobial product according to label or supplied instructions. This will be followed by the glove juice sampling procedure.

Each subject will complete this inoculation/wash procedure a total of 10 consecutive times, with a minimum of 5 and a maximum of 15 minutes between procedures. The glove juice sampling procedure will be performed after inoculation/wash cycles 1, 3, 7, and 10.

G. Wash Procedure

Following product application and hand-sampling, the subjects will be required to perform a supervised 1-minute hand rinse with 70% ethanol, with an air-dry, followed by a 4-minute wash with a 4% chlorhexidine gluconate (CHG) or 10% povidone-iodine solution, and a water rinse to remove any residual *E. coli* (ATCC #11229) from the hands.

H. Glove Juice Sampling Procedure

Following the prescribed wash, powder-free, loose-fitting sterile latex gloves are placed on the subject's hands. At the designated sampling times, 75 mL of sterile stripping suspending fluid without product neutralizers are instilled into the sampling gloves. The wrists are secured, and an attendant massages the hand through the glove in a uniform manner for 60 seconds. A 5.0 mL aliquot of the glove juice (dilution 10^0) is removed and serially diluted in sterile stripping suspending fluid with product neutralizers and Butterfield's phosphate buffer solution.

I. Bacterial Counts

Duplicate spread plates are prepared from each of these dilutions using MacConkey agar containing tested antimicrobial product neutralizers and incubated at 30 ± 2°C for approximately 48 hours. *Escherichia coli* (ATCC #11229) will produce purple colonies on MacConkey agar, and only those colonies should be counted. Those plates providing *E. coli* (ATCC #11229) colony counts between 25 and 250 should preferentially be utilized as the data source. The estimated number of viable microorganisms recovered from each hand is obtained from the following formula (11):

$$R = 75\left(\frac{\sum x_i}{n}\right)10^{-D}$$

where:

R = estimated number of bacteria
75 = amount of stripping fluid dispersed into each sampling glove
$\frac{\sum x_i}{n}$ = average of the duplicate agar plate counts
D = dilution level counted

Because the R-value represents an exponential mathematical distribution, statistical analysis should be conducted on a linearized data distribution. This is achieved by using an R' value in place of R, where $R' = \log_{10}R$.

J. Statistical Analysis

A pre-post experimental design is utilized to evaluate and compare the antimicrobial effectiveness [12,13]. For example, such a design for two test products and one reference product would appear as follows:

Preproduct Application Samples	Postproduct Application Samples
R(1) $O(1)_{BL}(1)_{BL}(1)_{BL}$	A(1) $O(1)_1 O(1)_3 O(1)_7 O(1)_{10}$
R(2) $O(2)_{BL}(2)_{BL}(2)_{BL}$	A(2) $O(2)_1 O(2)_3 O(2)_7 O(2)_{10}$
R(3) $O(3)_{BL}(3)_{BL}(3)_{BL}$	A(3) $O(3)_1 O(3)_3 O(3)_7 O(3)_{10}$

where R(1) = subjects randomly assigned to 1 of 3 products; A(1) = independent variables (1 is test product 1, 2 is test product 2, and 3 is reference product); $O(1)_i$ = dependent variables = microbial counts at baseline (BL) and after the ith product use (washes 1, 3, 7, and 10).

Prior to performing a statistical analysis, exploratory data analysis should be performed on the data. Stem-leaf ordering, letter value displays, and box plots are generated to assure the data collected approximate the normal distribution [14]. If this is the case, a series of Student's *t*-tests (adjusted for multiple comparisons) are conducted using the 0.05 level of significance for Type I (α) error.

K. Reliability

We have determined that, when using at least 18 subjects per product, the data variability or standard deviation (S) is $\pm$ 0.5 log.* Hence, 18 subjects are enough to ensure that Type I (α) error can be set at 0.05 and that Type II (β) error will not be excessive. Recall that Type I error is the probability of concluding the product is effective when it is not. This error, the most critical of the two, is controlled by setting α at the 0.05 level or less. Type II error (manufacturers' risk) is the probability of concluding that a product is not effective when it is. If any doubts exist as to maintaining the sampling variability at $\leq$0.5 log, increasing the subject sample sizes will be helpful.

V. OPTIONAL SKIN IRRITATION

It is important to understand the skin irritation potential of the product [15]. If it causes irritation to the hands, it simply will not be used. Detergents irritate the skin by damaging the stratum corneum, impairing its "barrier" functions, usually by removing normal skin oils [3]. Additionally, they may be toxic to living epidermis and the dermis cells. Increased levels of cytokines, which are associated with skin allergies and irritation, have been observed in skin lymphatic fluids, following skin exposure to sodium lauryl sulfate, an extremely common anionic detergent.

The most commonly observed irritation effects include (1) soap effect, where the skin appears shiny and wrinkled, (2) roughness, where the skin looks

---

* Using the formula equation [3]:

$$n \geq S^2 \left[\frac{(z_{\alpha/2} + z_{\beta})^2}{D^2} \right]$$

where n = sample size per product; S = known standard deviation of samples for $\log_{10}$ microbial populations per cm^2; $Z_{\alpha/2}$ = alpha level at 0.05 = 1.96 for two-tail test; Z_{β} = power of the statistic (80%) = 0.842; D = clinical difference of significance to be ruled out (20%). A 20% reduction from baseline of the control or reference product at a specific sampling time is considered adequate for detecting significance.

and feels rough, with fine scaling present, (3) redness, (4) swelling, and (5) cracks and fissues [16]. Usually these effects appear in combination at different sites on the skin and at different degrees of severity.

Although visual scoring of hand irritation indices are simple and generally are effective in providing important information concerning irritation potential, transepidermal water loss rate and skin moisture content (corneometer) measurements are better (more accurate and precise) for such evaluations. These measure the barrier function of the stratum corneum, a main indicator of skin irritation [16].

A. Skin Irritation Evaluations

A topical antimicrobial product can be usefully evaluated for its irritation potential, compared to that of competitors' products in a multiple product study. One can link the skin irritation and antimicrobial evaluations together or perform the skin irritation study as a free-standing evaluation. For the latter approach, the investigator must recruit a set number of human volunteers who meet protocol and the institutional review board (IRB) requirements for participating in the study. The subjects should be placed on a restricted ''conditioning'' products regimen for 7 days, just as in the antimicrobial efficacy study. This has the effect of bringing all subjects' skin conditions to a common state and eliminating the biasing influences of extraneous products on outcomes for the tested product(s). The subjects' hands can then be ''baseline'' graded for dryness, swelling, chaffing, rash, redness, cracking, fissures, etc. Depending on the study's intent, a visual examination or instrumented measurements of water loss and skin moisture content can be performed to provide these baseline values.

The subjects then use the product(s) in a standardized manner for 10–50 washes per day over the course of 1–4 days. Following every wash, every fifth wash, or some other predetermined standard interval, the hands are evaluated visually and/or instrumentally.

When using transepidermal water loss or corneometer instrumentation, standard parametric statistics—*t*-tests or ANOVA—can be applied to the data. However, nonparametric statistical models are more appropriate than parametric ones for analyzing data from visual grading, a subjective rating system [17]. Nonparametric statistics apply rank/order processes that do not utilize parameters (mean, standard deviation, and variance) in evaluating data and also have the advantage that data need not be normally distributed, as is required for parametric statistics [18]. Thus, when using small sample sizes such as may be encountered in pilot studies where the data distribution cannot be assured to be ''normal,'' nonparametric statistics are preferred [3].

Common nonparametric models include:

1. The Mann-Whitney statistic: This test is the nonparametric analog of the Student's *t*-test [19]. It is used to compare two product groups to one another. Unlike the parametric Student's *t*-test, which must assume a normal (bell-shaped) distribution, the Mann-Whitney statistic requires only that the sample data collected be randomly selected.
2. Kruskal-Wallis model: This is the nonparametric analog of a one-factor ANOVA model [16,20]. It is used to evaluate multiple groups in terms of one factor—for example, the comparative irritation effects of five different hand soaps.

VI. CONCLUSION

It is important, then, in manufacturing topical antimicrobial products for use in the food industry that they be tested modeling environmental conditions (e.g., organic fat load). This will assure that the products sold are effective in degerming the hands. Moreover, it is important to know the irritation potential of the product so it can be designed as not only effective, but also nonirritating to the skin of the user.

REFERENCES

1. DS Paulson. To glove or to wash: a current controversy. Food Quality 4:60–63, 1996.
2. DS Paulson. Foodborne disease: controlling the problem. Environ Health 5:127–132, 1997.
3. DS Paulson. Topical Antimicrobial Testing and Evaluation. New York: Marcel Dekker, 1999.
4. M Schaechter, BI Eisenstein. Mechanisms of Microbial Disease. 2nd ed. Baltimore, MD: Williams & Wilkins, 1993.
5. CW Van Way, III, CA Buerk. Surgical Skills in Patient Care. St. Louis: Mosby, 1978.
6. DS Paulson. A comparative evaluation of different hand cleansers. Dairy, Food Environ Sanit 14:524–528, 1998.
7. Food and Drug Administration. Tentative Final Monograph for Health Care Antiseptic Drug Products, 21 CFR, Parts 333 and 369:31402–31452, 1994.
8. Food and Drug Administration. Tentative Final Monograph for Health Care Antiseptic Drug Products, 21 CFR, Parts 333 and 369:31446, 1994.
9. D Polkinghorne. Methodology for the Human Sciences. Albany, NY: SUNY, 1983.
10. WJ Dixon, FJ Massey. Introduction to Statistical Analysis. 4th ed. New York: McGraw-Hill, 1983.

11. DS Paulson. A broad-based approach to evaluating topical antimicrobial products. In: JM Ascenzi, ed. Handbook of Disinfectants and Antiseptics. New York: Marcel Dekker, 1996:17–42.
12. CR Hicks. Fundamental Concepts in the Design of Experiments. 4th ed. New York: Holt, Rinehart & Winston, 1993.
13. J Neter, MH Kuter, CJ Nachtsheim, W Wasserman. Applied Linear Statistical Models, 4th ed. Chicago: Irwin, 1996.
14. PF Vellman, DC Hoaglin. Applications, Basics, and Computing of Exploratory Data Analysis. Boston: Duxbury Press, 1981.
15. DS Paulson. Statistical evaluations: how they can aid in developing successful cosmetics. Soaps/Cosmet/Chem Special 9:37–38, 1987.
16. DS Paulson. Skin irritation and what it means to you. Soaps/Cosmet/Chem Special 11:56–65, 1998.
17. MQ Palton. Qualitative Evaluation Methods. 2nd ed. Newbury Park, CA: Sage, 1990.
18. WW Daniel. Applied Nonparametric Statistics. Boston: Houghton-Mifflin, 1978.
19. WJ Conover. Practical Nonparametric Statistics. New York: John Wiley, 1980.
20. JD Gibbons. Nonparametric Methods for Quantitative Analysis. New York: Holt, Rinehart & Winston, 1976.

V
Consumer Products

The first chapter in this section, Chapter 21, presents an argument for the appropriateness of using the healthcare personnel handwash evaluation procedure—the one used in medical applications—to evaluate general consumer antimicrobial products that will not be used in the medical sector. Chapter 22 presents alternative ways for evaluating consumer topical antimicrobial products. Chapter 22 points to the concern that consumer topical antimicrobial products, to be effective, need antimicrobial properties, as well as skin care properties. The final chapter (Chapter 23) highlights characteristics important in the endeavor to develop consumer topical antimicrobial products.

21

Appropriateness of the Healthcare Personnel Handwash Test in Assessing Efficacy of Consumer Handwashing Products

Bruce H. Keswick, Kathy F. Wiandt, and Ward L. Billhimer
The Procter & Gamble Company, Cincinnati, Ohio

I. INTRODUCTION

The Healthcare Personnel Handwash Test (HCPHWT) was originally proposed to evaluate products specifically intended for use in healthcare settings [1] and has become the standard evaluation tool for consumer antimicrobial products. Consumers have many distinct and unique habits and practices that differ from those found in the healthcare setting. The purpose of this chapter is to discuss variables that can affect the performance of the test and its utility for evaluating handwashing products used by consumers.

II. THE HEALTHCARE PERSONNEL HANDWASH TEST

The HCPHWT was designed to model a hospital setting where hospital personnel come into contact with many patients and the frequency of handwashing is high following potential exposures to pathogenic bacteria. The purpose of the HCPHWT is to measure the reduction of transient bacteria after an initial contamination and the continuing reduction of the transient bacteria following multiple washes and contaminations.

The American Society for Testing and Materials (ASTM) published a Standard Test Method (EI174–94) for evaluating health care personnel handwash formulations [2]. It is based on the HCPHWT design proposed in the OTC monograph [1]. This method is generally recognized as the industry standard and is the method described in this chapter.

In the HCPHWT, the subjects' hands are contaminated with a suspension of *Serratia marcescens*. This species of bacteria produces a red pigment on agar that distinguishes it from the other bacterial flora on the hands. In this test, the subjects' hands are contaminated 11 times, treated with test product 10 times, and sampled 3 times to simulate a day's activities.

For the base count, the subjects' hands are contaminated with 5.0 mL of *S. marcescens* (10^7–10^9 organisms). Immediately following contamination, bacteria are sampled using a glove juice/plastic bag sampling procedure. The glove juice/plastic bag sampling procedure involves donning rubber gloves or affixing plastic bags to the subjects' hands and adding 50–75 mL of stripping solution containing adequate neutralizers to the glove or bag. The hands are massaged for 1 minute and an aliquot of the solution is removed for analysis.

To determine the efficacy of a test product, subjects' hands are recontaminated with *S. marcescens*. After completing the contamination step, subjects perform a wash procedure with the product being tested. Organisms on the subjects' hands are then removed using the sampling procedure within 5 minutes of treatment. Hands are sampled again after the subjects complete a total of 10 contamination/treatment regimens. Aliquots of the subjects' sampling solutions

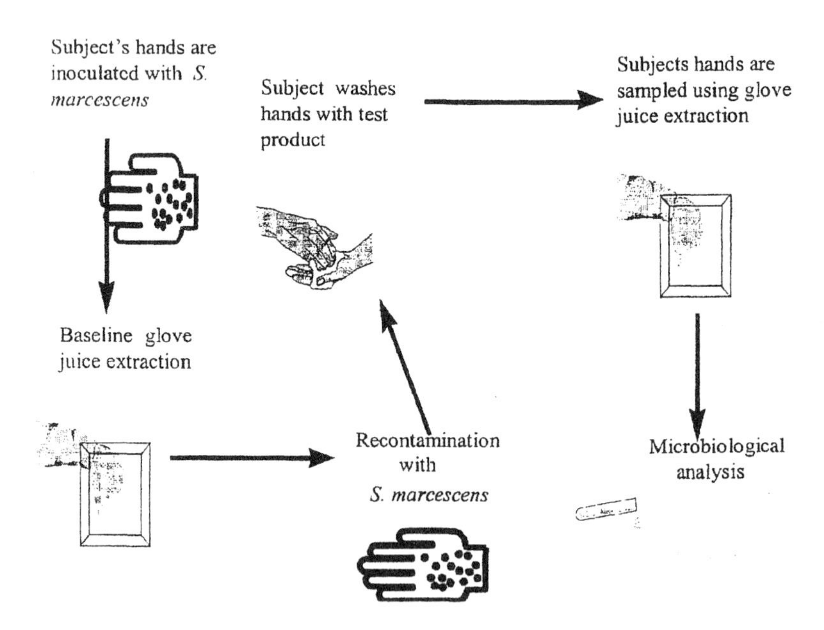

Figure 1 One-wash healthcare personnel handwash test.

Table 1 Log_{10} Reduction After First and Tenth Washes with an Antimicrobial Agent

Number of washes	Treatment	Baseline CFU[a] (log_{10})	Postwash CFU[a] (log_{10})	Difference[a] (log_{10})	Percent reduction
1	Hibiclens®	7.24 ± 0.10	4.31 ± 0.31	2.94 ± 0.29	99.88
10	Hibiclens®	7.24 ± 0.10	3.61 ± 0.35	3.63 ± 0.31	99.98

N = 12 subjects.
[a] Mean colony forming units (CFU) ± SD.

are diluted, plated, and incubated. Following incubation, the number of colony forming units (CFUs) is enumerated (Figure 1). Antibacterial activity is determined by comparing the number of bacteria removed from the hands after one and ten treatments with the product to the number of bacteria removed from unwashed hands.

Table 1 shows results of a HCPHWT conducted on a healthcare formulated product [3]. The data demonstrate that the bulk of bacteria are removed during the first wash.

The 1994 Tentative Final Monograph (TFM) proposed that the HCPHWT should be used to evaluate the effectiveness of antibacterial handsoaps [4]. For a product to be considered effective, it should produce a 2 log_{10} reduction within 5 minutes of the first wash and a 3 log_{10} reduction following the tenth wash. The need for increased removal of bacteria between the first and tenth wash was not explained. However, any increase in removal could demonstrate a residual effect of the soap.

III. VARIABLES THAT CAN AFFECT THE OUTCOME OF THE TEST

In a modification of the HCPHWT, certain variables such as the procedure for inoculating hands twice with bacteria, glove juice extraction sampling efficiency, cross-contamination of hands during the wash process, wash and rinse times, neutralization of the active ingredient, subungual areas, and air-drying can impact the outcome of the HCPHWT. Studies that have been conducted to develop an understanding of these parameters and their effects on the outcome of the HCPHWT will be described. All of these variables will be discussed in the context of using this method for evaluation of consumer products.

A. Inoculating Hands Twice with Bacteria

In a modification of the HCPHWT, hands are inoculated and sampled two times. The first inoculation and sampling are performed to determine a ''baseline'' count, and the second inoculation is performed prior to washing hands with the product and sampling to assess product efficacy. The efficacy of the product then is determined by comparing the counts after washing to the baseline count (first inoculation) and not to the actual number of bacteria placed on the hands immediately prior to washing with the product (second inoculation). It is assumed that the number of bacteria placed on the hands before washing is not significantly different from that of the first inoculation. Each inoculation step involves a degree of human error, which can potentially increase variability. Increased variability in the method can, in turn, decrease the sensitivity of the method and thus require an increase in the number of subjects needed to accurately assess the efficacy of a handwashing agent.

We devised a test to determine if the variability of the assay could be reduced by inoculating hands once with bacteria [4]. Eight subjects were enrolled in the study. The subjects conducted a practice wash with an unmedicated soap to clean their hands and learn the washing protocol. The subjects' hands were then contaminated with 5 mL of a suspension of *S. marcescens* (10^7–10^9 organisms). One hand of each subject was sampled to obtain a baseline specimen and the other hand was held in the air. The sampled hand was rinsed with water following the baseline sample and thoroughly dried. The subjects then washed their hands for 45 seconds with a bar soap containing an antimicrobial ingredient (1.5% 3,4,4′-trichlorocarbananilide, TCC). Both hands were then sampled using the plastic bag sampling procedure. The results from this study demonstrated that the antibacterial bar soap reduced the bacterial count by 2.5 log. The variability in the baseline counts was less than the previous study conducted according to the ASTM guidelines . Baseline recovery was 5.74 $\log_{10}$ with a standard deviation of 0.413. For the test in which the traditional recovery was conducted, the baseline counts were higher (7.94 $\log_{10}$), with a standard deviation of 0.603. In conclusion, inoculating hands once and performing baseline and test sampling on alternate hands reduced variability.

B. Inoculating Hands with a Large Volume

The HCPHWT suggests inoculation of the hands with 5 mL of the bacterial culture. Dosing the hands with such a large volume affects the consistency of bacterial loading from subject to subject and can result in loss of inoculum "down the drain." Experiments have been performed in which hands were dosed with three separate 1.5 mL aliquots (for a total of 4.5 mL) [5]. Each aliquot of the suspension was added to the subjects' hands, and they were instructed to rub the smaller aliquot on all surfaces (below the wrists) for approximately 20 seconds. The suspension was allowed to dry for approximately 30 seconds. This process was repeated two times. Following the final application, the hands were allowed to air dry for approximately 1 minute. Baseline count data from such studies

Table 2 Survey of Baseline Counts of HCPHWT Studies Conducted 1993–1997

Study	Inoculum dose method	$\log_{10}$ count (mean and standard deviation)
A	Single	7.51 ± 10.192
B	Single	8.00 ± 0.602
C	Single	7.94 ± 0.604
D	Single	7.60 ± 0.423
E	Single	6.64 ± 0.385
F	Split	7.60 ± 0.453
G	Split	7.10 ± 0.186
H	Split	6.93 ± 0.159
I	Split	7.34 ± 0.163
J	Split	7.15 ± 0.105
Range—Single Standard Deviation		0.192–0.604
Median—Single Standard Deviation		0.423
Range—Split Standard Deviation		0.105–0.453
Median—Split Standard Deviation		0.163

suggest that the variability is nearly one-third that observed in previous studies where the single dose method has been used. A survey of 10 HCPHWT studies conducted between 1993 and 1997 included five single-dose inoculation studies (Table 2). The survey revealed that the range of standard deviations for the $\log_{10}$ baseline counts from the single dose studies was 0.192–0.604 while the range of standard deviations for the $\log_{10}$ baseline counts from the split inoculum studies was 0.105–0.453. The median standard deviation for the single dose and the split inoculum group was 0.423 and 0.163, respectively [5]. ASTM has concurred with our recommendation and is currently in the process of revising the standard method to include inoculation with three 1.5 mL aliquots [6].

C. Extraction Efficiency and Cross-Contamination of Hands During the Wash

The glove juice method is considered to be an aggressive sampling technique; however, results of experiments described above suggest that it does not recover 100% of bacteria placed on the skin. We conducted a study to specifically determine the extraction efficiency of the glove juice/plastic bag sampling procedure [7]. In addition, the transfer of bacteria from one hand to the other as a result of the wash process was determined. Subjects first washed their hands with soap to remove bacteria and practice the protocol. Following the wash protocol, the subjects' hands were rinsed with ethanol. A sterile glove was then placed on one hand of each of the subjects. The glove was used to prevent exposing the skin surface of the hand to bacterial contamination. Next, hands were inoculated with 10^6 organisms of *S. marcescens*. The nongloved hand was sampled three successive times using the glove juice/plastic bag method, and removal of the marker bacteria was determined. For the next part of the study, the exposed hand was rinsed to remove any excess stripping solution. Both hands were then recontaminated (one hand was still covered with the glove). The glove was then removed. After being washed with a nonmedicated soap, both hands were sampled using the glove juice/plastic bag method.

The results from this study demonstrated that each subsequent sample from the bag juice method recovered bacteria . The bacterial recovery from each sample was 6.38, 4.08, and 3.54 log, respectively. The efficiency of the baseline recovery was 88.2%. The efficiency of the first sampling was 66.7%. Additional extractions did not appreciably increase the percent recovery (albeit the inoculum actually increased as the number of contaminations increased) . There is an opportunity to increase the bacterial recovery or identify a different recovery method for the HCPHWT. After the final contamination and wash, 3.56 and 2.18 log of bacteria were recovered from the nongloved and gloved hand, respectively. This indicates that bacteria are transferred from one hand to the other during handwashing, emphasizing the need for handwash products that prevent transfer and/ or have residual activity.

D. Wash and Rinse Times

The wash and rinse procedure outlined in the standard method recommends dispensing 5 mL of product or the amount specified by the manufacturer onto the hands [2,4]. Hands should then be lathered for 30 seconds and rinsed for 30 seconds. A parallel-designed study was conducted to determine the impact of handwashing time on subsequent germ removal [8]. Ninety subjects were enrolled into a preconditioning period. Those who returned to the test site were evaluated for participation in the study, qualified to participate in the study, and conducted a practice wash for 15 seconds and a rinse for 30 seconds. The subjects' hands were then rinsed with alcohol and contaminated with the marker organism *S. marcescens*. One hand was sampled via the glove juice method. Following baseline sampling, the subjects' hands were rinsed with water, alcohol, dried, and then recontaminated and washed with 5 mL of test product. The subjects lathered for 10, 20, or 30 seconds, or 1, 2, or 10 minutes. After being rinsed for 30 seconds, the subjects' hands were sampled as described above. The results are listed in Table 3. As shown, the number of bacteria removed by the glove juice method was not affected by lathering time.

These data suggest that the wash time over 10 seconds in the HCPHWT does not influence the efficacy of the hand soap. Rinse time, product amount,

Table 3 Log_{10} Reductions Provided in HCPHWT with Variable Lather Times

Lather time	Baseline CFU (log_{10})	Postwash CFU (log_{10})	Difference (log_{10})
10 sec	8.94	6.51	2.43
20 sec	9.16	6.89	2.27
30 sec	9.94	6.63	2.31
1 min	8.75	6.59	2.16
2 min	9.94	6.72	2.22
10 min	9.10	6.49	2.61
	8.97 ± 0.34[a]	6.64 ± 0.151[a]	2.33 ± 0.43[a]

N = 15 subjects for each time point.
Rinse time was held constant at 30 seconds.
CFU = Colony-forming units.
[a] Mean ± SD.

and other wash parameters may have a greater influence on the test than the actual ''wash'' times. The wash and rinse times examined in this HCPHWT are not reflective of the consumer wash experience consisting of about 7 seconds lathering and 8 seconds rinsing [9]. A second study was conducted using the HCPHWT incorporating a range of wash and rinse times. This study was conducted to assess the performance of plain soap and water in the HCPHWT under consumer-relevant wash and rinse times. Thirty subjects were enrolled into this study. Five subjects were used to evaluate a single test product (Babysan®) and one of six wash procedures (Table 4).

Subjects conducted a practice wash and rinse. The subjects' hands were then rinsed with alcohol and contaminated with the marker organism *S. marcescens*. The hands were sampled via the glove juice method. Following baseline

Table 4 Wash and Lather Time Procedures

Wash procedure	Lather time (sec)	Rinse time (sec)
A	5	2
B	5	8
C	5	16
D	15	2
E	15	8
F	15	16

sampling, the subjects' hands were rinsed with water, alcohol, dried, and then recontaminated and washed with 2 mL of test product. The subjects lathered their hands for 5 or 15 seconds and rinsed for 2, 8, or 16 seconds. The subjects' hands were sampled as described above. The results are listed in Table 5. As shown, the number of bacteria removed by plain soap was not affected by lathering time or rinse time. These studies demonstrate that equivalent bacteria are removed from the hands using different handwash procedures; however, while all the procedures removed approximately 2 $\log_{10}$ of bacteria, there were 5 log remaining on the hands postwash. The HCPHWT evaluates the reduction of bacteria from the hands but does not address the amount of bacteria left behind.

E. Neutralization

As outlined, the HCPHWT recommends neutralization and references the ASTM E1054-913 standard for ''Evaluating Inactivators of Antimicrobial Agents Used in Disinfectant, Sanitizer, Antiseptic or Preserved Products'' [10]. The ASTM method for neutralization gives general guidelines but does not specify steps needed to actually neutralize product in a handwash test. Furthermore, the HCPHWT indicates that neutralization may be accomplished by the 75 mL dilution factor without adding neutralizers to the stripping solution. Without the presence of neutralizers, the antimicrobial agent can continue acting in the samples up until bacteria are counted and falsely inflate the benefit provided by the antimicrobial agent. Benson et al. conducted an in vitro HCPHWT to determine whether neutralizers should be added to the stripping solution [11]. In this study, pieces of pigskin were contaminated with *S. marcescens* and washed 5 times with a 4% chlorhexidine gluconate handwashing product. When tests were performed with and without neutralizers in the dilution blanks but with adequate neutralizers in the stripping solution, there were no significant differences in numbers of bacteria surviving after five washes or each wash . When the tests were conducted without neutralizers in the stripping solution, significant differences were seen between the first and fifth washes. The skin sampled with neutralizing stripping solution achieved a 1.9 log reduction after the first wash and a 2.3 log reduction after the fifth wash. The pigskin sampled with the nonneutralizing sampling solution achieved a 3.2 log reduction after the first wash and a 6.1 log reduction after the fifth wash. The results of these studies indicate that adequate neutralization in the stripping solution is necessary to reflect the true activity of an antimicrobial ingredient.

F. Subungual Areas as Reservoirs for Bacteria

Several investigators have shown that the subungual regions of the hands serve as a reservoir for bacteria [12–14]. McGinley et al. measured the number and

Table 5 Log_{10} Reductions Provided in HCPHWT with Variable Lather Times and Rinse Times

Wash procedure		Sample	Baseline log_{10} (count)			Final log_{10} (count)		
Lather	Rinse	size	Mean	Median	Std. error	Mean	Median	Std. error
5	2	5	7.31	7.29	0.105	5.19	5.09	0.096
	8	5	7.38	7.35	0.039	5.07	5.06	0.158
	16	5	7.42	7.50	0.065	5.14	5.11	0.131
15	2	5	7.34	7.36	0.091	5.02	5.04	0.140
	8	5	7.43	7.44	0.038	4.74	4.66	0.078
	16	5	7.35	7.29	0.048	5.07	5.13	0.130

types of bacteria remaining on hands of healthcare workers after six consecutive washes [13]. The base of each digit (five sites), palmar surface (three sites), mid-dorsal surface (one site), interdigital spaces (four sites), outer surface of each fingernail (five sites), and the subungual spaces of each finger (five sites) were sampled. Serial sampling of each site yielded no significant differences in the bacterial recovery from the first to the sixth sample. The number of bacteria recovered from the subungual region was significantly higher than that of all other sites, suggesting that the subungual region was the source of continual bacterial shedding.

Leyden et al. studied the contribution of the subungual region to the continual shed of normal flora using the glove juice extraction method [14] . The nail beds on one hand were sealed with an acrylic polish and the nail beds on the other hand were not. The hands were serially scrubbed with a plain detergent six consecutive times. Whereas the bacteria recovered from the hands with sealed nail beds demonstrated a decay curve, the hands with ''open'' nail beds recovered bacteria continually. Furthermore, product treatments reduced the level of normal flora of the hands significantly when the subungual regions were sealed but not when they were open. Leyden et al. concluded that since the subungual flora was not likely to be a significant factor in the transfer of bacteria, the glove juice extraction method may not be appropriate for evaluating agents designed to degerm the skin surface.

The subungual experiments conducted by McGinley and Leyden demonstrates the impact of subungual areas on normal flora population estimates of the hands. We conducted a study to determine if the subungual areas serve as a reservoir for marker bacteria when using the glove juice extraction method to evaluate liquid soap products [15].

Eight subjects were enrolled into the study. The nail beds on one hand of each subject were sealed with acrylic nail polish, while those of the other hand were kept open. For the base count, the subjects' hands were inoculated with *E. coli*. Immediately following the contamination step, the organisms on the subjects' hands were sampled using a plastic bag sampling procedure. The subjects' hands were sampled two more times using this procedure to determine the efficiency of recovery of bacteria.

For the test procedure, subjects' hands were contaminated with the test organisms. Subjects treated their hands once with a liquid soap product. They washed their hands for 30 seconds and rinsed for 15 seconds. Afterward, hands were sampled three times using the plastic bag sampling procedure. Aliquots of the three sampling solutions were diluted, plated, and incubated. Following incubation, the number of CFUs was enumerated. Antibacterial activity was determined by comparing the number of bacteria removed from the hands after washing once with the test product to the number of bacteria removed from unwashed hands.

Sequential extractions from the hands continued to yield *E. coli* (Fig. 2).

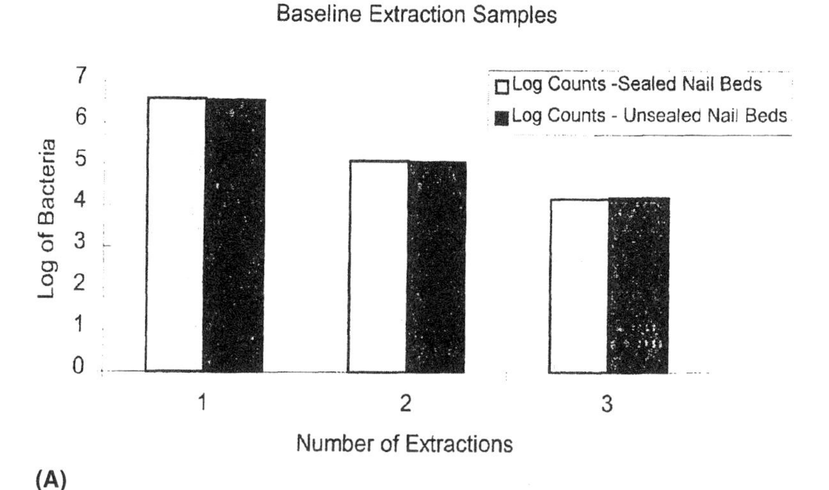

(A)

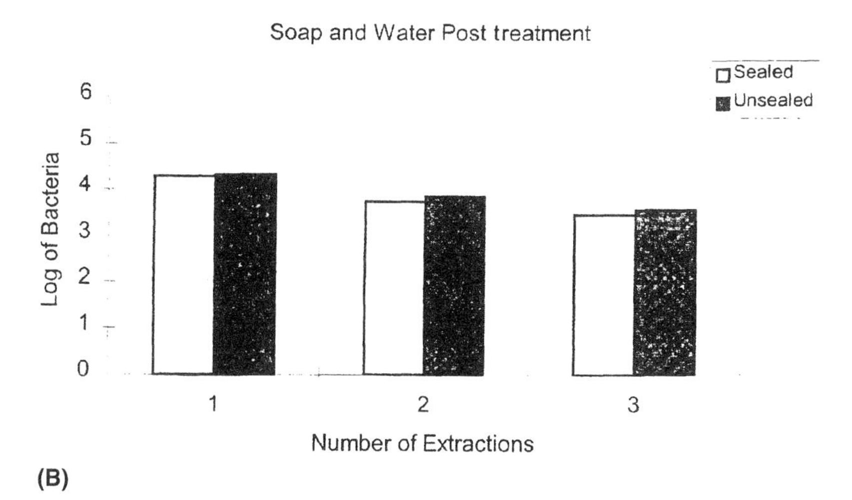

(B)

Figure 2 (A) Effect of sealing nail beds on recovery of transient bacteria: baseline comparison between sealed and unsealed nails. (B) Effect of sealing nail beds on recovery of transient bacteria: product comparison between sealed and unsealed nails.

However, fewer organisms were recovered with each additional extraction. The recovery technique was equally efficient for both sealed and unsealed nail beds, regardless of the number of extractions. Under the conditions of this study, no significant differences in bacterial counts from hands with sealed or unsealed subungual areas were observed after any of the extractions. In addition, sealing the subungual areas of the hands had no significant impact on the performance of any of the test products (regardless of the number of extractions performed), suggesting that, in contrast to normal flora, the subungual areas did not serve as a reservoir for marker bacteria when sampling using the glove juice method. Leyden et al. concluded that the glove juice extraction method continuously recovers normal flora from the nail bed regions and can impact evaluation of product performance. Our data suggest that for protocols using a marker organism, the subungual region does not impact the recovery of the organism or the evaluation of product performance.

G. The Effect of Air-Drying and Resident Time of Sampling

The effects of air-drying and desiccation on organisms can impact the final results of the HCPHWT. We conducted a small pilot study to determine the efficiency of the sampling process and to understand why many organisms remain on the hands [16]. Hands were contaminated with *S. marcescens* and allowed to air-dry for 4–8 minutes before sampling. Baseline counts were reduced by over 1 log in the 4- to 8-minute drying period, demonstrating that the time between collecting samples and conducting the test procedures can greatly influence the results of the HCPHWT. A standard air-drying time needs to be incorporated into the HCPHWT to provide consistency of the method. Such a standard should apply to the baseline as well as the posttreatment sampling.

IV. DRAWBACKS OF HCPHWT FOR ASSESSING EFFICACY OF HANDWASHING IN CONSUMERS

The HCPHWT was designed to assess the efficacy of products for use by healthcare workers, who wash their hands after frequently contacting potentially infected patients. The method is not ideal for assessing efficacy of consumer products because it does not encompass the myriad of sources of contamination in consumers or model their handwashing habits and frequency.

A. Source of Contamination

The Cosmetic, Toiletry, and Fragrance Association (CTFA) and the Soap and Detergent Association (SDA) have identified nine situations in which consumers have the potential to acquire bacteria that could lead to infection [17]. These include changing diapers, home hospice care, preparation of family meals, contact with pets, gardening or yard work, and travel and recreation. Many of these activities involve bacterial contamination in the presence of an organic load.

In consumers, the potential is high for transfer of bacteria to people from objects or other people. Jiang et al. demonstrated the rapid transfer of contaminants from objects to the hands in a daycare setting [18]. In this study, toy balls were introduced with DNA markers on them. The DNA markers were used to represent the spread of potentially pathogenic organisms. The study showed that within 24 hours of introduction of the probes, 62% of the hands of infants and teachers were contaminated. Hand touching was attributed as one of the major factors in the spread of the marker DNA, suggesting that bacteria are also spread in this manner.

The HCPHWT employs inoculation with a standard culture in a liquid (broth) medium, in the absence of an organic load. The HCPHWT guidelines specify the use of standard American Type Culture Collection (ATCC) organisms grown for 24 $\pm$ 4 hours [2]. However, wild strains of organisms transferred from consumer sources such as fecal matter may behave differently than a laboratory culture that has been maintained on standard media. The amount of organics present in the inoculum could increase survival of transient organisms by protecting them from cellular dehydration or by coating the cells and serving as a barrier from the active ingredients. Also, organic load could potentially inactivate an antimicrobial agent through protein binding or effectively blocking active sites on the molecule. While the broth culture in the HCPHWT provides some organic materials, it is at such a low level that it does not adequately model consumer-soiling situations where organic loads may be higher such as grease and feces from diapers, for example.

Broth inoculation is performed in the HCPHWT for standardization and logistical ease. However, placement of the organism in a high-moisture environment could potentially affect the product's active system by increasing solubilization and bioavailability. We conducted a pilot study to determine the impact of inoculum source on the effectiveness of an antibacterial product. Bacteria were transferred from a wet source, a dry surface, or a moist surface [19]. A technician wiped one of the subjects' hands for 15 seconds with an antibacterial hand product containing benzalkonium chloride prior to bacterial inoculation. The other hand was not treated to serve as a control. The palm surface of the first group of subjects was contaminated with 1.5 mL of a suspension of *E. coli*. The broth culture represented a wet inoculum source similar to the HCPHWT inoculum (high level wet). The fingertips of the second group of subjects were contaminated by rubbing them across a glass surface that had been treated with *E. coli*. The glass slide represented a dry surface. The third group of subjects contaminated their fingertips by rubbing them on an agar surface previously inoculated with *E. coli*. The agar plate represented a moist surface. The fingertips of the final group were inoculated by placing 10 μL of the *E. coli* suspension on each fingertip. This represented a wet source area–specific inoculum (lower level wet). An assigned fingertip on each hand was sampled within 1 minute of inoculation using

Table 6 Bacterial Recovery Immediately After Treatment

Method of inoculation	No treatment bacteria recovered ($\log_{10}$)	Hand wipe bacteria recovered ($\log_{10}$)	Standard deviation	Difference ($\log_{10}$)	*p*-value
1.5 mL of broth culture on whole hand	6.2	5.2	0.808	1.0	0.067
Wet transfer–placement of inoculum on the fingertip	6.4	6.1	0.808	0.3	0.5213
Moist transfer	4.3	3.2	0.808	1.1	0.058
Dry transfer	2.5	3.9	0.808	−1.4	0.0142

an agar imprint method [20]. Another fingertip was sampled 15 ± 1 minutes after inoculation to determine the residual effects of the product.

For sample analysis, the agar was removed from the imprint plates and was placed into a sterile centrifuge tube containing an extraction buffer. The sample was vortexed to adequately remove the organisms from the agar surface. Aliquots of the subjects' sampling solutions were diluted, plated, and incubated. Following incubation, the number of CFUs were enumerated.

The results of the immediate and the 15-minute post-treatment study are listed in Tables 6 and 7. The data indicate that moisture level of the inoculum source can influence bacterial counts and impact results of studies designed to test efficacy of an antibacterial product. Furthermore, the consistency of the results declines as subjects acquire bacteria from sources other than a broth culture. Because consumers are often exposed to bacteria by touching contaminated objects, use of a broth inoculate may not always be appropriate when assessing the

Table 7 Bacterial Recovery 15 Minutes After Treatment

Method of inoculation	No treatment bacteria recovered ($\log_{10}$)	Hand wipe bacteria recovered ($\log_{10}$)	Standard deviation	Difference ($\log_{10}$)	*p*-value
1.5 mL of broth culture on whole hand	4.3	2.2	0.790	2.1	0.0023
Wet transfer—placement of inoculum on the fingertip	3.6	2.8	0.790	0.8	0.1370
Moist transfer	1.3	0.4	0.790	0.9	0.1425
Dry transfer	2.8	2.2	0.790	0.6	0.3685

efficacy of products designed for consumers. The conditions of use for such products need to be considered as well to establish the appropriate exposure scenario.

B. Handwashing Habits

The HCPHWT models the behavior of healthcare workers, so it is based on the worker-patient relationship whereupon a worker contacts a patient (the bacterial/viral source) and then immediately washes his or her hands. Consumers typically are in contact with a wide variety of bacterial sources during the day and do not wash their hands after each encounter. They may not wash their hands if they do not perceive them as being dirty. Studies have shown that consumers wash their hands at certain intervention times (i.e., after using the restroom, preparing meat, wiping the kitchen countertop with a sponge, or removing laundry from the washer). While the U.S. Centers for Disease Control and Prevention (CDC) guidelines recommended that healthcare workers wash their lathered hands for at least 10 seconds followed by adequate rinsing [24], observation studies referred to earlier suggest that consumers typically wash and rinse their hands for a total of less than 8 seconds [9].

After washing their hands, consumers usually dry them with a towel, paper towel, or air-dryer. The HCPHWT does not incorporate a hand-drying step in the protocol. This allows for consistency within the method; however, omitting this from the protocol ignores a commonly performed step as part of the handwashing process. This parameter needs to be accounted for when evaluating efficacy of antibacterial products because towel drying will physically remove bacteria from the hands [21,22]. Ansari et al. compared the removal of rotavirus and *E. coli* from hands using three different hand-drying protocols and four different handwashing agents [22]. Fewer organisms were found on hands that were dried than on those that were not. Thus, when evaluating the efficacy and the benefits provided by antimicrobial handwashes, the method should incorporate a prescribed hand-drying technique. The incorporation of towel drying may introduce high levels of variability to a standardized method; however, typical behavior when consumers use the product should be part of any protocol in order to accurately ascertain the overall advantages from the wash product.

The type of product form to be evaluated in the HCPHWT can affect the outcome of the test. The HCPHWT was specifically designed to evaluate traditional rinse-off products for healthcare workers. This method may not be appropriate for other types of hand cleaners. Alternative products to soap and water such as hand sanitizers and towelettes are being marketed to consumers. These product forms are generally used when soap and water are not available. The appropriate way to evaluate the efficacy of ''leave-on'' alternative hand hygiene products is being debated. The European community is recommending the CEN hand rub method [23], while the United States continues to evaluate all antibacterial products using the HCPHWT. A hand product that can be used where there is no soap and water available may be beneficial to consumers. Therefore, it is important for investigators to use a method that models intended uses of alternate products and accurately evaluates their efficacy.

V. SUMMARY

The HCPHWT has been the traditional evaluation tool for antibacterial soap products in the United States. This is a controlled, clinical method with a historical database. Although it was intended to model in the healthcare setting, it has been used as the basis to support marketed claims for antibacterial soap products for both consumer and professional use. The above discussions demonstrate that the HCPHWT method does not provide control for many of the variables that can influence the performance of test products (especially consumer products). Consequently, use of this test may not adequately evaluate the efficacy of products intended for use by consumers. A need exists for additional test methods specifically designed to evaluate consumer products while modeling real-world conditions.

REFERENCES

1. Department of Health and Human Services. 21 CFR Part 333. Over-the-counter drugs generally recommended as safe, effective, and not misbranded. Fed Reg 43: 1210–1248, 1978.
2. Standard test method for evaluation of health care personnel handwash formulation. In:Annual Book of ASTM Standards, Volume 11.05, Standard Designation: El 174–94, 1994.
3. Data on file at the Procter & Gamble Company.
4. Department of Health and Human Services. 21 CFR Parts 333 and 369. Topical antimicrobial products for over-the-counter human use; tentative final monograph for health-care antiseptic drug products. Fed Reg 59:31402–31452, 1994.
5. Data on file at the Procter & Gamble Company.
6. American Society for Testing and Materials, personal communication.
7. Data on file at the Procter & Gamble Company.
8. Data on file at the Procter & Gamble Company, personal communication from Duane Charbonneau, Procter & Gamble Health Care.
9. Data on file at the Procter & Gamble Company.
10. Practices for evaluating inactivators of antimicrobial agents used in disinfectant, sanitizer, antiseptic or preserved products. In: Annual Book of ASTM Standards, Volume 11.05, Standard Designation: EI054–91, 1991.

11. L Benson, L Bush, D LeBlanc. Importance of neutralizers in the stripping fluid in a simulated healthcare personnel handwash. Infect Control Hosp Epidemiol 11:595–599, 1990.
12. J Haan. The source of the ''resident'' flora. The Hand 5:247–252, 1973.
13. KJ McGinley, EL Larson, JJ Leyden. Composition and density of microflora in the subungal space of the hand. J Clin Microbiol 26:950–953, 1988.
14. JJ Leyden, KJ McGinley, SG Kates, KB Myung. Subungal bacteria of the hand: contribution to the glove juice test: efficacy of antimicrobial detergents. Infec Control Hosp Epidemiol 10:451–453, 1989.
15. Data on file at the Procter & Gamble Company.
16. Data on file at the Procter & Gamble Company.
17. G Fischler, M Shaffer. Healthcare continuum: a model for the classification and regulation of topical antimicrobial wash products. Presented at joint meeting between CTFA/SDA, June 1995. Described in http://www.ctfa.org/viewpage.cfm?id=997.
18. X Jiang, X Dai, S Goldblatt, C Buescher, TM Cusack, DO Matson, LK Pickering. Pathogen transmission in child care setting studies by using a cauliflower virus DNA as a surrogate marker. J Infect Dis 177:881–888, 1998.
19. Data on file at the Proctor & Gamble Company.
20. S Namura, S Nishijima, KJ McGinley, JJ Leyden. A study of the efficacy of antimicrobial detergents for hand washing: using the full-hand touch plates method. J Dermatol 20:88–93, 1993.
21. DR Patrick, G Findon, TE Miller. Residual moisture determines the level of touch/contact-associated bacterial transfer following hand washing. Epidemiol Infect 119: 319–325, 1997.
22. SA Ansari, VS Springthorpe, SA Sattar W Tostowaryk, GA Wells. Comparison of cloth, paper and warm air drying in eliminating viruses and bacteria from washed hands. Am J Infect Control 19:243–249, 1991.
23. P Williamson, AM Kligman. A new method for the quantitative investigation of cutaneous bacteria. J Invest Dermatol 45:498, 1965.
24. JS Garner, MS Favero. Guideline for handwashing and hospital environmental control, 1985. Atlanta: US Department of Health and Human Services, Center for Disease Control, HHS Publication No. 99-1117, 1985.

22
Development of Methods to Evaluate Efficacy of Handwashing Products for Consumers

Bruce H. Keswick, Kathy F. Wiandt, and Ward L. Billhimer
The Procter & Gamble Company, Cincinnati, Ohio

I. INTRODUCTION

The purpose of this chapter is to discuss the need for and the development of methods to evaluate the efficacy of handwashing in the consumer product realm. Studies to assess how, where, and to what extent consumers acquire bacteria will be described. Examples will be provided that demonstrate how data derived from such studies can be used to estimate the level of risk associated with bacterial exposure in the nonhealthcare setting. Newly developed methods that are more reflective of the consumer's handwashing experience will also be described.

II. CONSUMER DATA THAT SUPPORT DEVELOPMENT OF NEW METHODS

The first step in developing a consumer-relevant handwashing model is to identify how consumers are exposed to and transfer bacteria. As shown below, the importance of developing a consumer-relevant model is underscored by the increasing potential for contamination due to societal changes and food importation and the ability of consumers to transfer large numbers of bacteria when conducting seemingly innocuous activities.

A. Impact of Societal Trends on Consumer Infection

Sattar et al. discussed the impact of changing social trends on infectious diseases in industrial nations [1]. These investigators observed that in the last 20 years, as healthcare costs have risen, acute-care hospital admissions have decreased dramatically. During this time period, the number of days per hospital stay decreased from 12 to 5 days. Consequently, the number of patients in home care and thus the potential for infectious disease spread in the home have increased.

Jackson and coworkers have found that up to 99% of foodborne infections in the developed world go unreported [2]. The Council for Agricultural Science and Technology estimates that in the United States, between 6.5 and 30 million infections per year are foodborne [3]. Increased importation and consumption of noncooked/minimally cooked fresh fruits and vegetables are associated with increased risk of illness and outbreaks.

Sattar and associates also hypothesize that social changes caused by dual income families may be responsible for spread of infectious diseases in industrialized countries [1]. Three fourths of all American children younger than 5 years old are in some form of childcare. This amounts to 13 million preschoolers and 6 million babies. This setting enhances transmission of pathogens between children who have not fully developed immunity to many infectious diseases. Increased mobility of the society and worldwide travel also translate to higher potentials for exposures to exotic diseases and the carriage of such diseases back home.

B. Sources of Consumer Contamination

In a study conducted by Scott et al., sites in over 200 homes were surveyed for bacterial recovery [4]. These investigators found that many sites in the home (particularly wet sites in the kitchen and bathroom) can harbor potentially pathogenic bacteria that may cause human infections. In an additional study, Scott et al. established the survival rates and the potential for cross-contamination of bacteria from laminated surfaces and cloths to fingertips [5]. The investigators found that bacteria could survive and be transferred to the hands up to 24 hours after contamination of the surface. This research demonstrates that consumers can encounter and transfer bacteria from their environment.

Rusin and coworkers have conducted three critical studies in order to understand bacterial acquisition and transfer potential to the mouth. The first of these studies measured microbial load on the hands following the day-to-day activities of ''typical'' consumers [6]. The hands of study participants were sampled after preparing a meal including a vegetable or fruit salad and meat; exiting a public restroom; cleaning the bathroom sink, toilet, and kitchen sink; doing laundry; or petting a household dog or cat. The hands of children returning from school or daycare were also sampled. Participating subjects conducted the activity and answered a survey designed to define various parameters associated with the activities.

With the exception of children returning home from school, participants disinfected their hands with alcohol before performing the designated activity. A record of handwashing was recorded following exiting a public bathroom. After the subjects completed the assigned activity, both hands were sampled by swabbing the palm side of the hands and hand size was recorded. Samples were collected, placed on ice, and returned to the laboratory for analysis. Samples were analyzed using standard microbiological procedures. The total number of bacteria and the number of coliforms and fecal coliforms were recorded. The summary statistics are listed in Table 1.

Rusin found the activities resulting in bacterial contamination of the hands to rank in descending order of contaminating potential as follows: (1) meal preparation, (2) cleaning the house, (3) petting a dog or cat, (4) returning home from school, (5) doing laundry, and (6) exiting a public restroom. The handling of raw chicken correlated with lower bacterial numbers than ground beef and other meats. The use of cleaning products with bleach did not result in lower bacterial loads on the hands after household cleaning. This may be due to contamination by cleaning with a kitchen sponge. However, hand contamination by handling laundry was reduced by the use of hot water, small washer loads, and detergent containing bleach. The petting of a dog resulted in only a 0.44 log greater bacterial contamination than petting a cat. Bathing of a pet did not reduce bacterial counts. There was a positive correlation between bacterial numbers and percent of time the animal spent outdoors.

These results suggest that everyday activities (especially meal preparation and household cleaning) can result in considerable contamination of the hands. The variation in the amount of bacteria acquired on the hands is very large. This

Table 1 Bacterial Recovery After Performance of Everyday Activities

Activity	Subjects (N)	Bacteria recovered ($\log_{10}$)[a]	Minimum ($\log_{10}$)	Maximum ($\log_{10}$)
Meal preparation	20	5.01 ± 0.33	2.1	7.9
Exiting a public restroom	21	3.53 ± 0.10	2.7	4.6
Household cleaning	20	4.50 ± 0.33	2.7	9.5
Laundry	20	3.99 ± 0.30	2.4	8.9
Handling a pet (dog or cat)	20	4.46 ± 0.21	3.2	7.8
Children returning from school or daycare	20	4.24 ± 0.27	3.2	8.9

[a] Mean ± SE.

variation is demonstrated by observing the minimum and the maximum log recovered on the hands conducting similar activities. While precautions such as handwashing or disinfection are important to interrupt cross-contamination in the home, potential microbial contamination hazards or ''hot spots'' may not always be obvious to the consumer. Consumers generally perceive a potential microbial risk associated with toileting and will wash their hands before exiting the restroom. The 20 participants exiting public restrooms in the Rusin study indicated they had washed their hands after using the toilet. The bacterial counts from their hands were relatively low (maximum $\log_{10}$ count 4.6). In contrast, handling wet, clean laundry does not present an obvious microbial risk to the consumer since they generally perceive the laundry to be clean and free of contaminants. However, Rusin demonstrated that, in fact, the microbial counts on the hands recently in contact with ''clean'' laundry can have nearly twice the level (maximum $\log_{10}$ count 8.9) observed on the hands after exiting a public restroom.

A second study conducted by Rusin et al. characterized the transfer efficiency of a pool of known microorganisms from different objects to hands [7]. The transfer rates from common objects that had been seeded with a known inoculum were established in these studies. Surfaces of the objects were prepared by inoculation with a pool of *Serratia rubidea*, *Micrococcus luteus*, and coliphage PDR-1. The study participants conducted an alcohol rinse and a soap-and-water control wash prior to handling the contaminated objects (fomites) listed below:

Sponge/Dishcloth—Subjects wrung out a sponge/dishcloth for 10 seconds.
Laundry—Subjects transferred a load of laundry to the dryer.
Faucet handle—Participants turned the faucet on and off two times.
Carrot—Participants cut a carrot into pieces.
Hamburger—Participants prepared four hamburger patties from one pound of inoculated hamburger meat.
Phone receiver—Subjects held an inoculated receiver as if answering a phone for 30 seconds.

Following transfer, the subjects' hands were air-dried for 1 minute. Hands were then sampled by swabbing the entire palm side of the hand, and the size of the hands was recorded. Samples were diluted, plated, and incubated using standard microbiological procedures. Following incubation, the number of colony-forming units (CFUs) or plaque-forming units (PFUs) was enumerated and the transfer efficiency for each organism from the object to the hand was determined. The results from this study are presented in Tables 2 through 4.

These data indicate that under controlled conditions, bacteria/bacteriophage are transferred to the hands from objects. In most cases, transfer efficiencies are extremely low (in terms of percent); however, the number of organisms that transfer to the hands from commonly contacted items are relatively high in

Table 2 *M. luteus* Transfer from Fomites to Hands

Type of fomite	Bacteria on fomite ($\log_{10}$)[a]	Bacteria recovered from the hand ($\log_{10}$)[a]	Transfer efficiency (%)[b]
Dishcloth	10.44	6.42	0.01
Sponge	9.58	5.50	0.01
Faucet	5.65	5.11	40.03
Carrot	9.05	5.84	0.07
Hamburger	9.79	5.23	0.02
Phone	6.13	5.71	41.81
Laundry—100%	9.73	5.70	0.04
Laundry—50:50	9.39	5.51	0.02

[a] Mean.
[b] Inverse log of bacteria recovered from hand ÷ inverse log of bacteria on fomite × 100.

some cases (i.e., 5–6 $\log_{10}$ bacteria). The study shows that consumers can come into contact with potentially infectious organisms while performing seemingly innocuous tasks. These organisms can cross-contaminate other sources, people, or the carriers themselves.

The final study conducted by Rusin et al. was to determine the transfer rate of bacteria from hands to the mouth [7]. Twenty healthy male and female subjects participated in the study. Two of the study participants' fingertips were dosed with a pool of *S. rubidea*, *M. luteus*, and PDR-1 coliphage (10^6 CFU/mL). The subjects placed one of their fingertips on the center of their lower lip for 10 seconds. The fingertip that touched the lip and the area of contact on the lower

Table 3 Coliphage PRD-1 Transfer from Fomites to Hands

Type of fomite	Bacteria on fomite ($\log_{10}$)[a]	Bacteria recovered from the hand ($\log_{10}$)[a]	Transfer efficiency (%)[b]
Dishcloth	9.85	5.47	0.01
Sponge	10.44	5.98	0.01
Faucet	4.90	4.22	33.47
Carrot	7.97	4.95	0.12
Hamburger	8.77	3.93	0.00
Phone	4.44	4.21	65.80
Laundry—100%	8.73	3.16	0.00
Laundry—50:50	8.34	2.27	0.00

[a] Mean.
[b] Inverse log of bacteria recovered from hand ÷ inverse log of bacteria on fomite × 100.

Table 4 *S. rubidea* Transfer from Fomites to Hands

Type of fomite	Bacteria on fomite ($\log_{10}$)[a]	Bacteria recovered from the hand ($\log_{10}$)[a]	Transfer efficiency (%)[b]
Dishcloth	10.34	4.94	0.00
Sponge	11.06	6.02	0.00
Faucet	5.60	4.74	27.59
Carrot	8.97	5.37	0.04
Hamburger	9.91	4.64	0.00
Phone	5.84	5.27	38.47
Laundry—100%	9.79	3.92	0.00
Laundry—50:50	9.01	3.16	0.00

[a] Mean.
[b] Inverse log of bacteria recovered from hand ÷ inverse log of bacteria on fomite × 100.

lip were sampled by swabbing. The additional fingertip that was inoculated was swabbed to determine the amount of original inoculum. After the sampling was completed, the samples were diluted, plated, and incubated using standard microbiological procedures. Following incubation, the number of CFUs and PFUs was enumerated and the transfer efficiency for each organism from the hand to the mouth was determined. As shown in Table 5, all three organisms consistently transferred from the fingertip to the lip.

This controlled laboratory study demonstrated that bacteria can readily transfer to the mouth at significant levels. There is evidence that survival of bacteria posttransfer on the hands provides multiple opportunities for oral exposure of bacteria. This study begins to quantify and establish rates in which bacteria acquired from the environment can transfer orally via the hands.

Table 5 Bacterial Transfer from Hands to Mouth

Organism	Inoculum placed on the finger ($\log_{10}$)[a]	Bacteria recovered from lip ($\log_{10}$)[a]	Bacteria recovered from fingertip after transfer ($\log_{10}$)[a]	Transfer efficiency (%)[b]
M. luteus	6.39	5.23	5.50	38.73
coliphage PDR-1	5.78	4.22	4.53	33.90
S. rubidea	6.66	4.73	5.06	33.97

[a] Mean.
[b] Inverse log of bacteria recovered from hand ÷ inverse log of bacteria on fomite × 100.

C. Summary

Experiments conducted by Rusin et al. [6,7], which identified the occurrence of bacterial load on the hands, the transfer efficacy from objects, and the transfer of microorganisms to the mouth provide insight into the impact of bacterial contact in the consumer realm. These studies provided information on the bacterial ''hazards'' associated with daily activities. Methodologies for evaluating consumer handwash products should try to dimensionalize the scenarios and high-risk situations for the consumer, such as everyday activities involving food preparation (as demonstrated with hamburger, carrot, sponge, dishcloth, and faucet) or incidental contact from typical contacts in society (i.e., phone use, using public restrooms, going to school or daycare, caring for a family pet, and home care).

The National Advisory Committee's Hazard Analysis and Critical Control Point (HACCP) program has been endorsed as an effective and rational means of assuring food safety from harvest to consumption [8]. The basic principles used to develop a HACCP plan include hazard analysis, verification procedures, critical control point identification, establishing critical limits, monitoring procedures, corrective actions, verification procedures, and record keeping and documentation. This same approach could be loosely applied to the identification and control of microbiological ''hazards'' encountered by consumers in their environment.

More research needs to be performed to better understand consumers' ''typical'' intervention points after contact with the environment. Unlike the healthcare setting, in which healthcare workers contact contaminated sources such as a patients and ''should'' wash their hands, consumers generally wash their hands only when they perceive them as being dirty. Consumers need to be better educated about how they can contact bacteria and when they should wash their hands. Many consumers may not wash their hands frequently enough to significantly reduce risk of bacterial transfer or infection. Increased use of antimicrobial products with persistent activity could reduce risk of contamination especially in situations where frequent handwashing is not practical or the use of alternative forms such as hand sanitizers allow for washing away from sinks.

III. PROMISING METHODS FOR ASSESSING THE BENEFITS OF HANDWASHING IN CONSUMERS

Literature published since the late 1800s describes the link between handwashing and the control of infections [9,10]. Good hygiene practices with antimicrobial products may be one way to counteract social trends that increase the potential for infections in consumers. As shown in Chapter 21, the use of the Health Care Personnel Handwash Test (HCPHWT) [11] to simulate consumer exposure to bacteria and evaluate the effect of antimicrobials may produce unreliable results. Methods used to demonstrate the benefits of antimicrobial soaps need to encompass the associated microbial risks that consumers encounter from conducting daily activities. The following are examples of methods that incorporate these risks.

A. Epidemiological Trials/Controlled Models of Deliberate Infection

The benefits of a handwashing intervention can be determined by conducting an epidemiological trial to show a reduction in illness. Several studies investigating the effects of handwashing on disease transmission have been conducted in environments that are conducive to the spread of disease [12–17]. Black et al. examined the effect of handwashing on diarrheal illness among daycare children [18]. Similar studies could be conducted to assess efficacy of consumer antimicrobial products. However, such studies may be costly and are often difficult to control.

Another approach to demonstrate the benefits of a handwashing product is using controlled clinical challenge models. These challenge models are conducted using a known organism to cause infection and a reduction in the disease serves as the endpoint. Such models have traditionally been used to evaluate the efficiency of vaccines [19]. More recently similar approaches have also been used to evaluate antiviral agents [20,21]. These studies generally involve fewer participants than an epidemiological trial and can be adequately controlled. Because challenge of human subjects with bacteria such as *Shigella* or *Salmonella* presents certain serious risks, safer bacterial models would need to be developed to demonstrate the efficacy of consumer products. Performance of a deliberate infection trial could be avoided by determining the removal and/or kill benefits provided by a product and understanding the attack rates and the infectious dose of a bacterium. These data could then be used to mathematically estimate risks and benefits.

B. Quantitative Microbial Risk Assessment

The efficacy of handwashing in reducing exposure to bacteria can be assessed using the Quantitative Microbial Risk Assessment (QMRA) model described by Haas and associates [22]. QMRA has successfully been used in a number of fields and is especially valuable when clinical testing is impractical. The general framework for conducting a risk assessment is to characterize the health hazards associated with exposure to the microorganisms(s), quantify the exposure, and determine the dose response for infection. An estimate of risk is made by employing an exponential risk model to dose-response and exposure data. The efficacy of a handwashing intervention is subsequently gauged by its potential to reduce that risk.

Rose et al. applied the QMRA methodology to estimate the effectiveness of antibacterial soaps in reducing skin infections caused by *S. aureus* [23]. Health hazards associated with exposure were primary skin infections including impetigo, folliculitis, furuncles, carbuncles, sweat gland infections, erysipelas, and erythema (primarily caused by *Staphylococcus* and *Streptococcus*). At-risk populations included children, the elderly, diabetics, and those with certain disease states such as atopic dermatitis. *S. aureus* was chosen to be modeled because of both relevancy and availability of data. A study by Singh et al., which determined the bacterial kinetics and infection rates of *S. aureus*, was used to assess the dose response [24]. The exposure assessment was determined by measuring *S. aureus* die-off and regrowth kinetics on skin after use of germicidal and control soaps. Risk of infection was then quantified by employing an exponential risk model to both growth curves integrated over a 24-hour exposure period.

There was a nearly 20-fold reduction in predicted skin infections when an antibacterial soap was used instead of nonantimicrobial control soap. This number was then compared to actual epidemiological studies, in which skin infections were reduced by almost twofold when using an antibacterial bar soap. Both the QMRA and the epidemiology study indicated that antibacterial soaps can reduce skin infections.

Rose et al. also used QMRA to assess reduction of risk of *Shigella* infection from diaper changing and disease transmission in daycare centers provided by handwashing with an antibacterial soap [25]. The exposures to the bacteria were based on oral fecal transmission data, and the dose-response assessment was based on infectivity data of *Shigella* provided by Crockett et al. [26] and results of HCPHWT studies conducted to quantify the reduction of *S. marcescens* after handwashing. The results of this QMRA model indicated that the use of a well-formulated antimicrobial soap might reduce the risk of infection 1000-fold compared to 100-fold for a control soap.

As shown by Rusin et al., a considerable number of bacteria can be transferred to hands by handling wet laundry [6]. The QMRA model also has been used to assess risk for exposure to *Shigella* by contaminated laundry [27]. Exposure data were obtained from studies designed to determine efficacy of bacterial transmission from laundry to hands and from hands to mouth [6,7]. The dose-response assessment was based on previously obtained infectivity data [26] and results of studies performed to determine the ability of detergents to reduce *Klebsiella* in laundry [27]. Based on these data, it was estimated that use of laundering agents with or without sanitizing agents reduced the risk of acquiring a bacterial disease through laundering by 99% or 90%, respectively.

C. Summary

Epidemiological trials have been performed that demonstrate the benefits of handwashing in the control of disease. Their regular use in assessing the efficacy of antimicrobials is limited by complex logistics, their cost, and potential variability. The QMRA has been utilized to demonstrate the benefits of handwashing in preventing skin infections with *S. aureus* and contamination from *Shigella* at daycare centers. It also has been used to assess the efficacy of detergents in eliminating bacteria from laundry. Currently, use of the QMRA is limited by the lack of exposure, transmission, and/or dose-response data. With adequate data, the QMRA could be a valuable tool to assess risk of bacterial contamination and efficacy of handwashing products in consumers.

IV. NEW METHODS THAT MODEL CONSUMER SCENARIOS OF CONTAMINATION AND HANDWASHING

In 1995, the Cosmetic, Toiletry, and Fragrance Association (CTFA) and the Soap and Detergent Association (SDA) developed the Health Care Continuum Model (HCCM) [28]. This model was presented to the Food and Drug Administration (FDA) as a means to classify and regulate topical antimicrobial products. The HCCM proposed six categories of classification based on use and health impact. The six categories include preoperative skin preparations, surgical scrubs, healthcare personnel handwashes, foodhandler handwashes, antimicrobial handwashes, and antimicrobial body washes. ''Antimicrobial handwashes'' are ''intended to help control the bacteria consumers acquire from the environment.'' The HCCM also identified nine situations in which consumers have the potential to acquire bacteria that could lead to infection. These are shown in Table 6.

These activities are associated with the potential for infections. Methods to assess the efficacy of handwashing in consumers should factor in the level of contamination, the potential for transfer from conducting various activities, and when and how consumers wash their hands.

A. Use of Agar Plates to Assess Potential for Transfer Pre- and Post-wash

In consumer situations, bacteria are able to transfer from the hands to other sources. The degree of transfer is not accurately determined by the glove juice sampling technique used in the HCPHWT, since hands are fully immersed into a buffer and massaged to remove bacteria. The actual transfer potential of bacteria from hands to an object can be determined using agar contact plates [29,30]. In this method, palmar surfaces of hands and/or fingers are pressed against plated agar. The amount of bacteria that transfers onto the agar is then quantified using image analysis. Namura and coworkers used a full-hand touch plate method to quantify the number of bacteria that remained on hands before and after scrubbing

Table 6 Potential for Various Activities to Cause Infection in Consumers

Activity	Examples of diseases transmitted (examples of organism transmitted)	Needs to be reflected in test methods
Diaper changing	GI (*Shigella*, rotavirus)	Varying levels and types of organic load
At-home sick care	ARI (rhinovirus)	Transfer potential of organisms to self or other sources such as food via hands
Attending daycare	Skin infections (*Staphylococcus*, *Streptococcus*)	
Meal preparation	Food poisoning (*E. coli*, *Salmonella*)	Multiple exposure to organisms
Gardening/yard work	Bacterial skin infections (*Streptococcus*, *Staphylococcus*), fungal skin infections (*Sporothrix schneckii*), GI infections (*Salmonella*, *E. coli*), ARI (rhinovirus)	Low frequency for handwashing "intervention" points
Travel and recreation		
Attending school or work	Skin infections (*Streptococcus*, Staphylococcus), GI infections (*Salmonella*, *E. coli*), ARI (rhinovirus)	
General contact with people		
Contact with pets	Skin infections (*Streptococcus*, *Staphylococcus*), GI infections (*Salmonella*, *Camphylobacter*)	

GI = Gastrointestinal illness; ARI = acute respiratory illness.

with antimicrobials commonly used to sterilize hands in hospitals [29]. In a study described by Charbonneau et al., this method was used to determine the efficacy of sanitizers in removing bacteria acquired by handling raw meat [30]. In both studies, reliable results were obtained with relatively few subjects.

B. Meal Preparation Test

We conducted two randomized, blinded clinical studies to determine the impact of handwashing with three different products on bacterial transfer under conditions found in a domestic kitchen [31]. In both studies, bacteria were transferred to the hands by handling ground beef that had been contaminated with 5×10^7 *E. coli*. For the first study, hands were sampled before and after being washed using the glove juice technique. In the second study, hands were sampled by placing a 25 cm^2 agar imprint plate on the palms for 10 seconds. The efficacy of each product was determined by comparing the mean CFUs (first study) or area of bacterial growth determined by image analysis (second study) before and after treatment. The results of the studies are summarized in Tables 7 and 8.

Both studies showed that organisms were consistently transferred from ground beef to the hands and that viable bacteria remained on the hands after washing. Results generated using the agar imprint study revealed that the remaining bacteria were transferable. According to the glove juice study, each product was equally effective in removing bacteria. By contrast, slight differences in efficacy were noted in the agar imprint study. Higher percentage reductions were observed in the glove juice test than the agar imprint study. This suggests that the glove juice method may not always accurately assess the potential for bacterial transfer by consumers and may overestimate the effectiveness of handwashing agents.

Table 7 Effect of Handwashing on Bacteria Transferred by Handling Ground Beef: Glove Juice Extraction

Test product	*N*	Baseline sampling[a]	Final sampling[a]	Change from baseline	
				Mean	% reduction
Healthcare personnel handwash	11	7.3 ± 0.20	4.0 ± 0.80	3.3 ± 0.89	99.9
Unmedicated soap	11	7.3 ± 0.24	4.0 ± 0.58	3.3 ± 0.57	99.9
Food service approved hand sanitizer/cleanser	11	7.2 ± 0.50	4.3 ± 0.60	3.0 ± 0.31	99.9

[a] Mean colony-forming units ± standard deviation (log_{10}).

Table 8 Effect of Handwashing on Bacteria Transferred by Handling Ground Beef: Agar Contact Plate Detection

Test product	*N*	Baseline[a] (pixels/mm)	Final[a] (pixels/mm)	Change from baseline	
				Mean[a]	% reduction
Healthcare personnel handwash	5	1735.1 ± 44.3	29.1 ± 15.3	1706.0 ± 47.7	98.3
Unmedicated soap	5	1754.1 ± 35.5	2.3 ± 0.9	1751.9 ± 35.1	99.9
Food service approved hand sanitizer/cleanser	5	1713.0 ± 13.9	64.7 ± 31.7	1648.3 ± 36.6	96.2

[a] Mean ± standard deviation.

In contrast to the traditional HCPHWT, the inoculation method used in this study involved an exposure scenario that is commonly encountered during meal preparation (i.e., high levels of bacteria associated with high organic loads). Under high organic (grease) conditions that a consumer might encounter during routine cooking, known effective active ingredients were only as effective as a well-formulated unmedicated product containing surfactants designed to disperse this type of organic material. Considering these results, it is worth exploring other consumer scenarios.

C. Potential of Hands to Cause Cross-Contamination

In the meal preparation test described above, we examined the ability of hands to transfer bacteria from meat to a surface (agar). The results showed that transferable bacteria remain on the hands after washing. The purpose of the next study was to determine the level of organisms that can be transferred from one food to another via the hands [32]. This method was designed to show the importance of consumers' hands in the transfer of bacteria in the foodhandling setting as well as the effectiveness of handwashing in reducing transfer.

Following a practice wash with a liquid-soap product, subjects handled a fresh, whole chicken for 45 seconds. After hands were air-dried for 1 minute, subjects handled a 50 g sample of irradiated-sterilized ground beef for 1 minute. They then decontaminated their hands by washing with soap and water and rinsing with alcohol. Subjects then handled another chicken. They lathered their hands for 30 seconds with one of the products being evaluated, rinsed for 15 seconds, and dried with a paper towel. Next, they handled another sample of sterile ground beef for 1 minute. Samples of ground beef taken before and after handling were placed into a sterile stomacher bag containing 450 mL of buffer

Table 9 Bacteria Transferred from Chicken to Sterile Ground Beef by Hands Before Washing

Subject number	Bacteria transferred ($\log_{10}$)
1	4.70
2	4.56
3	4.28
4	4.08
5	3.89
6	4.37
7	4.03
8	4.19
9	4.15
Mean ± SD	4.25 ± 0.24

SD = standard deviation.

and were mixed for 2 minutes. Samples of the solutions were serially diluted and plated in agar. Bacteria ($\log_{10}$) were enumerated after a 24-hour incubation period (Tables 9 and 10).

As shown in Table 9, bacteria were consistently transferred from the chicken to sterile ground beef by unwashed hands. These bacteria were natural contaminants of the chicken. The number of bacteria transferred to the beef was reduced (but not eliminated) by washing with a food service–approved hand sani-

Table 10 Effect of Washing Hands After Handling Chicken on Bacterial Transfer to Sterile Ground Beef

Subject number	Bacteria transferred after washing with food service approved hand sanitizer/cleanser ($\log_{10}$)	Subject number	Bacteria transferred after washing with plain soap and water ($\log_{10}$)
1	3.33	2	3.88
3	3.75	4	3.05
5	3.17	6	3.99
7	2.96	8	3.00
9	3.16		
Mean ± SD	3.27 ± 0.27	Mean ± SD	3.48 ± 0.45

SD = standard deviation.

tizer or plain soap and water after handling the chicken. The fact that reliable results were obtained with few subjects indicates that this methodology could be a powerful tool in assessing the benefits of handwashing in foodhandlers and consumers.

D. Residual Efficacy Test

Data from numerous tests show that bacteria remain on the hands after handwashing. Use of antibacterial agents that remain on the skin after washing may produce further reductions. Aly and Maibach have developed a method to evaluate the residual efficacy of topical antibacterial products against various skin pathogens [33]. This method involves inoculating the skin with bacteria, treating the site with an antimicrobial agent or a vehicle, and enumerating bacteria after 5 hours of occlusion. Finkey et al. and Scala et al. have used this method to evaluate the efficacy of antimicrobial soaps after repeated applications [34,35]. However, how quickly or how long the antimicrobial acts cannot be addressed using the standard method.

To determine the rate and duration of action of antibacterial soaps, we modified the conventional procedure by inoculating forearms with *S. aureus* either immediately or 24 hours after subjects had washed multiple times (seven or nine, respectively) over 3 days [36]. Surviving organisms were harvested after various intervals of occlusion (30 minutes, 2 hours, and 5 hours). In these experiments, subjects served as their own control by washing one forearm with an antibacterial soap bar containing 1.5% TCC and the other with an unmedicated, vehicle soap bar. Occluded sites were sampled for organisms using the cup-scrub technique [37,38]. In both studies, number of CFUs ($\log_{10}$) was determined after 48 hours of culture (Table 11). As shown, the antibacterial soap containing 1.5% TCC was significantly more effective than the vehicle in controlling the growth of *S. aureus* on the skin. The antimicrobial was effective in reducing bacteria for up to 5 hours. This effect was noted even after the bacterial challenge was given 24 hours after the final wash.

E. Consumer Realistic Residual Efficacy Test

Although the residual efficacy test described above is an improved means of assessing the residual effect of antimicrobials, it is not optimized for use in consumer situations where high levels of bacteria are contacted from sources that are not perceived to be contaminated and hands are not washed frequently. For instance, in the kitchen, bacteria are commonly transferred to hands by use of sponges [6,7]. After using sponges, consumers often do not wash their hands. Therefore, the kitchen sponge is an ideal source of bacteria for a consumer-relevant test.

Table 11 Residual Antibacterial Activity of Bar Soap Containing 1.5% TCC Versus Nonbacterial Bar Soap

Bacterial occlusion time[b]	Test product	Inoculation immediately after final wash[a]		Inoculation 24 hours after final wash[a]	
		Surviving CFU[c] (log_{10})	Difference (log_{10})	Surviving CFU[c] (log_{10})	Difference (log_{10})
30 minutes	Antibacterial soap	5.19 ± 0.42	0.47	4.75 ± 0.47	0.19
	Soap vehicle	5.65 ± 0.22	$p \leq 0.001$	4.94 ± 0.42	$p = 0.033$
2 hours	Antibacterial soap	4.39 ± 0.55	1.39	4.42 ± 0.56	0.63
	Soap vehicle	5.78 ± 0.15	$p \leq 0.001$	5.05 ± 0.24	$p \leq 0.001$
5 hours	Antibacterial soap	4.24 ± 0.87	1.49	4.39 ± 0.65	0.72
	Soap vehicle	5.72 ± 0.66	$p \leq 0.003$	5.11 ± 0.60	$p \leq 0.001$

[a] Time elapsed between final wash and inoculation.
[b] Occulsion time for inoculated treatment sites.
[c] Mean colony-forming units ± standard deviation.

We utilized a cellulose sponge that had been inoculated with *E. coli* and *S. aureus* (3.6×10^8 and 1.98×10^8 CFU/mL, respectively) to test the residual activity of two antibacterial soaps [39]. Prior to inoculation, the sponge (8 cm $\times$ 12 cm) was placed into a sterile stomacher bag and rehydrated for 4 hours with modified Letheen broth (Difco Manufacturing, Detroit, MI). This step was performed to neutralize preservatives in the sponge. The sponge was assayed for the level of bacterial contamination at the beginning and end of the study by removing an aliquot of liquid from the sponge and serially diluting and plating the sample.

Before gently pressing their fingers onto the sponge for 10 seconds, subjects either washed their hands three times (at least 1 hour apart) with a liquid test product or wiped them for 10 seconds with a towelette product. At time points of 10, 30, and 60 minutes after contamination, one fingertip of each hand was sampled using a fingertip extraction procedure in which a small test tube was affixed to the fingertip and inverted 20 times. Aliquots of the sample solutions were diluted, plated, and incubated using standard microbiological procedures. Following incubation (24 $\pm$ 4 hours), the number of surviving CFUs was determined (Table 12). As shown, both *E. coli* and *S. aureus* transferred to subjects' hands from the sponge. It appears that both products may be more effective against *E. coli* than *S. aureus*. However, this finding may also be explained by the fact that *E. coli* do not survive well on skin [40,41].

F. Continued Method Development

New methods that we have described are particularly well suited to study both the potential for bacterial contamination during meal preparation and how handwashing interferes with this process. However, additional methods need to be developed that are relevant for other exposure scenarios. When developing meth-

Table 12 Residual Effect of Two Antimicrobial Products on Bacteria Transferred to Fingers from a Sponge

Test product	N	Time (min)	*E. coli* ($\log_{10}$)[a]	*S. aureus* ($\log_{10}$)[a]
Antimicrobial towelette	12	10	3.6 ± 0.161	5.7 ± 0.046
		30	3.5 ± 0.161	5.7 ± 0.046
		60	3.1 ± 0.161	5.7 ± 0.046
Antimicrobial liquid soap	12	10	4.2 ± 0.161	5.8 ± 0.046
		30	3.7 ± 0.161	5.8 ± 0.046
		60	3.6 ± 0.164	5.7 ± 0.047

[a] Mean colony-forming units ± standard deviation.

Table 13 Needs for Assessing Efficacy of Handwashing Agents for Various Consumer-Related Activities

Activity	Method needs
Meal preparation	This method needs to incorporate a high inoculum level (6–8 $\log_{10}$) [6,7] and organism composition, similar to those found in food. Organic load needs to be present. Current methods, such as the meal preparation test and the sponge test, reflect these situations.
Diaper changing At-home sick care Attending daycare	Methods to reflect these situations need to incorporate inoculum levels of 5–8 $\log_{10}$ of bacteria [6,7] and organism composition, similar to those found in feces or other body excretions. Organic load needs to be present. Current methods do not encompass all of these situations.
Travel and recreation Attending school or work General contact with people	Methods to reflect these situations need to incorporate a moderate inoculum count 4–6 $\log_{10}$ [6,7]. Survey work to understand the organic load acquired during these activities needs to be conducted and incorporated into methods to reflect these activities.
Gardening/yard work Contact with pets	To develop a method that reflects these situations, survey work needs to be done to understand the level and type of bacteria. Further work needs be done to establish the typical soiling acquired from these activities.

odologies it is necessary to incorporate the organism type and soiling level that best fits the activity. As shown in Table 13, additional methods need to be developed to reflect situations that are encountered by the consumer.

One way to demonstrate the potential benefit of consumer antimicrobial products is to develop scenario-specific tests that are more reflective of situations, the activity, and the product use. One remaining question to be answered in order to evaluate products fairly and appropriately is if exposure is accurately represented by mimicking the level of bacteria and type of bacteria encountered or if subjects need to be exposed to the actual source of the bacteria.

V. SUMMARY

Methods other than the HCPHWT should be used to evaluate the efficacy of handwashing products intended for general consumers. Different product forms (i.e., wipes and sanitizers) and uses need to be evaluated in a method that has been developed for these specific forms and the situations in which they are used. We have developed methodologies that are appropriate for determining the efficacy of products in removing bacteria encountered during situations similar to meal preparation, where there are high levels of contamination and high organic loads. This does not cover all relevant scenarios. Other scenarios should also be considered to properly evaluate the performance of these consumer products under a wide variety of hand-soiling and bacterial levels. Clearly, the first step in developing methods to demonstrate product effectiveness in any situation would be to conduct studies to show that the surrogate method is a relevant representation of the situation.

Our experiments have shown that none of the products tested totally eliminated transferable bacteria from hands. This suggests that handwashing as currently practiced is not adequate and that antibacterial agents that produce a higher level of removal of bacteria and provide a residual effect need to be developed.

REFERENCES

1. SA Sattar, J Tetro, VS Springthorpe. Impact of changing societal trends on the spread of infections in American and Canadian homes. Am J Infect Control 27:S4–S21, 1999.
2. GJ Jackson. Principles and costs in the regulation of microbially contaminated foods. Southeast Asian J Trop Med Public Health 22S:3–9, 1991.
3. Council for Agricultural Science and Technology. Foodborne Pathogens: Risks and Consequences. Ames, IA: CAST, 1994.
4. E Scott, SF Bloomfield, CG Harlow. An investigation of microbial contamination in the home. J Hyg 89:279–293, 1982.
5. E Scott, SF Bloomfield. The survival and transfer of microbial contamination via cloths, hands and utensils. J Appl Bacteriol, 68:271–278, 1990.
6. P Rusin, S Maxwell, C Gerba. Quantitative evaluation of bacterial cross-contamination of the hands following everyday activities. American Society for Microbiology Poster Symposium, Los Angeles: May 2000.
7. P Rusin, S Maxwell, C Gerba. Comparative transfer efficiency of bacteria and viruses from common fomites to hands and from the hand to the lip. American Society for Microbiology Poster Symposium, Los Angeles: May 2000.
8. S Noteffilans, G Gallhoff, MH Zwietering, GC Mead. The HAACP concept: specification of criteria using quantitative risk assessment. Food Microbiol 12:81–90, 1995.
9. E Larson. A causal link between handwashing and risk of infection? Examination of the evidence. Infect Control Hosp Epidemiol 9:28–36, 1988.
10. BH Keswick, CA Berge, RG Bartolo, DD Watson. Antimicrobial soaps: their role in personal hygiene. In: RA Aly, KR Beutner, H Maibach, eds. Cutaneous Infection and Therapy. New York: Marcel Dekker, 1997: 49–82.
11. Standard test method for evaluation of health care personnel handwash formulation.

In: Annual Book of ASTM Standards, Volume 11.05, Standard Designation E 1174–1194.
12. BF Stanton, JD Clemons. An educational intervention for altering water-sanitation behaviors to reduce childhood diarrhea in urban Bangladesh. I. Application of the case control method for development of an intervention. Am J Epidemiol 125:284–291, 1987.
13. BF Stanton, JD Clemons. An educational intervention for altering water-sanitation behaviors to reduce childhood diarrhea in urban Bangladesh. II. A randomized trial to assess the impact of intervention on hygienic behaviors and rates of diarrhea. Am J Epidemiol 125:292–301, 1987.
14. O Leroy, M Garenne. Risk factors of neonatal tetanus in Senegal. Int J Epidemiol 20:521–526, 1991.
15. W Hlady, N Bennett, AR Samadi, J Begum, A Hafez, AI Tarafdar, JR Boring. Neonatal tetanus in rural Bangladesh: risk factors and toxoid efficacy. Am J Public Health 82:1365–1369, 1992.
16. MU Khan. Interruption of shigellosis by handwashing. Trans R Soc Trop Med Hyg 76:164–168, 1982.
17. JM Wilson, GN Chandler, Mushlihatun, Samiluddin. Handwashing reduces diarrhea episodes study in Lombok, Indonesia. Trans R Soc Trop Med Hyg 85:819–821, 1991.
18. RE Black, AC Dykes, KE Anderson, JG Wells, SP Sinclair, W Gary, Jr., MH Hatch, EJ Gangarosa. Handwashing to prevent diarrhea in day care centers. Am J Epidemiol 113:445–451, 1981.
19. L Kotloff, IP Nataro, GA Losonsky, SS Wasserman, TL Hale, DN Taylor, IC Sadoff, MM Levine. A modified *Shigella* volunteer challenge model in which inoculum is administered with bicarbonate buffer: clinical experience and experience for *Shigella* infectivity. Vaccine 13:1448–1494, 1995.
20. JO Hendley, IM Gwaltney, Jr. Mechanisms of transmissions of rhinovirus infections. Epidemiol Rev 16:242–258, 1988.
21. RB Turner, MT Wecker, G Pohl, TI Witek, E McNally, R St George, B Winther, FG Hayden. Efficacy of tremacamra, a soluble intercellular adhesion molecule 1, for experimental rhinovirus infection. A randomized clinical trial. JAMA. 281: 1797–1804, 1999.
22. CN Haas, JB Rose, CP Gerba. Quantitative Microbial Risk Assessment. New York: John Wiley & Sons, Inc., 1999.
23. JB Rose, CN Haas. A risk assessment framework for the evaluation of skin infections and the potential impact of antibacterial soap washing. Am J Infect Control 27:S26–S33, 1999.
24. RR Singh, AM Marples, AM Klingman. Experimental *Staphylococcus aureus* infections in humans. J Invest Dermatol 57:149–162, 1971.
25. JB Rose, CN Haas. Quantitative assessment of risk reduction from handwashing with antibacterial soaps. American Society for Microbiology Poster Symposium, Los Angeles, May 1999.
26. C Crockett, CN Haas, A Fazil, JB Rose, CP Gerba. Prevalence of shigellosis in the U.S.: consistency with dose-response information. Int J Food Microbiol 30:87–100, 1996.

27. LL Gibson, JB Rose, CN Haas. Use of quantitative microbial risk assessment for evaluation of the benefits of laundry sanitation. Am J Infect Control 27:S34–S39, 1999.
28. G Fischler, M Shaffer. Healthcare continuum: a model for the classification and regulation of topical antimicrobial wash products. Presented at joint meeting between CTFA/SDA, June 1997. Described in http://www.ctfa.org/viewpage.cfm?id=997. Federal Register, Washington, D.C.
29. S Namura, S Nishijima, KJ McGinley, JJ Leyden. A study of the efficacy of antimicrobial detergents for hand washing: using the full-hand touch plates method. J Dermatol 20:88–93, 1993.
30. DL Charbonneau, JM Ponte, HA Kochanowski. A method of assessing the efficacy of hand sanitizers: use of real soil encountered in the food service industry. J Food Prot 63:495–501, 2000.
31. WL Hillhimer, KF Wiandt, HH Keswick, JS Englehart, PH Neumann, GK Mulberry, AR Hrady. Transfer potential of transient bacteria upon contact after handwashing. American Society for Microbiology Poster Symposium, 1999.
32. Data on file at the Procter & Gamble Company.
33. R Aly, HI Maibach. In vivo methods for testing topical antimicrobial agents. J Soc Cosmet Chem 32:317–323, 1981.
34. MB Finkey, NC Corbin, LB Aust, R Aly, HI Maibach. In vivo effect of antimicrobial soap bars containing 1.5% and 0.8% trichlorocarbananilide against two strains of pathogenic bacteria. J Soc Cosmet Chem 35:351–355, 1984.
35. DD Scala, GE Fischler, BM Morrison, R Aly, HI Maibach. Evaluation of antibacterial bar soaps containing triclocarban. American Academy of Dermatology (poster), 1995.
36. WL Billhimer, CA Berge, JS Englehart, GY Rains, BH Keswick. A modified cup scrub method for assessing antibacterial substantivity of personal cleansing products. J Casmet Sci 52:369–375, 2001.
37. J Leyden, AM Kligman. Antimicrobials. In: AM Kligman, JJ Leyden, eds. Safety and Efficacy of Topical Drugs and Cosmetics. New York: Grune & Stratton, 1982: 289–309.
38. P Williamson, AM Kligman. A new method for the quantitative investigation of cutaneous bacteria. J Invest Dermatol 45:498–503, 1965.
39. Data on file at the Procter & Gamble Company.
40. CA Hart, MF Gibson, AM Buckles. Variation in skin and environmental survival of hospital gentamicin-reistant enterobacteria. J Hyg Camb 87:277–285, 1981.
41. CA Macintosh, PN Hoffman. An extended model for transfer of microorganisms via the hands: differences between organisms and the effect of alcohol disinfection. J Hyg (Lond) 92(3):345–355, 1984.

23

Importance of Skin Care Attributes in Developing Topical Antimicrobials

Daryl S. Paulson
BioScience Laboratories, Inc., Bozeman, Montana

Just casually looking on the shelves of discount, department, and grocery stores, one will find a plethora of personal care products claiming to have high skin-moisturizing abilities or, at least, to be nonirritating to the skin. Unfortunately, many of these claims are not grounded in valid data, and, worse, consumers over the long run will not use a product that irritates their skin.

If the maxim "80% of consumers buy from only 20% of the product lines" is correct, what do those 20% of product lines have that the other 80% fail to provide? These products offer consumers what they need, want, and demand. Therefore, capturing a greater market share, in most cases, is merely a matter of figuring out what consumers want and building a product that meets the criteria. This is clearly not a job just for marketing. Instead, it requires a unified, integrated systems approach to formulating, manufacturing, and marketing the product. This is a concerted effort in which different in-house groups work in a coordinated, effective manner—a system that requires formulators to blend a product with attributes desired by consumers; manufacturing personnel who accurately and precisely produce that product; and a marketing group that "gets the word out" regarding the product's ability to meet consumer wants [1]. Finally, consumers who use the product must find that it fits their needs.

Notice that this system relies on not one information perspective, but three—the physical attributes of the product, such as the ability to moisturize the skin and the ease of product use; consumers' subjective preferences (visual, tactile, fragrance, and tacit beliefs); and the beliefs, values, and perceptions concerning a product shared by target consumers. As long as the information system collects valid data from these three perspectives, incorporates them into the product development, and consumers realize it, it can be a winner, not just one of the many ''me, too'' products.

I. IMPORTANT FUNCTIONS

The following is a brief discussion of important functions formulators, manufacturers, and marketing personnel must address to produce a successful product. It is important that product formulators have the ability to develop the type of product desired by consumers, which is a challenge in itself. It is crucial first, however, that formulators know what consumers really desire.

Generally, the product must be formulated to hydrate the skin very quickly and have moisturizing effects lasting minimally several hours. The product must be physically and chemically stable under a variety of environmental conditions (e.g., extreme hot and cold climate conditions) and must not come ''out of solution,'' becoming visibly layered, or break down, leaving a ''funny'' odor on the body after use [2].

Hence, it is important that product stability testing, package integrity testing, and preservative effectiveness testing answer those questions. Finally, formulators need to assure that the raw ingredients they use are nonirritating or, better yet, hypoallergenic.

Manufacturers must produce, on a large scale, the products that formulators blend on a small scale—usually a challenging task. Ingredients must be guaranteed chemically pure, and the manufacturing process must be under strict quality control to assure that the product produced is what is expected.

While formulators are usually expected to formulate what marketing requests, that may not be the best approach. Marketing personnel often have a general idea of what consumers want, but to target consumers' wants more accurately, it is wise to go directly to potential consumers to find out from them. Two of the three informational perspectives can be collected through surveys (shared values) and product component preference testing (individual, subjective values). The third perspective (objective) is accomplished by evaluating the skin of human test subjects before, during, and after using the product. Collecting relevant and valid information from these three perspectives is critical, but the amount of data needed depends on the size of the market, what is already known about the market, market sales potential, and return-on-investment potential. The data collected from the three perspectives must then be integrated into the product design in order to meet consumer wants in shared values, individual, subjective preferences, and objective product performance requirements.

In one case, a company that manufactures a very high-end hand moisturizer did not observe these procedures. Instead, it performed an in-depth market survey throughout the U.S. to determine the shared values of the target consumer group. From this market survey, they learned what the target group, as a whole, valued. Additionally, product performance was evaluated through a number of skin moisturization evaluations employing human subjects using different moisturizer combinations. Based on these two sources of information, a high-end "Upper Manhattan" product was developed. It incorporated the information gained from the market survey accurately, and the product was very effective in rehydrating (moisturizing) the skin and also in retaining moisture over the course of several hours postapplication. But the product has not been a large success for the company.

Upon reviewing the collected data and performing an analysis, we discovered a very significant piece of missing information—the *individual subjective preferences*. So we sampled a panel of human subjects from a pool of target consumers and discovered that, upon repeated and prolonged use of the product, the consumer's subjective perception of using it was that it left a "greasy feeling" on the hands. The bottom line was that, while the product met the shared values of the target consumer group and the objective physical attributes desired, the personal subjective preferences of consumers were not met.

II. SUCCESSFUL PRODUCT DEVELOPMENT

Let us now examine in closer detail the three facets—objective product performance, subjective preference, and shared values—and what each entails.

With regard to objective performance, there are two general aspects of skin physiology that an effective moisturizing product must meet. The first is that it must rehydrate the skin by coating it with an oily film that holds in moisture. Second, it must continue to protect the skin over time by keeping the water moisture level on the skin's surface high enough to prevent significant dehydration [3]. A common way of meeting these goals is to add various nonirritating emollients to a number of potential moisturizing formulations and evaluate them using human volunteers in a small-scale pilot study. Two general approaches can be taken, depending upon whether the product is intended primarily to moisturize or rehydrate damaged skin or primarily to protect the skin from irritation from, for example, repeated hand washing. For moisturization studies, subjects with visible, measurable skin damage (e.g., redness and chaffing) can be recruited and enrolled in a study designed specifically to evaluate the degree to which different product formulations rehydrate the skin over time. In the second approach—protection against skin irritation—subjects with normal skill can be recruited and enrolled in a study where they wash repeatedly with irritating soaps (20–40 times per day over the course of several days) and treat the skin with different product formulations. The product formulations are evaluated based on their ability to prevent or, more likely, retard progression of skin irritation using statistical analysis.

The traditional way of evaluating skin irritation is through visual examination based on four indices: dryness (chafing), redness (erythema), swelling (edema), and rash, each at four severity levels. Often, however, only the first two indices—dryness and redness—are employed. Visual examination is very telling of skin irritation, but it does not evaluate the actual ability of the skin to hold moisture in, nor measure the actual moisture level of the skin. Two other methods, both instrumentational, are better for assessing these parameters by measuring actual skin moisture content and rate of transepidermal water loss [3].

When the barrier functions of the skin, mainly due to the covering of oils, is compromised, moisture leaves the skin at a higher rate than observed in normal skin. Measurement of transepidermal water loss, conducted using a tewameter, is not as straightforward as one might like, particularly because there are multiple reasons for variation in transepidermal water loss, including anxiety, stress, amount of fluid intake, alcohol consumption, smoking, diet (particularly salt intake levels), tiredness, sleep cycle changes, colds, flu, and emotional states. Hence, it is often very difficult to capture meaningful data using this measurement. It has been found, however, that sequestering subjects on-site at least 12 hours before testing and supplying a controlled diet, including liquid intake, can provide for much more reliable data in detecting transepidermal water loss from compromised and/or treated skin, but this is expensive to do. One way we have worked around this is to use adequately large sample sizes to achieve the statistical power required to differentiate products.

The total moisture content of the skin can be measured using a corneometer. We have found it to portray accurately the skin condition in many useful applications. For example, in studies designed to demonstrate the rehydration of damaged skin, it provides accurate measurements for evaluating a product's ability to "hold the water in," once the product has been applied, useful even with small sample sizes. The corneometer readings also tend to correlate well with visual skin grading/scoring, with the added benefit of providing quantitative measurements enabling rate and rate change estimates so that predictions can be made using parametric statistical models instead of relying on differences detected by qualitative visual grading and using nonparametric methods. This greatly enhances the power of the evaluation.

When using the tewameter and corneometer, we have found that several controls should be in place to provide the best results. First, the humidity of the environment where subjects are sequestered and where the measurements take place need to be held constant throughout the course of the study. Second, in order to reduce extraneous environmental nuisance variables, which can cause fluctuations, air drafts caused by movement of personnel and building heating/cooling equipment, the tewameter/corneometer and the subjects' hands, when measurements occur, need to be housed within a closed environmental system. We do this by employing a plexiglass environmental cabinet.

As previously stated, it is important that preference of individual consumers, a subjective measurement, be known. When the product is developed, individual consumers/users need to feel that it was made for them because, in actuality, it was. Ideally, consumers need to feel it is "their" moisturizer and identify the product as part of their world, perceptions very desirable for a moisturizing product company.

There are two effective and equally important ways of assessing subjective preference. The first method involves selecting panelists who match the target consumer population and having them participate in the preference testing of several prototypes of the product. The second method involves providing the product prototype to panelists; they then smell it, shake it, use it, etc., and describe what they like or do not like about the product.

Probably the most fundamental way of evaluating subjective attributes is to assemble a group of subjects that meet target consumer group characteristics and have them evaluate the product in terms of, for example, its ease of use, feel, clarity, color, and appearance. Table 1 provides a list of several important sensory attributes.

Please note that each of these perceptions is what psychologists call "surface structure questions" [1,4]. For example, if a person believes the product is too oily, the "too oily" phase is the surface or apparent structure. But underlying this surface structure is what is known as a deep structure, or the underlying

Table 1 Subjective Moisturizer Perceptual Attributes

Skin-softening ability
Degree of "oily feeling" after use
Fragrance characteristics
Perception while applying the product
Perception of skin before and after use
Moisturizing ability
What makes the product unique or special
Perception of what others will notice after use (e.g., fragrance, soft skin, etc.)
Perception of self after use (confidence level of being observed closely by partners, friends, strangers)
Perception of type of person who would use the product
Identification with the product

reason. ''Too oily'' from this deeper structural perspective can mean ''undesirable as a lover, less lovable, not good enough'' or whatever. Connecting surface structures to the deeper structures can provide a very strong motivational tool and can help assure that a product will be a success through using that information in subtly suggestive ads, packaging, etc.

The second way to gain valuable, often missed, preference testing information is to have panelists explore a prototype product and evaluate it without structured questions or rating scales, such as how it feels, how it goes on the skin, how it looks, etc. The key is to keep the evaluation process open-ended. This will help manufacturers get out of their preconceived ''product box'' and acquire novel information. To uncover greater information, the panelists can be interviewed with such questions as: How was it for you to use the product? What was your experience? How did you feel later about the product? How do you feel when you apply the product(s)? What did you first notice after applying it? How do you feel about it now?

The power of group perception goes a long way in assisting in the sale of a product, particularly if role models are seen using the product. For example, if Cindy Crawford is shown using a particular moisturizer, the surface message is ''to be beautiful and exciting like her, use this moisturizer.'' The implied, deeper message, however, is that a person will be more valued and more lovable for having beautiful skin, like the glamorous Cindy Crawford.

Additionally, the shared values of the group tend to position the consumer on the status hierarchy. For example, if one uses a discount, generic product, that will tend to promote the feeling of being a less glamorous person. However, if one uses a high-end cosmetic department store product, that will promote the image of a very sophisticated, well-groomed person.

Shared values tend to bind the limits of what is acceptable and valued to specific degrees. Using a moisturizing product that has a ''goaty'' fragrance, no matter how effective it is, will not be accepted in terms of shared group values. One that smells ''young, clean, and zesty'' will be highly valued, however, particularly by a culture obsessed with ''youth.'' Finally, shared surface and deep values can be magnified by advertising campaigns. For example, if a person perceives that she is taking better care of herself by using a specific moisturizer (a surface value), and if she feels more valuable, more loved, and/or more needed by her family (deep values), she will be motivated to purchase the product.

Figure 1 portrays the relationship between the three fundamental perspectives. A break anywhere in the circular flow will hinder the product's market success.

It is far easier to launch a product that conforms to consumer group needs and meets personal shared values and beliefs than to try to reeducate the market to accept other values. As time goes on, however, the product itself will influence the perceptions of groups, as well as personal users.

Generally, the initial concern in product development is to satisfy general

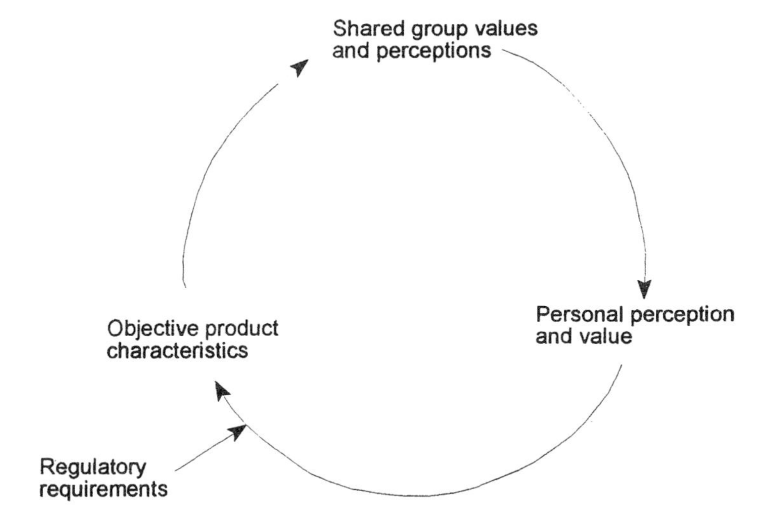

Integration of the Three Market Perspectives:

- Shared group values and perceptions
- Personal perception and value
- Objective product characteristics including regulatory requirements

Figure 1 Integration of the three market perspectives.

group values, beliefs, and perceptions of a product. The group sets general constraints as to what is and what is not acceptable. Within the group-constructed constraints are personal, subjective values, beliefs, and likes/dislikes. It is critical that one not view these three perspectives as separate. They are distinct, but not separate. They are interrelated and interdependent.

There is tremendous opportunity for marketing effective and desired skin-moisturizing products, but vision is needed. For example, markets wide open to moisturizers include food service/processing antimicrobial handwash formulations and consumer and medical topical antimicrobial products. The key is to stay close to targeted customers and listen, listen, listen.

REFERENCES

1. DS Paulson. Handbook of Topical Antimicrobial Evaluations. New York: Marcel Dekker, 1999.

2. HA Moskowitz. Cosmetic Product Testing. Vols. 1 and 2. New York: Marcel Dekker, 1984.
3. DS Paulson. Topical antimicrobials and skin irritation: the next step. Soap/Cosmet/Chem Special November 1998.
4. DS Paulson. Developing effective topical antimicrobials. Soap/Cosmet/Chem Special December 1997.

Part VI

Testing Methods

24
A Guide for Validation of Neutralizer Systems Used In Topical Antimicrobial Efficacy Evaluations

Christopher M. Beausoleil
BioScience Laboratories, Inc., Bozeman, Montana

I. INTRODUCTION

Many antimicrobial efficacy evaluations of topical antimicrobial products involve measurements of microbial population reductions at a specific time point after exposure to the product. To determine this accurately, the antimicrobial action of the product must be stopped at the time specified for sample, and it is for this action that neutralizer systems are employed. The validity of the neutralizer system must be established prior to performing the antimicrobial efficacy test. This concern for neutralizer validity has long been known, and a number of methods have been proposed for validating neutralizer systems [1–5]. Each of the methods focuses on two major concerns: (1) the neutralizer system must demonstrably neutralize the antimicrobial properties of the product, and (2) the neutralizer system must be proven nontoxic to the test microorganism(s). Few validation methods apply techniques of statistical analysis to the determination of their validity [5]. The purpose of this chapter is to incorporate statistical processes of analysis into a neutralizer validation system, as a means of providing more accurate and reliable outcomes.

Topical antimicrobial efficacy tests comprise test methods similar to those outlined in the Food and Drug Administration's Tentative Final Monograph for Healthcare Antiseptic Drug Products: time-kill kinetic studies; effectiveness testing of a surgical hand scrub; effectiveness testing of an antiseptic handwash or healthcare personnel handwash; and effectiveness testing of a patient preoperative skin preparation [6]. The validation method proposed here can be applied to these and modifications of these topical antimicrobial efficacy tests. Statistical methods will be used in the design of the validation study and applied to analysis of the data.

II. DESIGN OF THE VALIDATION STUDY

Application of statistical analysis to the neutralization data will permit a researcher to make reliable conclusions about the efficacy of a neutralizer system. The relevance of the statistical analysis depends entirely on the design of the experiment. The neutralization study design is grounded in four basic phases to be applied to one or more microorganism species to be tested.

Phase 1—A population of the challenge microorganism is exposed to a noninhibitory medium such as trypticase soy broth or a phosphate-buffered saline solution and then assayed for number of microorganism. This will establish the initial or baseline population of the challenge microorganism against which numbers resulting from other phases of testing will be compared.

Phase 2—The initial population of the challenge microorganism is exposed to the antimicrobial product at use-strength for a specific time and then assayed for number of surviving microorganisms. This is the positive control phase to establish that the product actually exhibits antimicrobial activity that must be neutralized.

Phase 3—The initial population of the challenge microorganism is exposed to the neutralizer system alone and then assayed for number of surviving microorganisms. This phase tests whether the neutralizing system is toxic to the test microorganism. Nontoxicity is demonstrated by no statistically significant difference between the counts from this phase and Phase 1.

Phase 4—The test product is dispersed into the neutralizer system, followed immediately by exposure to the initial population of the challenge microorganism. This phase tests whether the neutralizer system effectively neutralizes the antimicrobial activity of the product. Effective neutralization is demonstrated if the counts from this phase and Phase 1 are not significantly different.

III. PROPOSED NEUTRALIZER SYSTEM

There are three general approaches to neutralizing that can be utilized: to employ chemical inactivators that interact directly with the antimicrobial agent and elimi-

Table 1 Neutralizing Solution Ingredients

11.67 g Lecithin
100 mL Polysorbate 80
10.1 g Na_2HPO_4
0.4 g K_2HPO_4
5.0 Sodium thiosulfate pentahydrate ($Na_2S_2O_3$—$5H_2O$)
10 g Tamol
1 L Deionized water

Preparation: Suspend 11.67 g lecithin and 100 mL polysorbate 80 in 1 liter of hot deionized water; boil and stir for thirty (30) minutes. Add 10.1 g Na_2HPO_4, 0.4 g K_2HPO_4, 1.0 mL Triton ×-100, 5.0 g sodium thiosulfate pentahydrate ($Na_2S_2O_3$—$5H_2O$), and 10 g Tamol and mix thoroughly. Adjust final pH to 7.8–7.9 with 0.1 N sodium hydroxide or 0.1 N hydrochloric acid.

nate its killing action, dilution of the antimicrobial agent to a sublethal level, and membrane filtration to separate the antimicrobial agent from the microorganism physically [4,5,8]. The most commonly employed neutralizer systems incorporate the first two of these approaches—dilution of the antimicrobial product in a liquid medium containing chemical inactivators. Membrane filtration is also used but is time-consuming and very costly by comparison.

This chapter presents neutralization procedures for six topical antimicrobial compounds [9]: chlorhexidine gluconate (CHG), iodine (tinctures and povidone-iodine), alcohols (ethyl and isopropyl), triclosan, parachlorometaxylenol (PCMX), and quaternary ammonium compounds (QAC).

In the past, numerous formulas have been developed for neutralization of these compounds. Here, I present a single neutralizer system that is effective for all, there by simplifying the testing and analysis of products containing them. The neutralizing solution I suggest (Table 1) contains chemical inactivators suitable for each of the above six antimicrobial compounds [10–15], and these in conjunction with dilution will effectively neutralize products containing them. For triclosan, chemical inactivators will also be required in the plating medium, i.e., 0.5% (v/v) Tween 80 and 0.07% (w/v) lecithen.

IV. STATISTICAL ANALYSIS

If one formulates the neutralizer correctly, all that is left is a valid statistical comparison of the data. I find it expeditious to use a procedure presented by Paulson [16], which simplifies the statistical analysis into a series of six steps:

Step 1: Formulate the null hypothesis (H_0) and an alternative hypothesis (H_A); that is, formulate the claim and its negation to serve as the two alternatives being considered (''mutually exclusive'').

Step 2: Choose a sample size, n, and a risk factor, the probability of Type I error.

Step 3: Choose the test statistic to be used, as appropriate for the sample data.

Step 4: Formulate a decision rule; that is, decide what the evidence must show in order to either support or disprove the test hypothesis.

Step 5: Collect the sample data and perform the statistical calculations.

Step 6: Apply the decision rule and make the decision based upon statistical results.

For validation of the neutralizing system, the null and alternative hypothesis can be easily determined (Step 1). The null hypothesis (H_0) is that there is no significant difference between the initial microbial population recovered from Phase 1 and the test microbial populations recovered from the other phases. The alternative hypothesis (H_A) is that a difference does exist between the initial microbial population recovered from Phase 1 and the test microbial population recovered from the other phases.

Steps 2 and 3 should be considered together. An appropriate sample size can be calculated for the chosen test statistic, adjusting for the risk factors and using a basic understanding of the data to be collected (variability of the data). For the validation, a pooled, two-sample Student's t-test can be used to discern any differences between results of Phase 1 and the other four phases. The procedures for performing a pooled, two-sample Student's t-test are presented in many statistical books and will not be reviewed in this chapter [16,17]. The data collected are from a natural population and must be transformed to a linear scale for use in the Student's t-test by transforming the data to $\log_{10}$ values [18]. The chance of committing Type I error (α) should be small. That is, there should be little chance of rejecting a true null hypothesis, saying there is a significant difference between the two sets of data when in fact there is not. When validating neutralizing systems, a researcher must also take into consideration that a chance exists, too, of accepting a false null hypothesis, stating that there is no significant difference between the two sets of data when in fact there is. This is a Type II error (β). It would be disastrous if an ineffective neutralizing system were determined to be effective in the neutralization validation. For this test, the α and β will be small so that the chance of committing either type of error will also be small. With small α and β, the researcher will be able to accept or reject a neutralizing system with reasonable certainty. A common, accepted value of α for biological data is 0.05 (1 in 20 chance of committing Type I error). Because

Type II error is also important in this validation, a β of 0.10 (1 in 10 chance of committing type II error) will be used for the analysis.

An appropriate sample size, or number of replicates, can be calculated for the type of statistical test by using the α and β error, the minimal detectable difference between two test procedures, and the variability (standard deviation) of data determined from previous neutralization system validations. The statistical test chosen to detect if there was a significant comparative increase or decrease in microorganism populations is the two-tailed, pooled Student's *t*-test. Both and values have been determined, 0.05 and 0.10, respectively. The minimal detectable difference is the minimal difference between samples from two procedures that the researcher would consider as significant and would want to be assured of detecting. Minimal differences that have been published are 0.15, 0.20, and 0.30 $\log_{10}$ differences between data from Phase 1 and those from other phases [4,19,20]. The 0.15 $\log_{10}$ difference will be used for this validation, because it is the most conservative and is from a validation test that involves multiple samples (replication) and a statistical analysis [4]. The final requirement, variability of the data, will be difficult to establish, especially because many researchers will be performing this validation for the first time. If past data are unavailable, then an option is to use an excessive sample size (at least 10) and use the data from that validation to determine an appropriate sample size for future validation studies.

The required sample size for performing a pooled, two-sample Student's *t*-test is calculated from the formula [17]:

$$n \leq \frac{2S_p^2}{\delta^2}(t_{\alpha,(2),v} + t_{\beta,(1),v})^2$$

where:

n = sample size for each procedure
Sp = pooled sample standard deviation (variability)
δ = minimal detectable difference between the two tests
$t_{\alpha,(2),v}$ = critical value of Student's t distribution at $\alpha/2$ for v degrees of freedom
$t_{\beta,(1),v}$ = critical value of Student's t distribution at $\beta/1$ for v degrees of freedom
α = 0.05 ("/2 = 0.025 for a two-tailed test)
β = 0.10
$v = 2(n - 1)$

It was already mentioned that the sample size would be difficult to calculate in the absence of previous studies to draw on for data variability. Acceptable standard deviations can be calculated for any sample size by transforming the calculation and setting the sample size to various numbers.

$$S_p \leq \sqrt{\frac{n\delta^2}{2(t_{\alpha,(2),v} + t_{\beta,(1),v})^2}}$$

Table 2 Required Standard Deviations

Sample size (replicates)	Required standard deviation
15	0.122
14	0.117
13	0.113
12	0.108
11	0.103
10	0.098
9	0.092
8	0.086
7	0.079
6	0.072
5	0.064
4	0.055
3	0.043

Table 2 lists sample sizes from 3 to 15 and the corresponding standard deviations. For example, a researcher performing a validation using 15 replicates for each procedure has obtained a pooled standard deviation of 0.082. When the researcher performs another validation, as per Table 2, the sample size can be reduced to 8 replicates as long as the pooled standard deviation for that test remains at or below 0.082. Table 2 can also be used for instances when the variability of the data is small or nonexistent. If the researcher performing the validation using 8 replicates obtains data where variability is less than 0.082, then there is a possibility of the statistical test reporting a significant difference between procedures. In this case, and only in this case, if the difference between the two tests is less than the 0.15 (minimal detectable difference), then the researcher could conclude that there is no difference between the two procedures.

The decision rule, Step 4, is based on the significance level (″) of 0.05. If the t-test performed calculates a t value of *less than* the tabled t value at ″ = 0.05, then the null hypothesis, no detectable difference between samples, will be accepted. If the t-test performed calculates a t value *greater than* the tabled t value at ″ = 0.05, then the null hypothesis will be rejected—there is a significant difference between samples.

The final two steps can now be performed. For Step 5, perform the validation, collect the data, and perform the statistical analysis. Once that is complete, Step 6 then is performed, applying the decision rule established in Step 4 to the results of the statistical analysis.

V. VALIDATION METHOD

A. Apparatus, Reagents, and Materials

All specific conditions of the antimicrobial efficacy test must be duplicated in the validation. All equipment, types of media, incubators, and temperatures should be identical to those applied in the antimicrobial efficacy test. Even the technician who will be performing the antimicrobial efficacy test should also perform the validation of the neutralization system.

B. Test Microorganism(s)

The populations of the test inocula used in the validation should be kept to a minimum to allow plating directly from the neutralizing system with the highest concentration of antimicrobial product that will be used in the antimicrobial efficacy evaluation. A high-inoculum population would require diluting for enumeration. The diluting would prevent the researcher from determining if neutralization occurred immediately with the neutralizing system or from dilution of the product [4,5,8].

A validation of the neutralizing system must be performed with the microorganism(s) evaluated in the antimicrobial efficacy evaluation. Some evaluations involve testing of numerous microorganisms (time-kill evaluation) or microorganisms sampled from the skin of human subjects. In either case it would be impractical and nearly impossible to perform a validation for every microorganism. For antimicrobial efficacy evaluations such as the time-kill evaluation, representatives of gram-negative and gram-positive microorganisms and a yeast should at least be evaluated [6], preferably strains that are the most sensitive to the antimicrobial(s) being evaluated. For studies involving sampling of microorganisms from human subjects (surgical scrub or preoperative prep evaluations), *Staphylococcus epidermidis*, a common skin bacterium, should be used for the validation [21].

The procedures used to prepare the challenge suspension for the antimicrobial efficacy evaluation must be used to create the suspensions for the validation. Such methods involve creating a broth culture of the microorganism, transferring the broth culture to an agar plate, and using the plate culture to create a standardized population of the microorganism in a phosphate-buffered saline solution. The microorganism suspension must be diluted to a level that will permit plating directly from the test solutions. For example, dilute a challenge microorganism suspension to a concentration of 3.0×10^3–3.0×10^4 CFU/mL. This concentration is for validation studies that will use 0.1 mL of the diluted challenge microorganism suspension transferred to 9.9 mL of test solution, followed by plating of 1.0 mL of the test suspension. This will result in plates with colony counts of 30–300 CFU per plate.

C. Procedures

The following is a suggested method for performing a neutralizing system validation and is presented for the researcher to adapt to their specific antimicrobial efficacy evaluation. The procedures are intended for a neutralizing system using chemical inactivators, in combination with dilution of the antimicrobial product. The challenge microorganism suspension is appropriately 3.0×10^3–3.0×10^4 CFU/mL.

Phase 1—Test Microorganism Population Control

Add 0.1 mL of the test microorganism suspension to 9.9 mL of a noninhibitory solution and mix thoroughly. Immediately following mixing, plate 1.0 mL of the suspension in duplicate, using an agar medium appropriate for growth of the microorganism and the plating procedure to be used in the antimicrobial efficacy evaluation (i.e., pour-plating or spread-plating). Allow the suspension to stand for the maximum time that will be allowed prior to plating in the antimicrobial efficacy evaluation. After the exposure time has elapsed, plate 1.0 mL of the suspension, in duplicate, using a agar medium appropriate for growth of the test microorganism and the plating procedure to be used in the antimicrobial efficacy evaluation. Repeat this procedure until the number of samples required is completed.

Phase 2—Antimicrobial Product Control

Add 0.1 mL of the test microorganism suspension to 9.9 mL of the antimicrobial product and mix thoroughly. Immediately following mixing, plate 1.0 mL of the suspension, in duplicate, using the same medium and plating procedures as in Phase 1. Allow the mixture to stand for exposure period observed in Phase 1. After the exposure time has elapsed, plate 1.0 mL of the suspension, in duplicate, using the same medium and procedures as in Phase 1. Repeat this procedure until the number of samples required is completed.

Phase 3—Neutralizing System Control

Add 0.1 mL of the test microorganism suspension to 9.9 mL of the neutralizing solution and mix thoroughly. Immediately following mixing, plate 1.0 mL of the suspension, in duplicate, using the same medium and plating procedure used in the previous phases. Allow the suspension to stand for the same exposure period observed in the previous phases. After the exposure time has elapsed, plate 1.0 mL of the suspension, in duplicate, using the same medium and plating procedures as in the previous phases. Repeat this procedure until the number of samples required is completed.

Phase 4—Neutralizing System Effectiveness

Add 1.0 mL of the antimicrobial product to 8.9 mL of the neutralizing fluid and mix thoroughly. The antimicrobial product is added to the neutralizing solution because the act of neutralizing many antimicrobials requires a chemical reaction, and chemical reactions do take time. If the microorganism suspension added to the neutralizing fluid prior to adding the antimicrobial product, the product would likely be able to interact with the microorganism prior to being neutralized. Immediately following mixing of the neutralizing fluid and the antimicrobial product, add 0.1 mL of the challenge microorganism suspension to the 9.9 mL of neutralizing solution/antimicrobial product solution and mix thoroughly. Immediately following mixing, plate 1.0 mL of the suspension, in duplicate, using the same medium and plating procedure used in the previous phases. Allow the suspension to stand for the same exposure period observed in previous phases. After the exposure time has lapsed, plate 1.0 mL of the suspension, in duplicate, using the same medium and plating procedures as in previous phases. Repeat this procedure until the number of samples required is completed.

Prior to performing this validation, the researcher should first decide the appropriate dilution ratio of antimicrobial product to the neutralizing fluid. What has been presented is a 1:10 dilution of antimicrobial product to neutralizing fluid. For most antimicrobial efficacy evaluations, especially the time-kill evaluation, this is the highest concentration of product to neutralizer that must be validated. The researcher should take time to calculate the largest amount of antimicrobial product that will be neutralized during the antimicrobial efficacy evaluation and use that amount of product for Phase 4.

For antimicrobial efficacy evaluations involving human subjects, the 1:10 dilution ratio is far greater than the amount of product that would need to be neutralized during actual testing. For these evaluations, product can be applied to alcohol-treated skin of the subjects, removed with the relevant sampling procedure, and exposed to the neutralizing system, followed by addition of the test microorganism suspension, as stated for Phase 4. Phase 3 can also be modified in a similar manner, lacking only the application of the product. The test microorganism is then added to check for inhibition. Assaying the sampled neutralizing fluid without adding the test microorganism will provide a check for residual skin flora following the alcohol decontamination procedure. And note that for an in vivo neutralization study, all regulations pertaining to the protection of the human subjects must be followed [22,23].

D. Data Collection

All plating should be performed at least in duplicate. Incubate the plates from all procedures at the same temperature and for the same time, as detailed in the antimicrobial efficacy evaluation. Following incubation, remove the plates and count the colonies. For each replicate from each procedure, calculate the average colony-forming units per plate and perform a $\log_{10}$ transformation of the average. These $\log_{10}$ population values will be used in the statistical analysis.

E. Analysis of the Results

The outline of the statistical analysis has already been presented in Sec. IV. All comparisons will be performed using the two-sample Student's *t*-test.

First, compare the $\log_{10}$ population values obtained from the immediate and the maximum exposure time from the test microorganism population control (Phase 1). Further analysis cannot be performed if these populations are significantly different. A significant difference indicates that the maximum exposure time is too long and should be shortened for testing. If the populations from the maximum exposure time are greater than the immediate populations, then the microorganism populations had sufficient time to replicate. If the populations from the maximum exposure time are less than the immediate populations, then the microorganism population was losing viability.

After the initial comparison, compare the $\log_{10}$ population values from the other phases to the $\log_{10}$ population values from the immediate test microorganism population control (Phase 1). For a successful outcome, the immediate and the maximum-exposure-time $\log_{10}$ population values from the antimicrobial product control (Phase 2) should differ significantly from those of the test microorganism population control. This will demonstrate that the product was actually antimicrobial, and that the outcome of Phase 4 will be relevant.

The immediate and maximum-exposure-time $\log_{10}$ population values from the neutralizing system control (Phase 3) should not differ significantly from those of the test microorganism population control, thereby demonstrating that the neutralizing system was not toxic to the test microorganism.

The immediate and maximum-exposure-time $\log_{10}$ population values from the neutralizing system effectiveness (Phase 4) should not differ significantly from those of the test microorganism population control, thereby demonstrating that the neutralizing system was effective in neutralizing the antimicrobial action of the product.

VI. CONCLUSION

This chapter presented a statistically based method for validating neutralizing systems used in topical antimicrobial efficacy evaluations. This method should be performed prior to performing a topical antimicrobial efficacy evaluation so that the researcher can be assured that the neutralizing system to be used will, in fact, neutralize the antimicrobial products being evaluated. Failure to validate the neutralizing system will always put the results of the topical antimicrobial efficacy evaluation in question and could possible disqualify the evaluation from use in support of the efficacy of the product. By assuring neutralization, the researcher will always be able to proceed into testing confident of producing valid antimicrobial efficacy data.

REFERENCES

1. R Quisno, IW Gibby, MJ Foter. A neutralizing medium for evaluating the germicidal potency of quaternary ammonium salts. Am. J Pharm September: 320–323, 1946.
2. T Bergan, A Lystad. Evaluation of disinfectant inactivators. Acta Path Microbiol Scand Section B 80:507–510, 1972.
3. Food and Drug Administration. Topical antimicrobial drug products for over-the-counter human use; tentative final monograph for first aid antiseptic drug products; Proposed Rule, 21 CFR Parts 333 and 369. Fed Reg June 17:31401–31452, 1994.
4. United States Pharmacopeial Convention. Validation of Microbial Recovery from Pharmocopeial Articles. In: U.S. Pharmacopeia XXIV, NF 19, 2000.
5. SVW Sutton. Neutralizer evaluations as control experiments for antimicrobial efficacy tests. In: JM Ascenzi, ed. Handbook of Disinfectants and Antiseptics. New York: Marcel Dekker, 1996, pp. 43–62.
6. Food and Drug Administration. Tentative final monograph for health-care antiseptic drug products; Proposed Rule, 21 CFR Parts 333 and 369. Fed Reg June 17: 31401–31452, 1994.
7. RH Green. Sampling Design and Statistical Methods for Environmental Biologists. New York: John Wiley & Sons, Inc., 1979, pp. 25–64.
8. A Cremieux, J Freney, A Davin-Regli. Methods of testing disinfectants. In: S. Block, ed. Disinfection, Sterilization, and Preservation. Philadelphia, PA: Lippincot Williams & Wilkins, 2001, pp. 1305–1327.
9. G McDonnell, AD Russell. Antiseptics and disinfectants: activity, action, and resistence. Clin Microbiol Rev 12:147–179, 1999.
10. SM Bloomfield. Methods of assessing antimicrobial activity. In: SP Denyer, WB Hugo, eds. Mechanism of Action of Chemical Biocides: Their Study and Exploitation. Cambridge, MA: Blackwell Scientific Publications, Inc., 1991, pp. 1–22.
11. RA Robinson, RJ Osguthorpe, SJ Carroll, RW Leavitt, GB Schaalje, JM Ascenzi. Culture variability associated with the U.S. Environmental Protection Agency tuberculocidal activity test method. Appl Environ Microbiol 62:2681–2686, 1996.
12. WB Hugo. Disinfection mechanism. In: AD Russel, WB Hugo, GAJ Ayliffe, eds. Principles and Practice of Disinfection, Preservation and Sterilization. Oxford, England: Blackwell Scientific Publications, 1992, pp. 187–210.
13. SW Frantz, KA Haines, CG Azar, JI Ward, SM Homan, RB Roberts. Chlorhexidine gluconate (CHG) activity against clinical isolates of vancomycin-resistant *Enterococcus faceium* (VREF) and the effects of moisturizing agents on CHG residue accumulation on the skin. J Hosp Infect 37:157–164, 1997.

14. L Benson, L Bush, D Leblanc. The effects of surfactant systems and moisturizing products on the residual activity of chlorhexidine gluconate handwash using a pigskin substrate. Infect Control Hosp Epidemiol 11:67–70, 1990.
15. CA Lawrence. Inactivation of the Germicidal action of quaternary ammonium compounds. J Am Pharm Assoc (Sci. Ed.) 37:57–61, 1948.
16. DS Paulson. Topical Antimicrobial Testing and Evaluation. New York: Marcel Dekker, 1999, pp. 141–197.
17. H Zar. Biostatistical Analysis. Upper Saddle River, NJ: Prentice-Hall, Inc., 1999, pp. 122–136.
18. RR Sokal, FJ Rohlf. Biometry. San Francisco: W. H. Freeman and Company, 1981, pp. 419–421.
19. G Reybrouk. Efficacy of inactivators against 14 disinfectant substances. Zbl Bakt Hyg Abr Orig B 168:480–492, 1979.
20. RM Baird. Preservative efficacy testing in the pharmaceutical industries. In: MRW Brown, P Gilbert, eds. Microbiological Quality Assurance: A Guide Towards Relevance and Reproducibility of Inocula. New York: CRC Press, 1995, pp. 149–162.
21. MJ Marples. The Ecology of the Human Skin. Springfield, IL: Charles C Thomas, 1965, p. 582.
22. Food and Drug Administration. Code of Federal Regulations, Title 21, Food and Drugs, Part 50—Protection of Human Subjects, 1999.
23. Food and Drug Administration. Code of Federal Regulations, Title 21, Food and Drugs, Part 56—Institutional Review Board, 1999.

25
Testing Methodology of Preoperative Skin Preparation and Surgical Scrub as Over-the-Counter Drugs

David K. Jeng
Allegiance Healthcare Corporation, A Cardinal Health Company, McGraw Park, Illinois

I. INTRODUCTION

Almost a century has passed since Lister introduced the concept of asepsis with chemical antimicrobials. Today's aseptic technique, however, is not much different from that of Lister's era, namely, the application of preoperative skin preparation (POSP) on the body surface of the patients and the practice of surgical scrubbing (SS) and sterile gloving by surgeons and operating room (OR) nurses before surgery. However, evolution has taken place during recent decades in the synthesis of new antimicrobial chemicals and in the methods of application. The surgeons and OR nurses now have more antimicrobials of different chemical families to choose from than at any other time. The methods of qualification of POSP and SS are carefully governed by regulatory agencies, and the methods of application are described in detail by organizations such as the Association of Perioperative Registered Nurses (AORN).

Changes have accelerated in recent years in the design, formulation, and packaging of POSP and SS. For example, the concepts of polymer film-forming POSP with disposable applicator, brushless and waterless SS with emollient or lotion, and the combination of alcohol with other effective antimicrobials, etc., have been introduced into the POSP and SS market. To a certain extent, these new products changed the scenario of OR practice, as the old rules have gradually become insufficient to certify these new products and to govern their application.

The basic microbiological methodology proposed by the U.S. Food and Drug Administration (FDA) is relatively simple and straightforward. The bioassay methodology that measures the survival of microorganisms before and after treatment with test products, both in vitro and in vivo, is well known to the microbiological community and does not even need validation. Overnight, however, this same methodology has been found to be inadequate for evaluating, for example, water-insoluble products or products that are only partially water soluble because of incompatibility with water-based bioassay systems. Appropriate methods must be submitted to the regulatory agency to accommodate the new needs of testing.

This chapter focuses on issues involving testing methodology and provides some suggestions addressing the inadequacies. New methods, standards, and requirements, as well as certain simplifications of existing methods, are presented for discussion and consideration of future modification of the regulations in order to meet the testing needs of today, and those of tomorrow.

II. REVIEW OF THE POSP EFFICACY TEST REQUIREMENTS

The efficacy test requirements are described in the Federal Register Monograph published in 1994 [Tentative Final Monograph (TFM)] [1]. The following is a dissection and summary of the content of POSP efficacy test requirements in the TFM.

A. General Test Criteria

The general test criteria describe in vitro testing requirements for the antiseptic ingredient, the vehicle, and the final formulation. The proposed testing methods of Minimum Inhibitory Concentration study are described in the National Committee for Clinical Laboratory Standards (NCCLS) Manual M7, and the American Society for Microbiology time-kill study [2]. Researchers are encouraged to submit alternative test methods for approval.

1. Antimicrobial Spectrum of Activity: Minimum Inhibitory Concentration Study

Test Product. For products containing ingredients as listed in §333.412(a), (b), (c), (d), or (e) of the TFM, the undiluted final formulation, the vehicle, and the active ingredient are tested.

Test Method. A minimal inhibition concentration (MIC) method, as described in NCCLS-M7, is used.

Challenge Microorganisms. The antimicrobial spectrum of activity of the test products must be broad, versus 25 ATCC strains of 20 species of bacteria and 2 species of yeasts listed in §333.470(a)(1)(ii) of the TFM, plus 25 fresh clinical isolates of these same species of bacteria and yeasts. The challenge microorganisms include members of normal flora in humans, common environmental contaminants, or systemic pathogens.

2. Time-Kill Kinetic Study

Test Product. For products containing ingredients as listed in §333.412(a), (b), (c), (d), or (e) of the TFM, the active ingredient, the vehicle, and the final formulation are to be tested. A 10-fold dilution of the final formulation is required, but there is no dilution requirement for the active ingredient or the vehicle.

Test Method. Microorganisms are exposed to the test materials for proposed time intervals of 0, 3, 6, 9, 12, 15, 20, and 30 minutes. At the end of the exposure period, an appropriate neutralizing agent is used to quench antimicrobial activity chemically and/or by dilution. Microorganism populations before and after the treatment are enumerated to calculate the antimicrobial activity of the test product, expressed in a $\log_{10}$ reduction format.

Challenge Microorganisms. The same microorganisms specified for the MIC test are to be tested.

3. Test for Potential for Emergence of Microbial Resistance

This test determines the evolution of a point mutation by a sequential passage of a challenge microorganism through increasing concentrations of the antimicrobial in the culture medium or a survey of the published literature to determine whether resistance has been reported for the antimicrobial agent.

B. Specific Test Criteria

The specific test criteria relate to in vivo testing of final formulation on human volunteer subjects.

1. Test Product

For products containing ingredients as listed in §333.412(a), (b), (c), (d), or (e) of the TFM, the undiluted finished product is to be tested, along with a positive control.

2. Test Method

Establishment of a Test Subject List. An Institutional Review Board (IRB) is convened to review the study protocol for subject safety. After IRB approval of the study, sufficient human volunteers per product being tested are enrolled in the study, per acceptability criteria outlined in the protocol. The number of subjects required for testing is calculated as specified in the TFM, §333.470(b)(1)(iii)(F), and these are randomized to the products. All subjects are informed of the study descriptions, sign informed consents, and complete a washout period prior to participation in testing.

Test Organisms and Testing Sites. The skin normal flora of a dry skin site (usually the abdomen) and a moist skin site (usually the inguinal) are evaluated with a positive control product. A detergent scrub cup technique with appropriate neutralizing agent is used for sampling before and after treatment.

Sample Schedule. The baseline is obtained from untreated abdominal and inguinal sites at "0" time. Samples are taken from both treated testing sites of one third of the test subjects 10 minutes posttreatment, one third 30 minutes posttreatment, and remaining one third 6 hours posttreatment.

Efficacy Requirement. For products labeled according to §333.460(b)(1), a 2 $\log_{10}$ reduction of normal flora per cm^2 in abdominal testing site and 3 $\log_{10}$ reduction per cm^2 of inguinal testing site 10 minutes posttreatment are required. The number of the surviving organisms must not exceed the baseline for 6 hours following product application. For products labeled according to §333.460(b)(2), a 1 $\log_{10}$ reduction per cm^2 on a dry skin test site within 30 seconds of product application is required.

III. CRITIQUES ON THE POSP TESTING REQUIREMENTS

A. General Test Criteria: In Vitro Test Requirements

1. MIC Study

The results of MIC tests rely highly on the testing conditions such as testing media [3,4]. A comparison to the positive control provides relative information about antimicrobial activity. The MIC test in the monograph requires 25 strains of 20 species of ATCC bacteria and 25 of two species of ATCC fungi and equivalent fresh clinical isolates. The microorganisms specified represent the resident microbial flora most commonly encountered under actual use conditions of the test product, various common environmental species, and transient microbial flora most likely to be encountered by healthcare professionals in clinical settings [1]. The number of strians of the organisms used was not rationalized. The manufacturers may have practical difficulties collecting the required number of freshly isolated clinical strains. In addition, the list of the organisms, however, does not specify the use of antibiotic-resistant strains. Although no correlation between antibiotic resistance and antimicrobial resistance has been conclusively demonstrated, it is desirable to include antibiotic-resistant strains among the listed organisms, as this would entail no extra effort in conducting the test.

2. Time-Kill Kinetic Study

Product Requirements. The time-kill study requires testing a 10-fold dilution of the final formulation. The rationale for this requirement is not explained in the TFM, but it may be based on concern for a safety margin—that if the 10-fold dilution of the product performs well, the full-strength product would provide greater safety. However, the intended use of the finished POSP product is in its full, undiluted strength. In the entire ''use life,'' the POSP product is not applied in a diluted form. Although once dried on the patient's skin the product may be rehydrated and diluted by surgical irrigation solutions during surgery, this is only after the product has completed its mission as a POSP, as it is so defined. A 10-fold dilution of a full-strength, ready-to-use product may change physicochemical characteristics such as pH, viscosity, ionic strength, and the composite interactive relationship among ingredients in the formulation. The composite relationship may be vital for some designed antiseptic functions of the testing product.

The importance of the dilution issue is especially pronounced in testing of iodophor products. The antiseptic action of iodophor relies on the free iodine released from, for example, polyvinylpyrrolidone (PVP) polymer in an aqueous environment. A 10% PVP-I solution, for example, has 1% available iodine, but only releases approximately 1 ppm free iodine. However, it is the free iodine that is responsible for the antisepsis [5]. Dilution of PVP-I from 10- to 100-fold dilution controversially results in a greater concentration of free iodine and thus more effective antisepsis, creating a bias, which overestimates the true product efficacy [6,7].

Furthermore, there are new POSP products on the market using alcohol as the major active ingredient. Alcohol in concentrations from 60 to 90% (v/v) is an excellent antiseptic [8]. A 70% (v/v) alcohol-based POSP, once diluted 10-fold to 7% (v/v), will fail to show appreciable antimicrobial efficacy. Therefore, testing results would be biased to underestimate the true testing product efficacy. Such artificial bias is not desirable, whether true efficacy is over- or underestimated.

Exposure Time Requirement. The rationale for the timed exposure requirement of ''0, 3, 6, 9, 12, 15, 20, and 30 minutes'' in the TFM is ''because the time frames of greatest interest for antiseptic drug products intended for healthcare personnel handwash, surgical hand scrub, and patient preoperative skin preparation use are 1 to 30 minutes'' [§333.470(a)(1)(iv)(D)]. Since 1994, when the TFM was published, the efficiency and effectiveness of POSP product formulations and application designs have been significantly improved. A product that requires 10 minutes to eliminate the entire challenge inoculum of 10^5 or 10^6 cells/mL is hardly considered worthy in the marketplace. More often than not the testing product can reduce 5 $\log_{10}$ microorganisms or more within a minute, and it is not uncommon for some products to achieve that in 15–30 seconds. With demonstration of antiseptic action in less than one minute, to test the product over multiple extended exposure times makes little sense.

Requirements of Testing Organisms. Whereas the purpose of the MIC test is to demonstrate the broadness of the antimicrobial spectrum of a product, the purpose of the time-kill testing is to demonstrate how rapidly a product can kill at a specific concentration. In the analysis of time-kill efficacy, it is the bacterial resistance to the specific chemical, not the broad spectrum, that is important. If the test product can eliminate the most resistant organisms under testing conditions, other less resistant organisms likely would also be inactivated. The use of a biological indicator (BI) to monitor the quality of sterilization is a well-known concept in sterilization technology [9]. It uses the spores most resistant to a specific sterilization agent as an indicator in the sterilization process, with the rationale that if the most resistant spores are killed, the less resistant organisms are also eliminated. This concept may be adaptable to the evaluation of POSP and SS as follows.

VEGETATIVE BIOLOGICAL INDICATOR. Basically, all POSP product designs aim at a maximal antiseptic activity with minimum exposure time, although recent trends also include sustained activity. In sterilization technology, the BI system increases the safety margin of product sterility and improves cost-effectiveness.

Most antimicrobials are not designed for sporicidal activity, so if the concept of BI is adapted, a vegetative cell system must be used instead of spores. The vegetative biological indicator (VBI) should consist of lyophilized cultures instead of spore-bearing carriers. Instead of using a battery of ATCC organisms and fresh clinical isolates, a ''most'' resistant organism or a cluster of three to four ''most'' resistant organisms are lyophilized and passed only one passage in specific medium prior to use as VBIs. The VBI system is a useful alternative, especially to the fresh clinical isolates. Different strains of each clinical isolate are collected from hospital clinical laboratories by different repository laboratories with different procedures. The organisms, therefore, are not standardized cultures. The collection is not well characterized and is not maintained with standardized procedures or maintained by lyophilization. The VBI approach, with large-batch lyophilized cultures, not only makes the standardization of the testing organisms possible, but it is also cost-effective, beneficial especially to smaller companies that may be able to produce excellent products but cannot afford the extensive testing as currently required.

In order to apply the VBI system to the POSP efficacy evaluation, a database of vegetative organism resistance to specific families of antimicrobial chemicals must be established. Some of the organisms, such as *Enterococcus faecalis* and *Enterococcus faecium*, are known to be relatively resistant to iodophor, and *Pseudomonas* is more resistant to quaternary ammonium compounds, etc., but the choice of the organism does not have to be limited to human pathogens, as the emphasis of the VBI is the resistance to specific chemical agent. That is, they are simply indicator organisms with the most resistance to the testing chemical(s) or chemical group(s).

SAFETY ASSURANCE LEVEL. With the VBI, the safety of the efficacy can be determined by safety assurance level (SAL), a concept widely appreciated in sterilization technology. SAL requirements set up safety standards to be met by the industry. If a SAL requirement is 5, then a 5 $\log_{10}$ reduction must be achieved within the constraint of the testing conditions and time of exposure in order to meet the product safety requirement.

D VALUE. Another concept in sterilization technology, the D value [10], can also be ''borrowed'' for POSP evaluation. D value (time required to reduce 90% of challenge microorganisms under specific testing conditions) characterizes product efficacy.

The D value concept is in harmony with the philosophy of the TFM, which intends to provide information as to ''how rapidly the antimicrobial product produces its effect'' [§333.470(a)(1)(iv)]. The D value is also useful for product end users to compare the efficacy rating of the product. For example, a chlorhexidine gluconate (CHG) product reduces 1 $\log_{10}$ of a VBI in 15 seconds and eliminates all 5 $\log_{10}$ of the challenge microorganisms in 1.25 minutes, producing a straight slope characterized as D/15s − 5D/1.25m (D value = 15s; 5 $\log_{10}$ reduction of SAL of 5 in 1.25 minutes). This permits evaluation of comparative quality of various products at a glance. A code of D/15s − 5D/2m, on the other hand, would suggest that the chemical may not be as potent, thus a survival tailing results, although the initial reduction kinetics are the same as for the previous example. Furthermore, a formulation with a D value of one minute and elimination of 5 $\log_{10}$ in 5 minutes (D/1m − 5D/5m) is a relatively weak antiseptic. Figure 1 illustrates the D-value concept.

In summary, with application of a defined VBI, the SAL requirement, and the D value, the POSP efficacy can be characterized, the tests are simplified, and the relevant comparisons between products or with a standard are facilitated.

B. Specific Testing Criteria

The in vivo efficacy test uses an aqueous solution to retrieve the microorganisms that survive after the application of the test product. If the test product is not soluble in water or is only partially water soluble, complicating complete recovery of the surviving microorganisms, then the test method, unless validated, may

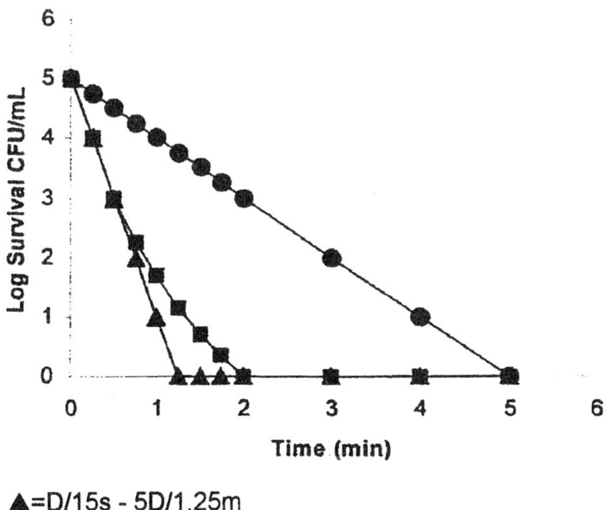

▲=D/15s - 5D/1.25m
■=D/15s - 5D/2m
●=D/1m - 5D/5m

Figure 1 D values.

become inadequate for accurate evaluation. The ineffectiveness in recovering surviving cells after the treatment may grossly overestimate the true efficacy of the product.

IV. PROPOSITION OF A NEW REQUIREMENT FOR POSP TESTS—THE REQUIREMENT OF METHOD VALIDATION

The information most urgently needed to fortify the test requirements described in the TFM is probably the validation of the methodology. In the manufacturing environment, it is compulsory that the methods be validated before implementation. For in vitro tests, TFM does not dictate the test method that must be used. In evaluating some products, such as the products that are not water soluble or only partially water soluble, the conventional methods may not be adequate. First, the products would precipitate in an aqueous in vitro assay system, making retrieval of the surviving microorganisms difficult or impossible, rendering an otherwise perfect method invalid for product evaluation. Therefore, regardless of the methods of choice, these methods must be validated before implementation. The same approach should also be used for in vivo tests. For example, the in vivo test method in the TFM requires using an aqueous solution to retrieve the surviving microorganisms after treatment. When some products form water-insoluble or only partially water-soluble films on skin after drying, the test method may be unable to retrieve the surviving microorganisms under the dry film, resulting in zero colony recovery and an erroneous conclusion that all test microorganisms were killed, hence 100% efficacy. For these products, alternative methods must be devised and validated. Below are some suggestions for validating the testing method for these products.

A. Methodology Validation Using Spore Challenge

Spores are not killed by the antimicrobials listed in §333.412 of the TFM under the testing conditions specified. Spore challenge, therefore, can be applied as a tool to validate a chosen method [11,12]. In a time-kill test, for example, by comparing for retrieval efficiency of the spores added to the test product and to saline, one could validate the efficiency of the chosen method in retrieval of surviving microorganisms. If the chosen method fails to retrieve spores efficiently, then it cannot retrieve vegetative cells either and, hence, is not a valid method to use. If the relative spore-recovery rate from the test product is less than 100%, then its potency should be discounted from the recovery rate in order to reflect the true efficacy. Table 1 shows, through method validation, how a traditional method described in the TFM effectively evaluates some formulations but is inadequate for others.

Table 1 shows that the very same bioassay method can recover inoculated spores from experimental formulation A, but not from B. The failure of spore recovery from formulation B indicates that the methodology is not suitable for evaluating B in an efficacy test when vegetative cells are used, because it would overestimate the true efficacy of formulation B.

Table 1 Method Validation Using Spore Challenge

	Log spore recovery/ mL ± STD[a] ($N = 10$)	p-value
Saline	4.143 ± 0.49	
Expt. formulation A (partially water-soluble)	4.158 ± 0.52	>0.05
Expt. formulation B (water-insoluble)	0.125 ± 0.02	>0.05

[a] 20 μL of *B. subtillis* spores were added to 10 mL partially water-soluble (A), water insoluble (B) experimental iodophor POSP formulations and saline. The spores are enumerated in an aqueous assay system.

Table 2 Methodology Validation: In Vivo Spore Recovery from Dry Film of Experimental POSP

Spore inocula	Log spore recovery/ cm^2 ± STD[a] (N = 10)	p-value
On bare skin	3.21 ± 0.21	
On skin, covered by partially water-soluble formulation A dry film	3.17 ± 0.23	>0.05
On skin, covered by water-insoluble formulation B dry film	0.21 ± 0.02	>0.05

[a] Ten microliters of *B. subtilis* spores in 70% alcohol containing log 3.45 spores were inoculated on human skin in an area approximately the size of a dime and allowed to dry. An experimental POSP was applied to the inoculated skin in an area approximately the size of a quarter and allowed to dry to a film. The spores were then recovered by the cup scrub method described in the TFM and enumerated. Spores inoculated on skin without application of POSP was used as a control. Efficiency of spore yield from bare skin is 3.21 × 100/3.45 − 93.04%.

Table 3 Free Iodine Study: Chemical Analysis of Free Iodine from Extracts of Dry Film of POSP Experimental Formulations

Sample	Free iodine from dry film extracts (ppm)		
	0–3	3–6	6–24 hours[a]
Formulation A (water-soluble)	1.24	5.70	<0.07[b]
Formulation B (partially water-soluble)	1.70	2.00	2.70
Expt. formulation C (water-insoluble)	<0.07[c]	<0.07[c]	<0.07[c]

50 mL of experimental POSP formulations were added to sterile bottles and dried with a gentle air flow while continuously rolled on a bottle roller. After completely dried, 50 mL saline was added to the bottles to extract free iodine for 3 hours. The extract was then transferred to a sterile storage bottle. Additional 50 mL fresh saline were added to POSP bottles, and the extraction procedure continued. Samples of 0–3, 3–6, and 6–24 hours were collected. Aliquots of samples were chemically analyzed for free iodine, and paired aliquots were tested for bacterial activity (see Table 4).

[a] Extraction time period.

[b] After 6 hours of extraction, all free iodine was extracted from water-soluble formulation B. No free iodine could be extracted from 6–24 hours.

[c] Detection limit = 0.07 ppm.

Free iodine control: PVP-I (7% in water) = 1.70 ppm.

Table 4 Free Iodine Study: Microbiological Analysis of the Extracts from Dry Film of POSP Experimental Formulations

	Log reduction/mL			
	S. aureus ATCC #6538		*E. coli* ATCC #11229	
Sample time	15	30	15	30 seconds
A—soluble 0–3 hr[a]	>5.68	>5.68	>5.81	>5.81
A—soluble 3–6 hr[a]	>5.68	>5.81	>5.81	>5.81
A—soluble 6–24 hr[a]	<1.00[b]	<1.00[b]	<1.00[b]	<1.00[b]
B—partially soluble 0–3 hr[a]	>5.68	>5.68	>5.81	>5.81
B—partially soluble 3–6 hr[a]	<5.68	>5.68	>5.81	>5.81
B—partially soluble 6–24 hr[a]	>5.68	>5.68	>5.81	>5.81
C—insoluble 0–3 hr[a]	<1.00	<1.00	<1.00	<1.00
C—insoluble 3–6 hr[a]	<1.00	<1.00	<1.00	<1.00
C—insoluble 6–24 hr[a]	<1.00	<1.00	<1.00	<1.00

An aliquot of free iodine extract from Table 3 was tested for bactericidal activity. A = water-soluble formulation, B = partially water-soluble formulation; and C = experimental water-insoluble formulation.

[a] Extraction time period.

[b] Free iodine was exhausted in water-soluble formulation B film in 6 hours. No free iodine could be extracted from 6–24 hours.

Spore challenge can also be used to validate test methods for in vivo evaluation. Table 2 shows the very same method, capable of recovering spores from the dry film on skin from one POSP experimental formulation, but not from the other. If vegetative cells were the targets of the evaluation, again, the surviving cells not recovered will be misinterpreted as killed. For formulation B of Table 2, therefore, the traditional method of the TFM is inadequate for efficacy evaluation. An alternative method must be used but should be similarly validated before implementation. Once validated, the exact same validated method should be used for efficacy testing without modification.

B. Efficacy Estimate Using Chemical Analysis

In correlation with the microbiological validation, chemical analysis of the active ingredient also becomes an important tool to validate the testing methodology. After all, it is the effective dose of chemical that kills the microorganisms. For most chemicals, the analysis of the active agent is relatively straightforward. However, for some chemicals, such as iodophors that contain elemental iodine bound to macromolecules such as PVP, the chemical analysis may be more involved. The iodine molecules bound to PVP are not effective antimicrobially unless they are released from PVP as free iodine [5]. Do water-insoluble or partially water-soluble iodophor POSP products, in liquid form or in dry film, release free iodine? Can an iodophor product, either in liquid or in dry film, without the chemical evidence of free iodine, demonstrate antimicrobial activity? Until recently, no studies have been available to examine the release of free iodine from the samples directly extracted from PVP-I POSP dry film in correlation with the antimicrobial activity [12]. Tables 3 and 4 demonstrate some aspects of the complicated nature of the iodophor formulations and provide some answers to the questions such as those posed above. Free iodine was extracted from the dry films of skin preparations, and the extracts were then tested microbiologically and chemically. The data in Tables 3 and 4 clearly show that the free iodine in the extracts found chemically is closely correlated to the antimicrobial activity. Such chemical analyses of antimicrobials under conditions of testing, therefore, can be additional tools to validate the test method.

V. REVIEW OF SS EFFICACY TEST REQUIREMENTS

A. General Test Criteria

In vitro testing requirements for SS products are similar to those for POSP—the antiseptic ingredient, the vehicle, and the final formulation. Researchers are encouraged to submit alternative test methods for approval by the FDA.

1. Antimicrobial Spectrum of Activity: MIC Study

Test Product. For products containing ingredients as listed in §333.412(a), (b), (c), (d), or (e) of the TFM, the undiluted final formulation, the product vehicle, and the active ingredient should be tested.

Test Method. A minimal inhibition concentration method, as described in NCCLS-M7, is used.

Challenge Microorganisms. The antimicrobial spectrum of activity of the test products must be broad, using 25 strains of 20 species of ATCC bacteria and two species of ATCC fungi listed in §333.470(a)(1)(ii) of the TFM, plus 25 strains of fresh clinical isolates of these same species bacteria and yeasts. The challenge microorganisms represent members of normal flora in humans, common environmental contaminants, or systemic pathogens.

2. Time-Kill Kinetic Study

Test Product. For products containing ingredients as listed in §333.412(a), (b), (c), (d), or (e) of the TFM, the active ingredient, the vehicle, and the final formulation are to be tested. A 10-fold dilution of the final formulation is required, but there is no dilution requirement for the active ingredient or the vehicle.

Test Method. Microorganisms are exposed to the test materials for proposed times of 0, 3, 6, 9, 12, 15, 20, and 30 minutes. At the end of the exposure period, an appropriate neutralizing agent is used to quench antimicrobial activity chemically and/or by dilution. Microorganism populations before and after the treatment are enumerated to calculate the antimicrobial activity of the test product, expressed in a $\log_{10}$ reduction format.

Challenge Microorganisms. The same microorganisms specified for the MIC test are to be tested.

3. Test for Potential for Emergence of Microbial Resistance

This involves determination of the evolution of a point mutation by a sequential passage of a challenge microorganism through increasing concentrations of the antimicrobial in the culture medium, or a survey of the published literature to determine whether resistance has been reported for the antimicrobial agent.

B. Specific Test Criteria

The specific test criteria relate to in vivo efficacy testing of final formulation, along with an active control, on human volunteer subjects.

1. Test Product

For products containing ingredients as listed in §333.412(a), (b), (c), (d), or (e) of the TFM, the undiluted final formulation is to be tested, along with a positive control.

2. Test Method

Establishment of a Test Subject List. An institutional review board is convened to review the study protocol for subject safety. After IRB approval of the study, sufficient human volunteers per product being tested are enrolled in the study per acceptability criteria outlined in the protocol. The number of subjects required for testing is calculated as specified in the TFM, §333.470(b)(1)(F), and these are randomized to the products. All subjects are informed of the study descriptions, sign informed consents, and complete a washout period prior to participation in testing.

Preparation. Appropriate scrubbing and handwashing using the methods described in the TFM.

Test Microorganisms and Efficacy Requirement. The normal skin flora of the hands is evaluated following use of the test product, and a positive control product is also tested.

Sampling Procedure. The glove juice method is used to evaluate the efficacy. Bacteria, pre- and posttreatment, are enumerated. Products are required to "reduce the number of bacteria 1 $\log_{10}$ on each hand within one minute of product use, and the bacterial cell count on each hand must not subsequently exceed baseline within 6 hours on the first day, produce a 2 $\log_{10}$ reduction of the microbial flora on each hand within one minute of product use by the end of the second day of enumeration, and produce a 3 $\log_{10}$ reduction of the microbial flora on each hand within one minute of product use by the end of the fifth day." The samples are retrieved and assayed in appropriate neutralizing agent after treatment.

VI. CRITIQUES OF SS TESTING METHODOLOGY

A. General Test Criteria: In Vitro Test Requirements

1. Time-Kill Kinetic Study

In practice, SS is used in conjunction with tap water. It is therefore appropriate to test SS finished product in vitro with a 10-fold dilution.

2. Neutralization

In testing some SS formulations containing emollients or lotions, it is often found that the neutralizing agent in the first of the 1:10 series dilution tubes may not be sufficient to neutralize the active ingredient completely. The neutralizing power of the neutralizing agent in the first 1:10 dilution tube is affected by the emollient or lotion materials in the formulation. In order to neutralize the residual chemicals completely, the product must be further diluted immediately into the second 1:10 dilution tube, and the survivor numbers must be enumerated from the second dilution tube and beyond. Table 5 compares the results of time in residence of inoculated bacteria in the first dilution tube in evaluating SS products. If the samples stayed in the first 1:10 dilution tube longer than 20 minutes, the cells were inactivated by residual antimicrobial activity not completely neu-

Table 5 Neutralization of SS Testing Product[a]

	CFU/mL	
Neutralization method	*E. coli*	*S. aureus*
Inoculum	3.54×10^6	6.92×10^5
First dilution tube for >20 minutes	$<1.00 \times 10^1$	$<1.00 \times 10^1$
First dilution tube for <15 seconds	3.52×10^3	6.39×10^2

[a] Surgical scrub formulation containing experimental parachlorometaxylenol (PCMX) scrub formulation with emollient and lotion. Bacteria were added to the formulation for 2 minutes and then assayed. At the end of incubation, the reaction materials were diluted 1:10 in specific neutralizing solutions and allowed to sit for appropriate time lengths and then assayed with a further 1:10 series of dilutions. Colony counts were taken from the second dilution tube instead of the first one.

tralized by the neutralizing agent. This would result in an overestimate of the true efficacy.

Furthermore, some of the new formulations may contain ingredients that cannot be neutralized by conventional neutralizing agents. For example, experimental formulations with alcohol, fatty acids, and compound containing zinc in lotion complex cannot be neutralized by conventional neutralizing agents containing lecithin and polysorbate 80. Some compounds with zinc are thought to be preservatives with sustained bacteriostatic activity. Attempts to neutralize zinc by a chelating agent such as EDTA in lotion systems was not successful (unpublished data). Even a dilution method described above cannot accurately assess the efficacy of the formulations. Therefore, unless a validated method of neutralization and bioassay is developed, the efficacy of the formulations could be overestimated.

B. Special Test Criteria

In clinical testing it is required that the performance of a SS product must improve during the course of testing. On the first day of a test, a 1 $\log_{10}$ reduction must be achieved in one minute, whereas in the second and fifth days, 2 $\log_{10}$ and 3 $\log_{10}$ reductions must be achieved in one minute, respectively. Among the antimicrobial chemicals, only CHG at appropriate concentration demonstrates such an immediate, persistent, and residual activity. The residual activity is thought to be due to skin absorption [13]. Frequent use of CHG would therefore result in an accumulative antimicrobial efficacy [14]. Iodophor also demonstrates a residual activity due to skin absorption [15], although a relatively short-lived one. Other chemicals, however, do not show appreciable residual activity. Therefore, there is no reason why the efficacy of the fifth day of those chemicals should be any different from that of the first day. With the current requirement, it is not surprising that many of the surgical scrub products containing antimicrobial actives other than CHG cannot achieve the required efficacy [13,16].

VII. CONCLUSION

The Tentative Final Monograph is a good guideline to regulate testing of POSP and SS products. The value of the document is apparent. As the title implies, the document is a tentative proposal, and, therefore, will be subject to modification over time to meet the needs of the industries, as deemed necessary. This chapter critically reviewed the testing methodology requirements of the TFM and suggested some solutions to remedy certain aspects of the testing methodology that may be found inadequate to meet current and/or future needs. Some of these recommendations may open the doors of discussion in the industry and technical community. It is hoped that such discussions may result in a perfected document to meet the change of the industries and the demand of the new era.

REFERENCES

1. Federal Register: 21CFR, Parts 333 and 369, Tentative Final Monograph for Healthcare Antiseptic Drug Products; Proposed Rule, 1994.
2. FD Schoenknecht, LD Sabath, C Thornsberry. Susceptibility tests: special tests. In: EH Lennette et al., eds. Manual of Clinical Microbiology, 4th ed. Washington, DC: American Society for Microbiology, 1985, pp. 1000–1008.
3. C Harris. MIC tests are not suitable for assessing antiseptic handwashes. J. Hosp. Infect. 13:95, 1989.
4. RW Lacy, A Catto. Action of povidone-iodine against methicillin-sensitive and -resistant cultures of *Staphylococcus aureus*. Postgrad Med J. 69 (suppl 3):S78–S85, 1993.
5. M Winicov, EL Winicov. Determination of free iodine and its significance in povidone-iodine solution. Proc Int Symp Povidone April 17–20: 186–192, 1983.
6. RL Berkelmann, BW Holland, RL Anderson. Increased bactericidal activity of dilute preparations of povidone-iodine solutions. J Clin Microbiol 15:635, 1982.
7. H Racckur. New aspects of mechanism of action of povidone-iodine. J Hosp Infect 6(suppl):13–23, 1985.
8. EL Larson, HE Morton. Alcohol. In: SS Block, ed. Disinfection, Sterilization, and Preservation, 4th ed. Philadelphia: Lea and Febiger, 1991, pp. 191–203.
9. TJ Macek. Biological indicators and the effectiveness of sterilization procedures. In: GB Phillips, WS Miller, eds. Industrial Sterilization. Durham, NC: Duke University Press, 1972, pp. 19–34.

10. CR Stumbo. Thermobacteriology in Food Processing, 2nd ed. New York: Academic Press, 1973, pp. 70–91.
11. DK Jeng, JE Severin. Povidone iodine gel alcohol: A 30-second, one time application preoperative skin preparation. Am J Infect Control 26(5):488–494, 1998.
12. DK Jeng. A new, water-resistant, film-forming, 30-second, one-step application iodophor preoperative skin preparation. Am Infect Control 29(6):370–376, 2001.
13. DS Paulson. Comparative evaluation of five surgical hand scrub preparations. AORN J 60(2):246–256, 1994.
14. AF Peterson, A Rosenberg, SD Alatary. Comparative evaluation of surgical scrub preparations. Surg Gynecol Obstet 146:63–65, 1978.
15. W Gortardi. Residual effects on the skin caused by povidone iodine preparations. Hyg Med 14:228–233, 1989.
16. DW Hubson, W Woller, L Anderson, E. Guthery. Development and evaluation of a new alcohol-based surgical hand scrub formulation with persistent antimicrobial characteristics and brushless application. Am J Infect Control 26(5):507–512, 1998.

26
Skin-Sampling Techniques

Barry Michaels
Georgia-Pacific Corporation, Palatka, Florida

I. INTRODUCTION

The human skin has been identified as an important source and reservoir of not only harmless resident microbial species, but also those responsible for various types of pathogenic conditions, processes, and infectious diseases. Various techniques have been developed to determine both qualitative and quantitative parameters regarding the microorganisms residing on and in the human skin [1,2]. In addition, testing of antimicrobial skin products sometimes requires that data be generated on efficacy against both nonresident transient skin contaminants and resident species. Yet all sampling methods have limitations.

The skin of the hands represents a specific anatomical area to which a variety of methods and perspectives have been applied to study contamination, soap product efficacy, and handwashing process effectiveness. Many consider the fingertip region of the hand an important area where transmission of bacteria occurs during food preparation or in health care environments [3–5]. This part of the hand has been an important focus of specific test sampling methods.

Methods used to sample the microbial status of human skin can be conveniently grouped into four main categories in which eight principal test methods are employed. Each of these methods may be performed with modifications, some of which link and integrate two or more of these methods. They include:

A. Methods Involving Skin/Stratum Corneum Removal

Biopsy/skin scraping
Adhesive tape stripping

B. Nondestructive Direct Sampling

Swabbing
Agar contact-plate
Glove/bag juice techniques

C. Washing Methods

Scrub method
Handwashing collection method

D. Other

Air sampling (for shedders)
Fingernail region

Sampling methods should be selected based on whether the sample is designed to represent the overall microbial status of the hand or a particular part of the hand. In antimicrobial effectiveness studies, hands are typically inoculated with up to 108 indicator bacteria. This is a higher number than might be encountered when examining normal flora or when looking for specific pathogenic species important for diagnostic purposes. In each of these cases, individual test methods may have specific advantages based on their detection characteristics.

II. METHODS INVOLVING SKIN/STRATUM CORNEUM REMOVAL

A. Biopsy/Skin Scraping

The use of a keratotome [6–9], designed only to remove stratum corneum, yields slightly better results than tape stripping. An underlying defect in these methods is that repeat sampling of the same site is difficult and skin physiology is changed in the process of sample collection; thus, samples from adjacent sites are required, increasing population variability. Biopsy of skin tissue has been used to determine microbial counts. In cases where skin pathology is encountered, more definitive information is required [10–12]. Early skin microbiology experiments frequently employed cadaver skin for experimental purposes. Techniques have been described utilizing pickled dehydrated animal skins to test antibacterial activity of soap products [13]. Woodroffe [9] used small pieces of skin taken from freshly killed pigs' feet to measure the substantiality of antimicrobial compounds. Although convenient, the shortcomings of this technique are apparent, as it does not involve living human skin with an intact biofilm [14]. Ultimately, the value of the data derived through use of skin samples obtained at autopsy is questionable [15], and the primary use of biopsy is for diagnostic or treatment purposes [12,16,17]. Occasionally, biopsies are used to validate results from other sampling methods [18]. Unfortunately, with skin sampling methods that involve removal of stratum corneum, the dynamic aspects of skin are severely affected or lost, making comparative studies difficult.

B. Adhesive Tape–Stripping Technique

The tape-stripping method initially described by Rockl et al. [19,20] appears to be a valuable and versatile tool for providing information on the distribution of bacteria in and on the skin [19–21]. The tape method has been used as a reliable inexpensive diagnostic tool for quantitative determination of, for example, *Pityrosporum* yeast [22], *Candida* spp. [23], *Enterobius* spp. [24], and *Trichophyton* spp. [25]. The tape method was found to be more sensitive than swab testing, skin scraping, and direct impression techniques for the determination of quantitative and distribution data of *Malassezia* spp. on canine skin [26].

Whereas the tape-stripping method aids in enumerating bacterial colonies, the handwashings collection and scrub methods result in a dispersal of bacterial colonies [27]. Use of tape stripping has allowed for exploration of the microbial anatomy of normal human skin by employing plasmid profiles through use of agarose gel electrophoresis [28]. Results of this technique suggest that the reservoir for normal resident skin flora is located below the stratum corneum, perhaps in hair follicles and ducts of sebaceous glands. Tape stripping has also been used to characterize the effectiveness of microelectrical currents at reducing microbial counts in subsurface skin layers as compared to controls [29].

Care should be taken to utilize sterile adhesive tape, available for this type of sampling [30], as adhesive tape has been identified as a potential source of pathogenic bacteria in hospitals [31,32]. By sequential stripping with cellophane tape, 15–25 keratinized layers of stratum corneum can be sampled for the presence of subsurface coagulase-negative staphylococci [33]. Adhesive tape used to collect and identify gunshot powder residue has also been used to extract human DNA for matching identification and attribution of specimen [34]. High-performance liquid chromatography (HPLC) has been used to measure urocanic acid in sun-exposed and shielded skin [35], as well as free long-chain sphingoid base distribution in layers of stratum corneum [36].

Various methods have been investigated to quantify the amount of stratum corneum removed by tape stripping [37]. Use of differential transparency measurements of sequentially stripped stratum corneum has been used to develop a profile of stripped layers [38,39]. Weighing and quantification of the sodium hydroxide soluble protein fraction, along with spectrophotometric protein analysis of the tape, has been used successfully to provide information regarding layer composition obtained through sequential strippings [40,41]. Because of furrows in the skin, it has been found that cells derived from different layers may end up on one tape strip [42].

Tape stripping is damaging to the skin and produces a certain degree of trauma to the epidermal layers. The degree of trauma has been measured by laser-doppler imaging [43], changes in capacitance measurements [44,45], and ionic mobility [46]. The most evident physiological changes concern impairment of barrier function as measured by transepidermal water loss (TEWL) [47–50]. TEWL can increase up to 40-fold that of baseline values, leading to tape-induced dermatitis [47]. Damage caused by tape stripping has been found to increase prostaglandin E_2 and interleukin-1α over threefold [51], initiate mitotic bursts, and increase cellular migration rates as a regenerative response [52–55]. Topical drug delivery has been investigated in skin with the penetration barrier removed by tape stripping [56]. If this method is employed utilizing the same skin area repeatedly to obtain baseline readings and determine treatment effects, as is occasionally described [57], then results may be compromised. The use of adjacent skin surfaces for comparison of baseline readings and skin surface treatments can provide acceptable data, although statistically less precise than other methods due to increased sample variability [14].

III. NONDESTRUCTIVE DIRECT SAMPLING

A. Swabbing

Swabbing techniques have been extensively utilized and are subject to a great many variations [27,58–61]. This method permits sampling, quantitative evaluation, and identification of microorganisms on only small areas of skin. Surface organisms and a variable proportion of subsurface organisms are obtainable when using this technique [62]. This method consists of swabbing a defined area for a measured period of time and counting bacteria by either streaking or dispersing microorganisms in suspending fluids. Swabbing time, pressure, and swab head moisture level all affect microbial recovery [63]. Variability in counts is also affected by site selection [27], with palm, dorsum, fingertips, wrist, and finger sides all having differing microhabitats. This shortcoming has been partially compensated for by either pooling swabs from each of these areas or sequentially swabbing these areas [64].

Moist swabs have been found to ''pick up'' more bacteria than dry swabs, with pick-up being significantly influenced by swab sampling pressure. Since heavier pressures pick up more microorganisms than light pressures, pressure consistency during execution is important when this method is used for comparative purposes. Moistening fluid may consist of either sterile water, saline, 1% peptone, or tryptone soya nutrient broth. Nonionic detergents may also be added to maximize dispersion, with pH adjusted to provide optimal release [65] from surfaces. Calcium alginate, rayon, and cotton swabs have each been found to have slightly different pick-up characteristics [27]. Choosing a liquid moistening agent seems to be less important than the pressure applied during sampling [27].

B. Agar Contact-Plate Method

This is a simple and inexpensive method that yields accuracy and good reproducibility [62,66]. It is commonly used for fingerprint impressions [5,67–74]. Hand dorsum impressions [75], palm impressions [73], and finger and palm impressions [76] have also been used. This method is good for the screening of gross microbial hand contamination, particularly during hospital infectious disease outbreaks [62].

Fingertip cultures have been used successfully to evaluate the effectiveness of various surgical scrub techniques involving the fingertip to elbow scrub [77]. Here, the fingertip is used as an indicator for measuring the success of the entire regimen. The fingertip count has been used to sample for fecal coliform bacteria, and thus is an indicator for fecal–oral transmission potential [78] and for evaluation of antiseptic creams against these types of contaminants [79]. When used for testing effectiveness of antimicrobial products on fingertip contamination, suitable quenching agents are required in agar [14]. With the contact plate method, limitations include that it is not quantitative, as colonies may be very difficult to count and identify due to their proximity to each other. In testing the effectiveness of antimicrobial compounds on marker organisms, selective or differential culture media is often used to distinguish them from normal flora.

A variation of this is the finger streak method, in which the fingertip is drawn across an agar plate, providing better distribution of colonies to obtain counts and species characterization [80–82]. A drawback of the finger streak method is the inconsistency in microbial counts due to variations in skin sweating, in which inaccurate conclusions may be obtained from test results [79,83]. Dry swabbing seems to yield results similar to those obtained using the contact-plate method. In some instances, the contact-plate method has been found superior to other methods due to the increased sensitivity for detection of low numbers of the target organism [84].

C. Glove/Bag Juice Techniques

The glove or bag juice technique is a variation of earlier work that employed samples from used gloves [72,85,86] or glove fluid [87,88] for relative data on microbial counts and population structure. Eventually, stripping liquids were placed in gloves for gloved hand samples [89,90]. The bag juice method is a simple variation utilizing a polyethylene or stomacher bag in place of the glove [91,92]. The hand is placed in a loose glove or bag with 50–75 mL of stripping liquid containing nonionic surfactants. A laboratory worker then massages the gloved or bagged hand for a period of one minute. The fingers, fingertips, and palm (but not the back of the hand) are rubbed through the wall of the bag or glove, after which time a quantity of liquid is removed, diluted, and plated [2,62].

Studies performed comparing the glove juice, swab, and finger contact-plate (finger press) methods to obtain microbial counts on the entire hand have concluded that at higher levels of accuracy, lower levels of variation are found with the glove juice method when high initial inoculum levels are used [76]. It should be noted that at low inoculum levels (200 CFU), all three methods yielded similar results. This method was codified by American Society for Testing Materials [93] and further adopted by the FDA for use of proof of antimicrobial soap efficacy [94]. Some investigators prefer bags rather than gloves because of more uniformity in sample mixing, whereas others contend that gloves provide a better ability to remove bacteria from individual fingers. Despite the overall acceptance of this method for skin sampling, complaints of inaccuracy due to variations caused by the degree of skin sweating or small glove punctures have been made [14,83]. The glove juice technique [95] has had wide acceptance for testing antimicrobial hand soap, surgical scrubs, and healthcare personnel handwash products, as evidenced by the adoption by the American Society for Testing Materials [96,97]. This method has also been accepted by the FDA as the standard for healthcare personnel handwash and surgical scrub products [98,99].

IV. WASHING METHODS

A. The Cup Scrub Method

The removal of bacteria from skin, an important aspect of wash and scrub procedures, is dependent on mechanical factors such as the degree of scrubbing [10] and/or chemical aspects [14]. Surfactant action coupled with optimal alkaline pH has been shown to favor effective bacterial removal [100–102]. This method utilizes a Teflon spatula or rubber policeman within a premoistened (with skin-stripping solution) enclosed area [93]. A cup or ring (1.5–4.0 cm in diameter) is held down on to the skin surface while the Teflon spatula is used to rub the skin. The stripping solution is drawn off, diluted, and plated. This method seems to have been first described by Colebrook and Maxted [103] and modified by Williamson and Kligman [65], Story [104], and later Shaw [27]. Bacteria on skin surfaces are well dispersed, and counts tend to identify individual cells rather than colonies or clumps that may be picked up using the impression plate method or tape stripping [27]. Although the so-called scrub cup or more appropriately named cylinder-sampling method [2] has the potential to increase microbial recovery rates over swab testing and is lacking in reproducibility, it has been used satisfactorily in antimicrobial efficacy studies [105]. Because of the need for two laboratory technicians to execute this test method, its usefulness is limited. This technique has also been used to obtain microbial counts and flora samples in cattle where ballotini beads were used to break down squame and bacterial microcolonies [106]. It should be noted that prolonged shaking with beads can decrease bacterial viability.

By combining swab and scrub test methods on the same site, differential identification of both surface and subsurface microorganisms can be obtained [60]. A study by Chevalier [18] compared results of the scrub method to results of aerobic counts obtained from biopsies with remarkable similarity in data sets. From detailed studies using the scrub test method, the percentage of different species and types can be calculated for various parts of the body [107]. Due to the aggressiveness of this method, skin damage inevitably occurs and some studies of the human skin, under dynamic conditions, are precluded.

Another variation of this technique is rubbing fingertips against the bottom of a Petri dish or scrubbing fingertips against glass beads contained in a bowl or tube [108,109]. Both methods yield similar results and were found useful for finger contamination sampling.

B. The Handwashings Collection Method

The handwashings collection method was first described by Price [10] and later practiced by others [110–112] and eventually modified by Cade [113,114]. It consists of scrubbing the skin with a brush for a predetermined period of time and rinsing the area into a known volume of liquid. Each washing is held in a separate wash basin. Microbial counts from serial scrubbings are performed and cumulative serial counts are used to chart decreases in microbial counts with each washing. The plotted data are fit to a standard curve to provide a total number of bacteria on hands.

This method is time consuming, with up to one week needed for full recovery of skin populations. Because a certain percentage of bacteria is removed with each scrubbing, controlled data can be obtained by limiting scrubbing time. Wash fluid can be collected after wash periods either with or without the use of a scrub brush [72,85,115–120].

The modified Cade handwashing procedure uses handwashing in five sterile polyethylene wash basins or large zip-lock freezer storage bags [1]. Only samples from washes 1 and 5 are sampled and plated, saving time over the full Price or Cade technique.

Fingertip washing has also been used [121–123]. Fingertip washing or fingertip impression techniques work well for identification of contact contamination and potential dissemination of specific microbial species [72].

V. OTHER METHODS

A. Air Sampling

Air sampling has been identified as a useful technique to measure shedding of bacteria attached to loose skin squames [124,125]. Certain individuals shed significant quantities of bacteria (especially staphylococci) into the air and can be a danger in hospital areas where immune impairment or lack of immune system development places patients at risk [126]. Air samplers impinge microorganisms onto selective agar surfaces where, after removal from impinger, they can be identified following growth [127]. Standardized methods have been developed which monitor microbial counts from persons in small rooms during various time periods at different levels of activity [66,124]. Bacterial counts have been shown to increase after handwashing [128]. As shedding of bacteria in conjunction with skin squames has been associated with up to an 18-fold increase due to handwashing with particularly drying soap products [129], air sampling is seen as a valuable tool for evaluating these effects.

B. Fingernail Region–Sampling Techniques

One problem with both the handwashings collection method and the glove or bag juice method is the fact that microorganisms residing in the fingernail region are counted and characterized as part of the whole hand count. While this has been described as a drawback of the glove juice method, it is by no means thorough enough to be used for testing of sanitizers designed for disinfection of the fingernail region. This region with the subungual space and nail folds is a common site of transient contamination and colonization. It is a microhabitat very different from other skin surfaces found on the hands. Since over 90% of the bacteria on the hands are found in this region [3,4], then counts taken using both of the above-mentioned sampling techniques run the risk of obscuring the finer details of microbial ecology taking place on skin surfaces. Where samples from multiple sites of the hands show bacterial counts from 102 to 103 per area, subungual spaces may contain numbers up to 2 orders of magnitude higher [3,4,130]. In addition to the inability of the glove juice method to adequately remove marker bacteria from that region, only two sampling sites (each hand) are available, whereas a specifically designed test method can provide 10 sites [131]. This allows possible testing of a larger number of products with multiple sampling sites with theoretically fewer volunteers. The method developed by Mahl employs an electric toothbrush [131]. Efficacy of this method has been described as close to 100% when using *Bacillus subtilis* spores. Individual toothbrush heads can be used for each fingernail, as used heads can be sanitized by use of 70% ethyl alcohol for 10 minutes or disinfected with 1000 ppm freshly reconstituted hypochlorite solution for 10 minutes followed by rinsing in sterile distilled water and air-drying or steam autoclave or ethylene oxide. This method has been formalized by American Society for Testing Materials [132] with use of either *S. marcescens* or *B. subtilis* spore suspension. The nail region of each finger is inoculated with 0.02 mL of marker organism and allowed to dry in front of a small electric fan. After treatment with antimicrobial product, the electric toothbrush is used for a 1-minute period using 7 mL of collecting fluid in separate petri plates.

REFERENCES

1. DS Paulson. A proposed evaluation method for antimicrobial hand soaps. Soap/Cosmetics/Chemical Specialties June:64–67, 1996.
2. DS Paulson. Topical Antimicrobial Testing and Evaluation. New York: Marcel Dekker, 1999.
3. JB Hann. The source of the ''resident'' flora. Hand 5:247–252, 1973.
4. KJ McGinley, EL Larson, JJ Leyden. Composition and density of microflora in the subungual space of the hand. J Clin Microbiol 26:950–953, 1988.
5. AZ Sheena, ME Stiles. Efficacy of germicidal handwash agents in hygienic hand disinfection. J Food Prot 45:713–720, 1982.
6. R Castroviejo. Electro-keratotome for the dissection of lamellar grafts. Am J Opthalmol 47:226, 1959.
7. H Blank, EW Rosenberg, I Sarkany. An improved technique for obtaining uniformly thin sheets of skin. J Invest Dermatol 36:303, 1961.
8. H Baxby, RCS Woodroffe. The location of bacteria in skin. J Appl Bacteriol 28: 316, 1965.
9. RCS Woodroffe. A respirometer technique for extimating germicidal activity on skin. J Hyg Cambr 61:283, 1963.
10. PB Price. The bacteriology of normal skin; a new quantitative test applied to a study of the bacterial flora and the disinfectant action of mechanical cleansing. J Infect Dis 63:301–318, 1938.
11. R Marks, ND Ramnarain, B Bhogal, NT Moore. The erythrasma microorganism in situ: studies using the skin surface biopsy technique. J Clin Pathol 25[9]:799–803, 1972.
12. MH Brownstein, AD Rabinowitz. The invisible dermatoses. J Am Acad Dermatol 8[4]:579–588, 1983.
13. LJ Vinson, EL Ambye, AG Bennett, WC Schneider, JJ Travers. In vitro tests for measuring antibacterial activity of toilet soap and detergent bars. J Pharm Sci 50: 827, 1961.
14. BM Gibbs, LW Stuttard. Evaluation of skin germicides. J Appl Bact 30[1]:66–77, 1967.
15. EJL Lowbury. Removal of bacteria from the operation site. In: Skin Bacteria and Their Role in Infection. New York: McGraw-Hill, 1965.
16. GP Wormser, G Forseter, D Cooper, J Nowakowski, RB Nadelman, H Horowitz,

I Schwartz, SL Bowen, GL Campbell, NS Goldberg. Use of a novel technique of cutaneous lavage for diagnosis of Lyme disease associated with erythema migrans. JAMA 268[10]:1311–1313, 1992.

17. AB Fleischer Jr, SR Feldman, RE White, B Leshin, R Byington. Procedures for skin diseases performed by physicians in 1993 and 1994: analysis of data from the National Ambulatory Medical Care Survey. J Am Acad Dermatol 37(5 Pt 1):719–724, 1997.
18. J Chevalier, GM Mercier, A Cremieux. Evaluation of a standard scrubbing method for the recovery of aerobic skin flora. Ann Inst Pasteur Microbiol 138[3]:349–358, 1987.
19. H Rockl, E Muller. Studies on the localization of bacteria in the skin. Arch Klin Exp Derm 209:13, 1959.
20. DM Updegraff. A cultural method of quantitatively studying the microorganisms in the skin. J Invest Dermatol 43:129, 1964.
21. DM Updegraff. Methods for determining the distribution of bacteria in the skin. J Am Oil Chem Soc 44[8]:481–483, 1967.
22. JR Wikler, P de Haan, C Nieboer. The 'tape-method': a new and simple method for quantitative culture of *Pityrosporum* yeasts. Acta Derm Venereol 68[5]:445–449, 1988.
23. RS Barnetson, LJ Milne. Skin sampling for *Candida* with adhesive tape. Br J Dermatol 88[5]:487–491, 1973.
24. MA Kvitko, AN Robster, RG Pavlenko. Experience in using an adhesive tape in an examination for enterobiasis. Med Parazitol (Mosk) 48[6]:72–73, 1979.
25. LJ Milne, RS Barnetson. Diagnosis of dermatophytoses using vinyl adhesive tape. Sabouraudia 12[2]:162–165, 1974.
26. RA Kennis, EJ Rosser, NB Oliver, RW Walker. Quanitity and distribution of *Malassezia* organisms on the skin of clinically normal dogs. J Am Fet Med Assoc 208(7):1048–1051, 1996.
27. CM Shaw, JA Smith, ME McBride, WC Duncan. An evaluation of techniques for sampling skin flora. J Invest Dermatol 54:160–163, 1970.
28. E Brown, RP Wenzel, JO Hendly. Exploration of the microbial anatomy of normal human skin by using plasmid profiles of coagulase-negative staphylococci: search for the reservoir of resident skin flora. J Infect Dis 160(4):644–650, 1989.
29. L Bolton, B Foleno, B Means, S Petrucelli. Direct-current bactericidal effect on intact skin. Antimicrob Agents Chemother 18(1):137–141, 1980.
30. JA Fairclough, CG Moran. The use of sterile adhesive tape in the closure of arthroscopic puncture wounds: a comparison with a single layer nylon closure. Ann R Coll Surg Engl 69(3):140–141, 1987.
31. DM Berkowitz, WS Lee, GJ Pazin, RB Yee, M Ho. Adhesive tape: potential source of nosocomial bacteria. Appl Microbiol 28(4):651–654. 1974.
32. DA Redelmeier, NJ Livesley. Adhesive tape and intravascular-catheter-associated infections. J Gen Intern Med 14(6):373–375, 1999.
33. JO Hendly, KM Ashe. Effect of topical antimicrobial treatment on aerobic bacteria in the stratum corneum of human skin. Antimicrob Agents Chemother 35(4):627–631, 1991.

34. C Torre, S Gino. Epidermal cells on stubs used for detection of GSR with SEM-EDX: analysis of DNA polymorphisms. J Forensic Sci 41(4):658–659, 1996.
35. K Shibata, Y Nishioka, T Kawada, T Fushiki, E Sugimoto. High-performance liquid chromatographic measurement of urocanic acid isomers and their ratios in naturally light-exposed skin and naturally shielded skin. J Chromatogr B Biomed Sci Appl 695(2):434–438, 1997.
36. N Flamand, P Justine, F Bernaud, A Rougier, Q Gaetani. In vivo distribution of free long-chain sphingoid bases in thehuman stratum corneum by high-performance liquid chromatographic analysis of strippings. J Chromatogr B Biomed Appl 656(1):65–71, 1994.
37. Y Yamamoto, T Yamamoto. Volume of the epidermal stratum corneum stripped with adhesive tape. Iyodenshi To Seitai Kogaku 13(6):360–361, 1975.
38. F Klaschka, M Norenberg. A device for the measurement of the horny layer transparency. The stripping method. Arch Dermatol Res 254(3):313–325, 1975.
39. F Klaschka, M Norenberg. Individual transparency patterns of adhesive-tape strip series of the stratum corneum. Int J Dermatol 16(10):836–841, 1977.
40. E Marttin, MT Neelissen-Subnel, FH De Haan, HE Bodd'e. A critical comparison of methods to quantify stratum corneum removed by tape stripping. Skin Pharmacol 9(1):69–77, 1996.
41. F Dreher, A Arens, JJ Hostynek, S Mudumba, J Ademola, HI Maibach. Colorimetric method for quantifying human stratum corneum removed by adhesive-tape stripping. Acta Derm Venereol 78(3):186–189, 1998.
42. RG van der Molen, F Spies, JM van't Noordende, E Boelsma, AM Mommaas, HK Koerten. Tape stripping of human stratum corneum yields cell layers that originate from various depths because of furrows in the skin. Arch Dermatol Res 289(9): 514–518, 1997.
43. HN Mayrovitz, SG Carta. Laser-doppler imaging assessment of skin hyperemia as an indicator of trauma after adhesive strip removal. Adv Wound Care 9(4):38–42, 1996.
44. A Triebskorn, M Gloor, F Greiner. Comparative investigations on the water content of the stratum corneum using different methods. Dermatologica 167(2):64–69, 1983.
45. J Palenske, VB Morhenn. Changes in the skin's capacitance after damage to the stratum corneum in humans. J Cutan Med Surg 3(3):127–131, 1999.
46. YN Kalia, F Pirot, RO Potts, RH Guy. Ion mobility across human stratum corneum in vivo. J Pharm Sci 87(12):1508–1511, 1998.
47. H Hofman, H Maibach. Transepidermal water loss in adhesive tape induced dermatitis. Contact Dermatitis 2(3):171–177, 1976.
48. RC Scott, GJ Oliver, PH Dugard, HJ Singh. A comparison of techniques for the measurement of transepidermal water loss. Arch Dermatol Res 274(1–2):57–64, 1982.
49. H Tagami, K Yoshikuni. Interrelationship between water-barrier and reservoir functions of pathologic stratum corneum. Arch Dermatol 121(5):642–645, 1985.
50. PG van der Valk, HI Maibach. A functional study of the skin barrier to evaporative water loss by means of repeated cellophane-tape stripping. Clin Exp Dermatol 15(3):180–182, 1990.

51. DM Reilly, MR Green. Eicosanoid and cytokine levels in acute skin irritation in response to tape stripping and capsaicin. Acta Derm Venereol 79(3):187–190, 1999.
52. OP Clausen, T Lindmo. Regenerative proliferation of mouse epidermal cells following adhesive tape stripping. Micro-flow fluorometry of isolated epidermal basal cells combined with 3H-TdR incorporation and a stathmokinetic method (colcemid). Cell Tissue Kinet 9(6):573–587, 1976.
53. OP Clausen, E Thorud. Perturbation of cell cycle progression in mouse epidermis prior to the regenerative response. J Invest Dermatol 75(2):129–132, 1980.
54. E Proksch, J Brasch, W Sterry. Integrity of the permeability barrier regulates epidermal Langerhans cell density. Br J Dermatol 134(4):630–638, 1996.
55. J Welzel, KP Wilhelm, HH Wolff. Skin permeability barrier and occlusion: no delay of repair in irrritated human skin. Contact Dermatitis 35(3):163–168, 1996.
56. C Gunther, A Kecskes, T Staks, U Tauber. Percutaneous absorption of methylprednisolone aceponate following topical application of Advantan lotion on intact, inflamed and stripped skin of male volunteers. Skin Pharmacol Appl Skin Physiol 11(1):35–42, 1998.
57. HH Woodworth, PM Newgard. Measurement of Skin Contamination. California: Stanford Research Insititute, 1963.
58. ME McBride, WC Duncan, JM Knox. Physiological and environmental control of gram negative bacteria on skin. Br J Dermatol 93:191–199, 1975.
59. R Aly, HI Maibach. Effect of antimicrobial soap containing chlorhexidine on the microbial flora of skin. Appl Environ Microbiol 31:931–935, 1976.
60. CA Evans, RJ Stevens. Differential quantitation of surface and subsurface bacteria of normal skin by the combined use of the cotton swab and the scrub methods. J Clin Microbiol 3:576–581, 1976.
61. VJ Tucci, AM Stone, C Thompson, HD Isenberg, L Wise. Studies of the surgical scrub. Surg Gynecol Obstet 145(3):415–416, 1977.
62. EL Larson, MS Strom, CA Evans. Analysis of three variables in sampling solutions used to assay bacteria of hands: type of solution, use of antiseptic neutralizers, and solution temperature. J Clin Microbiol 12(3):355–360, 1980.
63. JA Ulrich. Technics of skin sampling for microbial contaminants. Health Lab Sci 1:133, 1964.
64. JM Dunsmore. The effect of hand washing on the bacteria of skin. Aust J Dairy Techonol 27:137–141, 1972.
65. P Williamson, AM Kligman. A new method for the quantitative investigation of cutaneous bacteria. J Invest Dermatol 45(6):498–503, 1965.
66. JA Ulrich. Dynamics of bacterial skin populations. In: Skin Bacteria and Their Role in Infection. New York: McGraw-Hill, 1965.
67. R Berman, R Knight. Evaluation of hand antisepsis. Arch Environ Health 18:781–783, 1969.
68. NE Dewar, DL Gravens. Effectiveness of septisol antiseptic foam as a surgical scrub age. Appl Microbiol 26(4):544–549, 1973.
69. A Gross, L Cofone, MB Huff. Iodine inactivating agent in surgical scrub testing. Arch Surg 106(2):175–178, 1973.
70. G Reybrouck, H van de Voorde. The meaning of the results of four national disin-

gectant testing techniques (author's transl). Zentralbl Bakteriol (Orig B) 160(6): 541–550, 1975.
71. PC Galle, HD Homesley, AL Rhyne. Reassessment of the surgical scrub. Surg Gynecol Obstet 147(2):215–218, 1978.
72. G Reybrouck. Handwashing and hand disinfection. J Hosp Infect 8:5–23, 1986.
73. HE Eitzen, MA Ritter, ML French, TJ Gioe. A microbiological in-use comparison of surgical hand-washing agents. J Bone Joint Surg (Am) 61(3):403–406, 1979.
74. ML Rotter. Hygienic hand disinfection. Am J Infect Control 5(1):18–22, 1984.
75. A Gross, WJ Selting, DE Cutright, SN Bhaskar. Evaluation of two antiseptic agents in surgical preparation of hands by a new method. Am J Surg 126(1):49–52, 1973.
76. DS Paulson. Evaluation of three microorganism recovery procedures used to determine handwash efficacy. Dairy Food Environ Sanita 13:520–523, 1993.
77. C Poon, DJ Morgan, F Pond, J Kane, BR Tulloh. Studies of the surgical scrub. Aust NZ J Surg 68(1):65–67, 1998.
78. JV Pinfold, NJ Horan, DD Mara. The faecal coliform fingertip count: a potential method for evaluating the effectiveness of low cost water supply and sanitation initiatives. J Trop Med Hyg 91:67–70, 1988.
79. J Murray, RM Calman. Control of cross-infection by means of an antiseptic hand cream. Br Med J i:81, 1955.
80. HG Smylie, CU Webster, KL Bruce. pH index and safer surgery. Br Med J ii:606, 1959.
81. HG Smylie, JR Logie, G Smith. From Phisohex to Hibiscrub. Br Med J 4(892): 586–589, 1973.
82. GA Ayliffe, K Bridges, JR Babb, HA Lilly, EJ Lowbury, J Varney, MD Wilkins. Comparison of two methods for assessing the removal of total organisms and pathogens from the skin. J Hyg (Lond) 75(2):259–274, 1975.
83. LW Stuttard. Release of bacteria from surgeon's hands. Br Med J i:591, 1961.
84. J Brown, LW Wannamaker, P Ferrieri. Enumeration of beta-haemolytic streptococci on normal skin by direct agar contact. J Med Microbiol 8(4):503–511, 1975.
85. EJ Lowbury, HA Lilly. Gloved hand as applicator of antiseptic to operation sites. Lancet 2(7926):153–156, 1975.
86. A Cremieux, H Guiraud-Dauriac, J Duval, G Otterbein, J Guilbaud, B Epardeau, A Guillemart. Multicenter study in actual situations of the activity of an antiseptic in the surgical washing of hands. Pathol Biol (Paris) 32(5 Pt 2):599–603, 1984.
87. A Rosenberg, SD Alatary, AF Peterson. Safety and efficacy of the antiseptic chlorhexidine gluconate. Surg Gynecol Obstet 143(5):789–792, 1976.
88. AF Peterson, A Rosenberg, SD Alatary. Comparative evaluation of surgical scrub preparations. Surg Gynecol Obstet 146(1):63–65, 1978.
89. MW Casewell, I Phillips. Hands as route of transmission for *Klebsiella* species. Br Med J 2:1315–1317, 1977.
90. R Aly, HI Maibach. Comparative study on the antimicrobial effect of 0.5% chlorhexidine gluconate and 70% isopropyl alcohol on the normal flora of hands. Appl Environ Microbiol 37(3):610–613, 1979.
91. CA Bartzokas, JECorkill, T Makin. Evaluation of the skin disinfecting activity and cumulative effect of chlorhexidine and triclosan handwash preparations on hands

artificially contaminated with *Serratia marcescens*. Infect Control 8(4):163–167, 1987.
92. E Larson, K Mayur, BA Laughon. Influence of two handwashing frequencies on reduction in colonizing flora with three handwashing products used by health care personnel. Am J Infect Control 17(2):83–88, 1989.
93. ASTM E 1173–87. Standard Test Method for Evaluation of a Pre-operative Skin Preparation. Philadelphia: American Society for Testing Materials, 1987, pp. 729–731.
94. Food and Drug Adminstration (FDA). Topical antimicrobial drug products for over-the-counter human use: tentative final monograph for health-care antiseptic drug products. Fed Reg 59(116):31402–31452, 1994.
95. RN Michaud, MB McGrath, WA Goss. Application of a gloved-hand model for multiparameter measurements of skin-degerming activity. J Clin Microbiol 3(4): 406–413, 1976.
96. ASTM E 1327–90. Standard Test Method for Evaluation of Health Care Personnel Handwash Formulations by Utilizing Fingernail Regions. Philadelphia: American Society for Testing Materials, 1995, pp. 1–4.
97. ASTM E 1115–86. Standard Test Method for Evaluation of Surgical Hand Scrub Formulations. ASTM Standards Materials Environ Microbiol 1:201–204, 1987.
98. Food and Drug Adminstration (FDA). OTC topical antimicrobial product, drug and cosmetic product monograph. Fed Reg 39(179):33102–33141, 1974.
99. Food and Drug Adminstration (FDA). OTC topical antimicrobial product monograph. Fed Reg 43(4):1210–1249, 1978.
100. L Arnold. Relationship between certain physico-chemical changes in the cornifield layer and the endogenous bacterial flora of the skin. J Invest Dermatol 5:207–223, 1942.
101. H Blank, MH Coolidge. Degerming the cutaneous surface. I. Quaternary ammonium compounds. J Invest Dermatol 15:249, 1950.
102. P Williamson. Quantitative estimation of cutaneous bacteria. In: Skin Bacteria and Their Role in Infection. New York: McGraw-Hill, 1965.
103. L Colebrook, WR Maxted. Antisepsis in midwifery. J Obstet Gynaec Br Commonw 40:966, 1933.
104. P Story. Testing of skin disinfectants. Br Med J ii:1128, 1952.
105. J Chevalier, A Cremieux. Comparative study on the antimicrobial effects of Hexomedine and Betadine on the human skin flora. J Appl Bacteriol 73(4):342–348, 1992.
106. DH Lloyd. Evaluation of a cup scrub technique for quantification of the microbial flora on bovine skin. J Appl Bacteriol 56(1):103–107, 1984.
107. A Cremieux, JL Cazac. Evaluation of normal aerobic skin flora (author's transl) Ann Microbiol (Paris) 131B(1):59–68, 1980.
108. EJ Lowbury, HA Lilly. Disinfection of the hands of surgeons and nurses. Br Med J 2:1445–1450, 1960.
109. ML Rotter, W Koller. Test models for hygienic handrub and hygienic handwash: the effects of two different contamination and sampling techniques. J Hosp Infect 20(3):163–171, 1992.

110. WD Pohle, LS Stuart. The germicidal action of cleaning agents—a study of a modification of Price's procedure. J Infect Dis 67:275–275, 1940.
111. EF Traub, CA Newhall, RJ Fuller. Studies on the value of a new compound used in soap to reduce the bacterial flora of the human skin. Surg Gynec Obstet 79:205, 1944.
112. CV Seastone. Surg Gynec Obstet 84:355, 1947.
113. AR Cade. Antiseptic soaps a simplified in vivo method for determining their degerming efficiency. Soap Sanit Chem 26:35–38, 1950.
114. F Heiss. Cade-test with a new soap. Asthet Med (Berl) 18(12):223–232, 1969.
115. H Reber, M Muntener, K Neck, U Lips. Test methods for surgical hand disinfection (author's transl). Zentralbl Bakteriol (Orig B) 160(6):601–627, 1975.
116. J Charrel, MJ Gevaudan, MN Mallet, A Blancard, P Gevaudan. Evaluation de pouvoir bactericide de huit antiseptiques destines au lavage de type chirurgicale des mains. Rev Epidemiolo Sante Publique 25:297–313, 1977.
117. JD Jarvis, CD Wynne, L Enwright, JD Williams. Handwashing and antiseptic-containing soaps in hospital. J Clin Pathol 32(7):732–737, 1979.
118. ME Reverdy, A Martra, J Fleurette. Effectiveness of 9 soaps and/or antiseptics on hand flora after surgical-type washing. Pathol Biol (Paris) 32(5 Pt 2):591–595, 1984.
119. C Savage, H Leclerc, ME Reverdy, J Fleurette, H Guiraud-Dauriac, A Cremieux, B Joly, R Cluzel, G Garrigue. Multicenter study of the antimicrobial activity of antiseptics on the hands. Pathol Biol (Paris) 32(5 Pt 2):581–584, 1984.
120. ME Stiles, AZ Sheena. Efficacy of germicidal hand wash agents in use in a meat processing plant. J Food Prot 50(4):289–295, 1987.
121. M Rotter, W Koller, G Wewalka. Povidone-iodine and chlorhexidine gluconate-containing detergents for disinfection of hands. J Hosp Infect 1:149–159, 1980.
122. DGHM (German Society for Hygiene and Microbiology). Richtlinien für die Prüfung und Bewertung chemischer Desinfektionsverfahren. Zentralbl Bakteriol Hyg I Abt Orig B 172:528–556, 1981.
123. OGHMP. (Austrian Society of Hygiene, Microbiology and Preventive Medicine). Richtlinie vom 4. November 1980 zur Prüfung der Desinfektionswirkung von Verfahren für die chirurgische Händedesinfektion. Hyg Med 6:4–9, 1981.
124. RR Davies, WC Noble. Dispersal of bacteria on desquamated skin. Lancet ii:1295–1297, 1962.
125. RP Clark. Skin scales among airborne particles. Hygiene 72:47, 1974.
126. EA Mortimer, E Wolinsky, AJ Gonzaga, CH Rammelkamp. Role of airborne transmission in staphylococcal infections. Br Med J 1:319–322, 1966.
127. G Kraidman. The microbiology of airborne contamination and air sampling. Drug Cosmet Ind 116:40–43, 1975.
128. PD Meers. Intravenous infusions: the potential for and source of contamination. In: PD Phillips, PD Meers, PF D'Arcy (eds.) Microbiological Hazards of Infusion Therapy. Lancaster, PA: M.T.P. Press, 1976, p. 59.
129. PD Meers, GA Yeo. Shedding of bacteria and skin squames after handwashing. J Hyg Camb 81:99–105, 1978.
130. A Gross, DE Cutright, SM D'Alessandro. Effect of surgical scrub on microbial population under the fingernails. Am J Surg 138(3):463–467, 1979.

131. MC Mahl. New method for determination of efficacy of health care personnel hand wash products. J Clin Microbiol 27(10):2295–2299, 1989.
132. ASTM E 1174–87. Standard test method for evaluation of health care personnel hand wash formulations. ASTM Standards Materials Environ Microbiol 1:209–212, 1987.

27

The Need and Methods for Assessing the Activity of Topical Agents Against Viruses

Shamim A. Ansari
Colgate-Palmolive Company, Piscataway, New Jersey

Syed A. Sattar
University of Ottawa, Ottawa, Canada

I. INTRODUCTION

Topical agents intended for the decontamination of skin have routinely been examined for their antibacterial properties, and indeed this has often been assumed to be all that is necessary. Viral contamination of skin has not been widely recognized, and, therefore, neither is the need for topical antiviral agents. While pathogenic viruses are not a part of the normal resident microflora of the body, they are shed for varying periods by those infected with them. Shedding of virus generally begins prior to the onset of clinical symptoms and lasts for several days, or occasionally weeks, after recovery. It is also important to note here that most cases of viral infections remain asymptomatic while silently shedding infectious viruses into their surroundings and can, therefore, be a source of serious disease for others.

Symptomatic viral infections remain a leading cause of morbidity and mortality [1]. Indeed, the relative significance of viral infections has been increasing as we successfully combat common bacterial diseases. Many societal changes are also making us more vulnerable to attack by viruses [2], and in the absence of safer chemotherapy and vaccines, measures such as handwashing remain an effective means of prevention against many types of viral pathogens.

Since all viruses are obligate parasites, suitable living hosts in the form of either intact animals, embryonated eggs, or cell cultures are essential to detect and measure their viability in the laboratory. This need for a living host adds considerably to the expense and time when assessing the activity of topicals against viruses.

Topicals are applied chiefly to the hands for antisepsis, especially in clinical settings, and because hands are considered to be major vehicles for virus spread, this chapter will concentrate on the importance of virus spread by hands. This does not preclude the activity of similar agents applied to other areas of skin but simply recognizes the predominance of hands as vehicles for most enteric and respiratory viruses that are not predominantly airborne.

This chapter will critically assess the published information on the methods available as well as the conditions and criteria needed to assess the virus-eliminating activity of antiseptics, particularly handwash agents, and then discuss why and which viruses may be suitable as surrogates to determine the activity of handwash agents against human pathogenic viruses. The reader is referred to a recent review on the basic properties of viruses and their relative significance as human pathogens [3].

II. VIRUSES AND THEIR SPREAD BY HANDS

Infectious virus particles have been recovered from naturally contaminated hands of caregivers, from fomites, and from environmental surfaces [4]. The actual amount of virus discharged from infected individuals varies considerably, depending on the type of infecting agent and the stage of the infection. On contaminated hands, viruses can survive from a few minutes to at least several hours [4]. Although virus survival on many types of inanimate surfaces and objects is frequently much longer [4], viruses, which are particularly sensitive to drying, may survive better on the skin than when dried onto surfaces, depending on the ambient relative humidity.

To initiate an infection, sufficient numbers of infectious particles of a given virus must first enter the body of a susceptible host at an appropriate site. Since virus particles are discharged from an infected host in various body secretions and excretions, transmission of viral infections requires direct or indirect contact with such contamination and the deposition of its required dose at a suitable portal of entry.

What is the smallest number of viable virus particles needed to infect a susceptible host? This question cannot be answered categorically, but limited studies with human and animal hosts have shown that this number may be as low as 1–10 infectious units [5]. Whether or not this is universally true, it would be prudent to assume the infectious dose of viruses to be quite small.

Human hands could act as vehicles for many types of viruses [4], and, by corollary, regular and proper decontamination of hands could reduce the risk of spread of such infectious agents. But the link between hands and the spread of viral infections is based mainly on circumstantial evidence and on limited experimental studies using human subjects [4]. This lack of direct evidence is not surprising in view of the general difficulties in working with viruses, the seasonal nature of most viral infections, as well as our inability to distinguish between simultaneous spread of a particular infectious agent by hands and other vehicles in a given setting.

We know even less about the relative importance of hand decontamination in interrupting the spread of viral infections under field settings because of obvious problems in designing and conducting such investigations. Other major impediments in this regard have been the absence of a proper regulatory framework and the lack of availability of proper quantitative protocols to determine the relative effectiveness of handwashing agents in the removal and or inactivation of viruses on human hands. Only in recent years have standardized test methods become available to generate information on the ability of viruses to survive on human hands [4], to be transferred to and from hands during casual contact [4], and the potential of topical agents to rid hands of viral pathogens [4,6].

Hands are among the most obvious surfaces to become contaminated by infected individuals; this is true whether the contamination is of self or a caregiver. The nature and extent of such contamination will depend on the site of infection, the degree and nature of the discharge medium from the host, as well as the personal habits of the infected individual and the hygienic facilities available. The degree of contamination can vary widely. For example, some enteric viral infections can produce a profuse and almost explosive diarrhea, which may be difficult to contain. Addressing such an infection in wards or facilities for bedridden or mentally handicapped patients can be quite difficult.

Regular and frequent interactions between hands and their surroundings suggests that transfer of contaminating virus can occur readily between the contact points. Such transfer of infectious virus to and from hands upon casual contact with objects or other animate or inanimate surfaces can be demonstrated to occur readily in experimental settings [4,6,7]. Several studies have shown that clean hands can readily become contaminated when objects or surfaces with infectious virus on them are touched or handled [6]. The reverse has also been shown to be true. Transfer of a rhinovirus was observed in 15 of 16 trials in which a plastic surface, contaminated 1–3 hours previously, was touched by a volunteer [8]. Individuals with acute rhinovirus colds were shown to deposit infectious rhinovirus particles on objects they touched [9]. Infectious rhinovirus particles could be recovered from fingertips of volunteers who handled objects such as doorknobs previously touched by virus-contaminated (donor) hands, and rhinovirus transfer also has been shown to occur by direct hand-to-hand contact [10].

Studies using human subjects have also established that self-inoculation with rhinovirus- and rotavirus-contaminated fingers can lead to infection [11,12]. Whether or not sufficient virus can be exchanged and subsequently acquired by the susceptible host obviously depends on a range of factors. Virus transfer can be promoted by moist conditions, as well as by increased friction and pressure [13]. Presumably, increased frequency of contact will also promote virus acquisition and transfer; for example, a caregiver with frequent contacts among daycare participants may inadvertently transfer viruses from one child to another, possibly simply by hand contact. Although significant numbers of viral particles can be transferred when contamination levels are high, generally, the percentage of virus transferred during experimental contacts has been shown to be fairly low. This suggests that the further along the chain of contacts the susceptible host is from the point of primary contact with the virus, the lower the risk that an infection will result, and vice versa.

III. IMPORTANCE OF HAND ANTISEPSIS

Topical antimicrobials are used for a variety of purposes in infection control, and there is also increasing recognition of their potential as prophylactics against sexually transmitted viral infections [14,15]. However, their principal application in dealing with viruses is in "hygienic handwashing," a term that refers to decontamination of hands to eliminate transient microflora. Consequently, this chapter will focus on methods to evaluate the virus-eliminating activity of chemicals and formulations used primarily for the decontamination of the intact skin of hands. With the exceptions discussed below, we are unaware of any attempts to standardize methods to assess the virucidal potential of topicals applied on mucous membranes; the limited information available in this regard comes from the use of experimental animal models [16].

Although the main focus of this chapter is the carriage, transfer, and decontamination of viruses on intact skin, it is necessary to mention the possibility of inapparent parenteral exposure to viruses through compromised skin. This is especially important for caregivers when their skin becomes damaged through frequent handwashing and when they may be exposed to bloodborne viruses. Whether or not hand decontamination could prevent virus infection in these circumstances will depend on the exact conditions prevailing on a case-by-case basis.

IV. DESIGNING AND PERFORMING TESTS AGAINST VIRUSES

No test for antiviral activity can predict the effectiveness of products used in the field, but properly designed tests can assess the potential of topical products to rid hands of contaminating viruses. As detailed below, several factors (Table 1) are crucial in the proper assessment of the activity of topicals against viruses.

A. Test Virus(es)

As yet, there are no generally recognized surrogates for testing the activity of topicals against viruses, and this encourages the practice of testing a given formulation against as many viruses as possible and listing them all on the product label. This is particularly true in the United States, where label claims can only be made against individually tested viruses. Such an approach (1) makes product development unnecessarily expensive and time-consuming, (2) encourages the use of the pathogenic viruses themselves (e.g., HIV or hanta viruses), which are unsafe to handle and may cause undue risk of laboratory-acquired infections, (3) results in the listing of easy-to-kill (enveloped) viruses, such as HIV, on product labels, thus gaining an unjust market advantage, (4) encourages label claims against viruses (e.g., influenza viruses), which may not be amenable to control through the use of chemical germicides, and (5) makes product compari-

Table 1 Factors Important in Assessing the Activity of Topicals Against Viruses

Culture and infectivity assay of test virus(es)
Test virus(es) to be used
Nature of in vitro carrier to be used
Volunteers to be selected for testing
Nature and level of soil loading
Diluent, if required, for the test product
Time used for the initial drying of the inoculum
Contact between virus and test germicide
Neutralization of virucidal activity
Procedure for the elimination of cytotoxicity
Method for quantitating virus infectivity
Number of test and control carriers/volunteers
Number of product lots to be tested
Product performance criterion
Essential controls

sons difficult because of the use of nonstandardized viral strains and variations in test protocols.

Such testing should be conducted using properly selected surrogates provided the test conditions are rigorous enough. As can be seen from Table 2, certain types of viruses possess many of the characteristics desired in a surrogate, and it should be possible to select one or two of them for testing the capacity of hygienic handwashing in eliminating viruses from experimentally contaminated hands. With the anticipated eradication of poliomyelitis in the next 3–5 years, there are moves already underway to phase out the experimental use of all types of polioviruses [17].

In some investigations on the interruption of transmission of viral infections by chemical germicides, the hands of volunteers were contaminated with viruses naturally shed in body fluids [9,11]. Such an approach is feasible only when the virus titer is high and the clinical specimen containing it has been thoroughly screened for other extraneous and potentially harmful agents. However, the use of virus-positive clinical specimens is unsuitable in standard test protocols because such material can vary greatly not only in the infectivity titer of the virus but also in its soil loading.

Some studies on chemical germicides against viruses have used bacteriophages (or phages), which are viruses of bacteria that are relatively inexpensive and easy and safe to work with [18–20]. However, although such testing alone would not be sufficient for product registration and marketing purposes, it could provide for the manufacturer an inexpensive and rapid method of screening a large number of potential formulations.

B. Culture and Infectivity Assay of Test Virus(es)

The use of experimental animals in testing the activity of chemical germicides in general against viruses is considered neither necessary nor desirable [21]. Established and well-characterized cell lines are the recommended host system. In some cases, such as when working with adenoviruses [22], it is advisable to carry two cell lines: one for the preparation of virus pools and the other to assay for its infectivity. The use of embryonated eggs may be needed in rare cases, such as for the production of high-titered pools for influenza viruses.

In working with chemical germicides, an accurate measure of the degree of loss in *virus infectivity* is essential. Therefore, assay systems that detect and/or measure the presence of viral proteins, nucleic acids, or enzymes without determining viral infectivity are considered unsuitable [21]. It must also be remembered here that no matter what host system is used, inability to detect infectious virus particles in it does not necessarily mean the complete absence of infectivity for the natural host.

C. Nature and Level of Soil Loading

In nature, viruses are always shed in an organic matrix and such ''organic or soil load'' can interfere with the activity of a germicidal chemical either by interacting with it and reducing its effective concentration or by preventing its access to the target virus through physical protection. Therefore, any proper test against viruses must simulate the presence of such soil. In practice, there are wide variations in the nature and levels of substances used, and, as far as we are aware, no substance or a combination thereof can be regarded as a universal soil load. Bovine serum (5–10%) is commonly used for this purpose, but it is relatively expensive, not readily available, and may contain specific or nonspecific virus inhibitors. While substances such as feces [23] have been used in testing the virucidal activity of chemical germicides, they are inherently variable and thus unsuitable as a soil load for standardized test protocols to assess the potency of the germicidal chemicals. They may, however, have some value in secondary test methods to further examine the potential for field effectiveness.

We have recently developed a soil load that appears to be suitable for working with a variety of test organisms, including viruses [24]. It consists of mucin in combination with low molecular weight (peptides) and high molecular weight protein (albumin) mixture. The concentrations are designed to provide a challenge approximately equivalent to 5–10% serum in testing for virucidal or other types of germicidal activity.

D. Diluent for Test Germicide

Many topicals require the addition of water to prepare their use dilution, but product labels often do not clearly specify the type of water to be used for this purpose. Normally, one would use tap water for this purpose, but label claims for germicidal activity may be based on distilled water as the product diluent. Current regulations in the United States assume that distilled water is used and do not require that this be stated on the label. Consumers are usually unaware of the need to use distilled water.

Even when distilled water is meant to be used, most users do not have ready access to it, and in the vast majority of cases tap water becomes the practical diluent; germicides with marginal virucidal activity may work when diluted in distilled water, but not when tap water is used as the diluent [25].

Even though tap water may represent a stronger challenge, its quality, as well as the nature and levels of disinfectants in it, varies both temporarily and geographically. In view of this, water with a standard level of hardness in it (e.g., 200–400 ppm $CaCO_3$) makes for a more desirable diluent in tests for virucidal activity.

Table 2 Viruses Relevant in Hand Antisepsis and Possible Surrogates for Testing Activity Against Viruses

Virus(es)	In vitro infectivity assay method?	Safe for skin?	Survival on hands?	Potential for spread by hands?	Suitability as a surrogate?	Comments
Adenoviruses	Yes	Yes	Good	Yes	Yes	Many types of adenoviruses are safe and relatively easy to work with in the lab; however, they may be less resistant to chemical germicides than other nonenveloped viruses such as rotaviruses.
Bacteriophages	Yes	Yes	?	No	?	Carefully selected phages may be suitable only in the preliminary screening of topicals against human pathogenic viruses.
Caliciviruses (Norwalk agent)	No	Yes	?	Very high	No	Human caliciviruses cannot be grown in the lab, but some animal caliciviruses can be.
Hepatitis A virus	Yes	Yes	Very good	Very high	Possible	Relatively resistant to inactivation by many germicides used as topicals; vaccination of personnel handling the virus is recommended.

Herpesviruses	Yes	No	Poor	High	No	Very fragile viruses with low resistance to many chemicals.
Poxviruses	?	No	?	?	No	Generally difficult to work with in the lab and also require specialized facilities for handling and containment except vaccinia virus.
Papilloma viruses	No	No	?	High	No	Human papillomaviruses cannot be grown in the lab, while some animal papillomaviruses may be cultured and quantitated with some difficulty.
Enteroviruses (e.g., polioviruses)	Yes	Yes	Good	Not known	Possible	Although the vaccine strains of polioviruses are very safe, the use of all polioviruses will soon be phased out in view of the anticipated eradication of poliomyelitis; whereas a coxsackie- or echovirus may be used instead, their safety will be a concern.
Respiratory syncytial virus (RSV)	Yes	Yes	Very poor	High	No	Very fragile viruses with low resistance to many chemicals.
Rhinoviruses	Yes	Yes	Good	High	Possible	Relatively safe and easy viruses to work with in the lab.
Rotaviruses	Yes	Yes	Very good	Very high	Yes	Relatively safe and easy viruses to work with in the lab.

E. Carriers

Topicals are intended for use on human skin. Therefore, they should be tested on a suitable surface. Many studies on topicals have used hard surface carriers, or even suspension tests, and extrapolated the results on product potency to human skin. We do not believe this to be a valid approach because skin may present a greater challenge than a hard inanimate surface [26] and a much greater challenge than suspension tests. Mucous membranes may present an even greater challenge.

F. Replicates for Test and Control Carriers

The expensive and labor-intensive nature of work with viruses limits the number of replicates in tests for virucidal activity. Nevertheless, enough replicates must be included to make the results statistically meaningful. This requires some knowledge of the degree of reproducibility of the assay methods, and because viruses require a host system, the results tend to be inherently more variable than is observed for bacteria, for example. In general, methods that determine virus plaque- or focus-forming units are more accurate than most-probable-number (MPN) assays. Each measure of reduction in virus infectivity by a germicide is obtained by comparison with controls not exposed to test formulation. Therefore, it is crucial to include sufficient numbers of such controls to obtain an accurate mean value against which each test carrier can be assessed.

G. Drying of Virus Inoculum

Unlike testing on inanimate surfaces, the virus inoculum on an animate surface such as skin may never be totally dry. Therefore, the objective is to challenge the inoculum by waiting until it looks visibly dry, keeping in mind the fact that viruses differ in their ability to withstand drying [27]. It is also essential to include a control that measures the virus titer after drying and to use it as the baseline to calculate the loss in virus titer after exposure to the test or control solution.

H. Time for Virus-Germicide Contact

The handwashing guidelines from the U.S. Centers for Disease Control and Prevention (CDC) recommend that caregivers lather their hands for at least 10 seconds [28]. Surveys in hospitals have found caregivers to spend, on an average, no more than 8 seconds washing their hands [29]. While contact times of 15–60 seconds have been used in recent studies on handwash agents [30], we believe that the contact between the challenge virus and the test formulation for in vivo testing of such products should not be longer than 10–15 seconds. This will require some skill and organization on the part of the experimenter to properly apply such short contact time in the testing. Contact times longer than these in a test protocol would be meaningless in predicting the performance of a hygienic handwash agent in the field. In case a surgical scrub or preoperative skin prep is to be tested against viruses, an appropriate extension in the contact time may be justified.

I. Neutralization of Virucidal Activity

For accurate and meaningful results, the virucidal activity of the test formulation must be arrested immediately and effectively at the end of the contact time without the process itself killing the virus or enhancing cytotoxicity. This can be achieved by either the addition of a neutralizer or by dilution of the virus-germicide mixture, or by a combination of both [31,32]. Whichever approach is adopted, its effectiveness must be properly validated before the test results can be accepted.

The following are salient among the many factors to consider in choosing a suitable chemical neutralizer:

1. No one substance or a combination thereof is known that can successfully neutralize all known ingredients of germicide formulations.
2. If used, a neutralizer must be shown to be free of any deleterious effects on the test virus(es) and also must be proven to be devoid of cytotoxicity and interference with virus infectivity assay.
3. Any steps to arrest virucidal action must not dilute the eluates such that the detection and quantitation of infectious virus in them becomes difficult.
4. The selected neutralizer(s) must act quickly to avoid any increase in the contact time between the test virus and the germicide.
5. It must be readily available and standardizable.
6. It must not alter virus infectivity or host-cell susceptibility.

While a 100-fold dilution of the virus-germicide mixture soon after the end of the contact time has proven effective in dealing with most types of germicidal chemicals [32], this procedure requires that the volume of the diluent be kept relatively small to allow for the titration of most of the eluate.

In some studies using carrier tests, the dried virus inoculum is scraped from the surface of the carriers prior to the addition of a neutralizer [33]. This effectively can turn the carrier test into a suspension test. Therefore, the neutralizer must be added to the carriers and or the eluting fluid before virus is eluted/ scraped.

J. Elimination of Cytotoxicity

The need for living hosts for the detection and measurement of virus infectivity introduces the serious and sometimes difficult-to-deal-with issue of cytotoxicity. The considerations involved in the successful elimination of cytotoxicity are somewhat similar to those enumerated above for neutralizing virucidal activity.

In many instances, a given chemical may or may not be effective against the test virus while being toxic to the cells employed to detect residual virus infectivity remaining after treatment. Such cytotoxicity can seriously interfere with the interpretation of test results, and can often be assessed simply by microscopic examination for visual cell damage of the cell culture system for up to 24 hours after addition of the residual virus-disinfectant mixture. Such apparent absence of cytotoxicity can be misleading because host cell monolayers may appear to be undamaged but be unable to support virus replication. Even when toxicity appears to be visibly removed, subtle effects on the cells, and potentially on their ability to support virus replication, may remain. Failure to culture infectious virus and virus kill can be confounded, and this needs to be examined through a low-level virus challenge to test the functionality of the assay system. Proper demonstration of a lack of cytotoxicity to the detection system will include separate determinations that low levels of virus can still be detected in untreated cell cultures after the virus itself has been exposed to the residual level of test product remaining at the end of the test—positive virus control—and, similarly, that prior exposure of the cell cultures to residual levels of disinfectant product does not interfere with the detection of low numbers of untreated viruses—positive cell control. Moreover, germicides with fixative properties can effectively kill host cells without detaching them or producing any apparent damage to them. In such cases it may be possible to observe cytotoxicity in a dilution that was not detected in the undiluted test sample. Cell culture monolayers, which are the host system of choice for virus infectivity assays, are also most affected by cytotoxicity, while whole animals and embryonated eggs are likely to be more resilient in this regard.

Gel filtration [34,35] or centrifugation [36,37] of virus-germicide mixtures may be effective in the removal of cytotoxicity, but such steps invariably extend the contact of the virus with the test germicide by several minutes or more and put into question the accuracy and relevance of claims for virucidal activity of topicals. A 10- to 100-fold dilution of the virus-germicide mixture at the end of the contact time is one simple and potentially widely applicable approach to reducing cytotoxicity [32]. This approach, however, requires relatively high-titered pools of the test virus and may not work on its own for chemicals that are highly cytotoxic. For highly cytotoxic formulations, simple dilution of the reaction mixture may be insufficient [38], and the use of additional steps such as gel filtration [35,39], dialysis [40], and ultrafiltration [41–43] may be required. As stated above, the most serious limitation to these approaches is that they all inevitably extend the contact time between the virus and the test formulation.

K. Quantitation of Virus Infectivity

Any good test for virucidal activity must incorporate a measurement of infectivity of the challenge virus. This is essential to determine the level of loss in viral infectivity for the host system used. In many tests for virucidal activity, only very small fractions (often only 1–10%) of the test sample eluates are titrated for infectious virus, and the absence of infectivity in the host system may be taken as evidence of the formulation's effectiveness. The limit of detection for the assay system is rarely quoted, but when no infectious virus is detected, the results should be quoted as "less than" the theoretical detection limit. For a higher level of confidence in the results, it is desirable to titrate most or all of the test sample eluates. In order to save time and materials, ultrafiltration can be used to reduce the volume of the eluates using filters that allow very little protein binding.

L. Number of Product Lots to be Tested

For a higher level of confidence in the test results, it is considered necessary to evaluate more than one product lot of a given germicide for virucidal activity, and several standards require at least three lots to be tested for government registration purposes.

M. Other Essential Controls

The need for cell cultures for tests with viruses requires the incorporation of controls over and above those necessary for working with bacteria. In addition to cell culture controls to demonstrate lack of contamination and virus controls to demonstrate a functioning assay system, the level of input virus and the loss in virus infectivity upon the drying of the inoculum on the carrier need to be measured. In some cases, these measurements are only made once and the data used with a series of tests. However, it is recommended that, for proper accuracy, such measurements should be included with every test due to inherent variations in cell cultures and viruses. When there is a need to separate virus kill from simple mechanical removal of the test virus during the test, it is recommended that a control also be included to determine the mechanical removal of the test virus; standard hard water should be used for this purpose in place of the test topical product. For any claim of topical antisepsis, reduction in the virus titer on treatment with the test product must be substantially higher than that obtained with the standard hard water treatment alone.

An often-ignored complication of titrating viral infectivity, particularly in cell culture systems, is that germicide residues, even in diluted eluates, may increase or decrease the susceptibility of the host cells to the test virus. In case of decreased susceptibility, the host system could overestimate the activity of the tested germicides by not being able to detect the presence of low levels of infectious virus in the inoculum (see above discussion of cytotoxicity). An increase in the level of infectivity could also occur due to either unmasking of more viral receptors on the host cell surface or deaggregation of viral clumps. For the results to be considered valid, controls must, therefore, be included to rule out the presence of such interference. The best way to approach this for a cell culture host system is to first expose the cell monolayer to a noncytotoxic level of the test germicide and subsequently challenge the cells to the test virus diluted to give countable infectious foci such as plaques. If the number of infectious foci in such preexposed monolayers is not statistically significantly different from that in the monolayers treated with a control fluid, the product can be assumed to be free from such interference.

N. Product Performance Criteria

The true relationship between the germicidal activity of a product and its ability to prevent the spread of infections in the field is not known and remains difficult to determine. Therefore, performance criteria for potency testing of germicidal chemicals are a matter of policy and practicality rather than being based on sound public health science. Nevertheless, such arbitrary criteria have a long history and provide a valuable means of registering suitable products. In tests for bactericidal activity, it is generally feasible to measure viability reductions of 5–6 $\log_{10}$. When working with viruses, it is more usual to aim for 3–4 $\log_{10}$ reductions in infectivity titer on hard surfaces. For topical products, it is often not practical to set such high performance criteria. However, some agents such as alcohol are widely documented to routinely produce kills of greater than 2 $\log_{10}$.

Product performance criteria are normally set by regulatory agencies, and they may differ from one jurisdiction to another. Some national or international standards also specify the level of reduction in the viability titer of the test organism for the evaluated product to meet the requirements of that standard. The Canadian General Standards Board [44], for example, requires products for use on environmental surfaces and medical devices to show a >3 $\log_{10}$ reduction (beyond the level of cytotoxicity) in the level of infectious virus to meet its requirements. Certain regulatory agencies also specify that, in addition to showing the specified level of $\log_{10}$ reduction in the infectivity titer, no infectious virus be detectable in the highest dilution of the virus-germicide mixture titrated. However, in North America, no standards exist for virucidal inactivation by topicals, and further discussions are needed to establish practical criteria whereby topical products can be considered to be safe and effective against viruses.

V. METHODS TO ASSESS THE ACTIVITY OF TOPICALS AGAINST VIRUSES

There has been considerable progress in this area of research in the past decade, and several standard (Table 3) and experimental procedures are now available to assess the activity of topicals against viruses. The following is an analysis of the relative advantages and disadvantages of the general classes of published protocols.

A. Suspension Tests and Testing Using Inanimate Carriers

In many instances, the activity of a hygienic handwash agent is determined using a simple suspension test [45] or inanimate carriers (Table 4). In most suspension tests, one part of the virus is mixed with nine parts of the handwash agent and the mixture held at 20°C or room temperature (22 ± 2°C) for as long as 10 minutes before being titrated for infectious virus.

In carrier tests, the virus inoculum is first applied to and dried on an inanimate carrier such as metal, plastic, or glass, and the product under test is then placed on the dried inoculum and left to act on it at room temperature for 1–10 minutes [23,32]. At the end of the contact time, the virus-product mixture is eluted from the carrier and titrated for infectious virus. As a rule, the carrier test is more stringent when compared to a suspension test, because the test product must act on the virus in the dried inoculum, and products that pass the suspension test may or may not pass the carrier test.

Neither the suspension nor the in vitro carrier methods give results that are truly predictive of a handwash agent's activity on human skin. Furthermore, these tests are generally conducted at ambient temperatures, whereas the temperature of the skin is around 33°C. The contact time of even one minute is too long for testing handwash agents, because most caregivers apply them on their hands for no more than 4–8 seconds [29], even though 10 seconds is the time recommended by the CDC in their handwashing guidelines [28]. Also, the level of moisture on it is quite different from that encountered on nonporous inanimate surfaces, and the surface topography of the skin gives a level of protection for applied virus that cannot be simulated by inanimate surfaces.

Table 3 Standard Methods Currently Available to Assess the Activity of Topicals Against Viruses

Organization	Title of standard	Number	Purpose	Year adopted
CEN	Virucidal Suspension Test for Chemical Disinfectants and Antiseptics Used in Human Medicine	DRAFT	A **suspension** of the test virus, with or without a soil load, is mixed with the test agent and the mixture held at 20°C for the desired contact time. The mixture is then titrated for infectious virus.	Under review by CEN/TC 216
ASTM	Standard Test Method for Determining the Virus-Eliminating Effectiveness of Liquid Hygienic Handwash Agents Using the Fingerpads of Adult Volunteers	E 1838-96	**Fingerpads** are contaminated with the test virus and allowed to dry. The dried inoculum is exposed to test product. Virus is then eluted with or without a water rinse. Eluates and controls are titrated for infectious virus.	Adopted as a standard in 1996
ASTM	Standard Test Method for Evaluation of Handwashing Formulations for Virus-Eliminating Activity Using the Entire Hand	E 2011-99	Target virus is placed on the hands and spread over the **entire surface of both hands** and allowed to dry. The hands are washed with the test product. The virus is eluted from fingertips. Controls and eluates are titrated for infectious virus.	Adopted as a standard in 2000

CEN = Comité Européen de Normalisation, Brussels, Belgium. ASTM = American Society for Testing and Materials.

Table 4 Chronological Listing of In Vitro Studies on the Activity of Topicals Against Viruses

Topical(s) tested	Viruses tested	Ref.
	Suspension test	
Chlorhexidine	Herpes hominis, poliovirus type 2 (Sabin), adenovirus type 2	46
Aqueous iodine	Rhinovirus	47
Alcohols	Rotavirus, astrovirus, echovirus 11	48
Betadine, hexol, hibitane, hibiclens, ethanol	Rotavirus SA11	49
Povidone-iodine, chlorhexidine digluconate	Poliovirus type 1 (Sabin)	41
Listerize	Herpes simplex virus types 1 & 2	50
Zilactin®	Herpesvirus type 1	51
Chloroxylenol (PCMX)	HIV-1	52
Povidone-iodine, ethanol, isopropanol, chlorhexidine, hydrogen peroxide	Poliovirus 1 (Sabin)	53
Chlorhexidine gluconate, povidone-iodine, carbanilide, benzylkonium chloride	Vaccinia orthopoxvirus, herpes virus KOS 1	38
Bleach, Lysol, povidone-iodine, hibiclens, osyl, ethanol, Listermint	Respiratory syncytial virus	54
Levermed HDI (a hand disinfectant gel)	HIV strains RF & IIIB	37
Povidone-iodine, chlorhexidine gluconate, benzylkonium chloride, alkyl diaminoethylglycine hydrochloride	Adenovirus, mumps virus, rotavirus, coxsackievirus, herpesvirus, rubella virus, measles virus, influenza virus, HIV	55
Chloroxylenol, benzalkonium chloride, cetrimide/chlorhexidine	Coxsackievirus, adenovirus type 25, HSV-1, polio virus type 1 (Sabin), coronavirus	45
Dishwashing detergents, antibacterial hand soaps	Respiratory syncytial virus	56
Antimicrobial hand soaps	Vaccinia virus	43
Benzylkonium chloride, hydrogen peroxide, acetic acid	Adenovirus type 5	57
	Carrier test using stainless steel disks	
Products containing ingredients such as ethanol, povidone-iodine, chlorhexidine gluconate, etc.	Coxsackievirus B3, HPIV-3, HCV 229E, adenovirus type 5	58

B. Tests Using Human Subjects

Table 5 presents a summary of in vivo studies for testing the activity of topicals against viruses. Such protocols are based mainly on the use of either the whole hand [59,60,64], fingertips [69], or fingerpads [64,68].

1. The Whole-Hand Method

In the whole-hand method [64,74], generally 0.5 mL of the test virus is placed on the palm surface of the hands, and the hands are rubbed together to spread the contamination. The hands are then allowed to air-dry. To obtain the base titer of the virus at the end of this procedure, 20 mL of an eluent is poured over the hands while they are being rubbed together to recover as much of the virus as possible. To test a handwash agent, the contaminated and dried hands may be wetted in water to simulate pretreatment rinse and then receive 0.5–5 mL of the test product on the palm surface of one of the hands, and the hands are rubbed together for between 10 seconds and several minutes to simulate the normal lathering procedure. They are then washed in water, dried, and the virus eluted either from the entire surface of both hands or by dipping only the fingertips in ≥20.0 mL of an eluent. Depending on the level of cytotoxicity of the product under test, the eluate may require the removal/neutralization of cytotoxicity prior to virus titration in cell culture.

In our view, the whole-hand method has the following weaknesses when used to test handwash agents against viruses:

1. There is potential for virus wash-off during the pre- and posttreatment tap water rinse.
2. The virus initially placed on one of the hands is spread over the entire surface of both hands during the application of the handwash agent and at the time of sampling, and if only fingertips are immersed in the recovery medium, this leads to virus recovery from a small fraction of the contaminated surface. This is particularly important in dealing with formulations with strong detergent activity but weak virucidal activity, because they spread the inoculated virus over a much wider area of the hand during the lathering procedure.
3. The volume (about 20 mL) of the recovery medium is too large to allow for the detection of infectious virus in all or most of it without a virus concentration step.
4. In any given sitting, the volunteer can be used to test either the control or only one of the products against a single type of virus.
5. Incorporation of the paper towel–drying step as an integral part of the test procedure makes it difficult to account for the true extent of virus elimination by the handwash agent itself.

Table 5 Chronological Listing of In Vivo Studies on the Activity of Topicals Against Viruses

Test surface	Viruses tested	Topical(s) tested	Ref.
Skin	Rhinovirus	Aqueous iodine	47
Whole hand and fingertips	Adenovirus, poliovirus, coxsackie, echo, influenza A, vaccinia etc.	Desderman	59
Swine skin model, fingerpads	Porcine enterovirus type 3, polioviruses type 1,2,3 (Sabin)	Derma Klenz I, Clean 'N Smooth, vegetable oil based soap, ethanol	60
Fingerpads	Rhinovirus	Glutaric acid	61
Wrists	Bacteriophage (X174)	70% ethanol	62
Human gloved fingertips, cadaver skin	Cytomegalovirus	Chlorhexidine gluconate, povidone-iodine, soap	63
Fingerpads, whole hand	Rotavirus	Alcohols, chlorhexidine gluconate, PCMX	64
Fingertips	Adenovirus type 3	Welpas™ (alcoholic disinfectant)	65
Whole hand	Poliovirus type 1, echovirus 11 & 12, reovirus type 3, papovavirus SV40, adenovirus type 2, coxsackievirus type 4, vaccinia virus	Alcohols, antiseptic products (V, VP 1, Desderman, Betaisodona)	66
Fingertips	Polio 1, bacteriophages (MS2, K 1–5)	Alcohols, povidone-iodine, soap	67
Fingerpads	Hepatitis A virus, poliovirus type 1 (Sabin)	Handwash agents	68
Fingertips	Bovine rotavirus	Povidone-iodine, triclosan, chlorhexidine digluconate, alcohols	69
Guinea pig skin	Herpesvirus type 1	Ivory soap	70
Whole hand, fingerpads	Poliovirus type 1 (Sabin)	Alcohol-containing products, non-medicated soap	71
Porcine skin	Bacteriophage MS2 & 06	Chlorhexidine, hydrogen peroxide, 70% alcohol	26
Ex vivo human skin and umbilical cord	Herpesvirus type 2 and adenovirus type 4	Chlorhexidine gluconate and benzalkonium chloride	72

6. At the end of the experiment, it is difficult to ensure the proper decontamination of the hands of the volunteers because of the large surface area and hard-to-reach interdigital and subungual spaces.
7. Relatively large volumes (≥0.5 mL) of high-titered virus pools must be used here, and this makes the method expensive and limits its use and its ability to detect high levels of virus inactivation.
8. In most cases, viral contamination is picked up by the palm surface of the hand; therefore, the use of the whole hand may be quite relevant to testing the bactericidal activity of surgical scrubs and products for preoperative skin antisepsis [76], but not for dealing with viruses.

2. The Fingertip Method

In the fingertip method [70], 20 μL of the virus suspension is placed on the palm surface of each fingertip and rubbing together of opposing fingers for 40 seconds to spread the inoculum. The fingers are then dried for 80 seconds and moistened under running tap water. The test product (5 mL) is poured into the cupped hands and rubbed for 30 seconds over the entire surface of both hands simulating normal handwashing. The hands are then rinsed for 15 seconds in running tap water and dried for 15 seconds using paper towels. For virus recovery, the tips of fingers and thumbs are immersed and rubbed together for 1 minute in a bowl containing 20 mL of a cell culture medium and glass beads. The recovery medium and controls are titrated for infectious virus. This protocol is subject to most of the same limitations as described above for the whole-hand method.

3. The Fingerpad Method

Keeping the above-mentioned factors in mind, we developed the fingerpad protocol [64,68] to avoid many of the cited difficulties. It has also been used by other investigators [26,70] and is now a standard of ASTM [74]. Table 6 lists the types of viruses that have been tested using this method. Figure 1 gives the main steps in the procedure.

In the fingerpad method test areas on the fingerpads of washed and decontaminated hands of the test subject(s) are demarcated by pressing them over the mouth of an empty plastic vial. Test virus (10 μL), with a suitable soil load, is then placed at the center of the demarcated area on each of the thumbs, and the inoculum on the thumbpads is immediately eluted in 1 mL of a suitable eluent to serve as the virus "input" control. A similar inoculum is placed on each of the other digits, and the fingerpads are allowed to dry for 20–25 minutes under ambient conditions until the inocula are visibly dry. In determining the length of the drying period on the fingerpads, it is valuable to understand the kinetics of the virus decay. Experiments with a relatively stable virus (rotavirus) suggest

Table 6 Quantitative In Vivo and Ex Vivo Methods Developed at the University of Ottawa and Their Application in Studies on the Survival and Germicide Inactivation of Viruses

Virus tested	In vivo (fingerpads)	Ex vivo (human skin)	Ex vivo (umbilical cord)
Rotavirus (Wa)	+	−	−
Rhinovirus type 14	+	−	−
Adenovirus type 4	+	+	+
Herpesvirus type 2	−	+	+
Hepatitis A virus (HM-175)	+	−	−
Influenza virus type A (PR8)	+	−	−
Parainfluenza virus type 3	+	−	−
Poliovirus type 1 (Sabin)	+	−	−
Coxsackievirus type 3	+	−	−
Feline parvovirus	+	−	−

that 60 minutes may be a more appropriate drying time than 20 minutes, because the percentage of virus transfer is higher at 20 minutes and the virus may not be properly dry. However, many viruses can decay quite rapidly on human skin, and after 60 minutes very few infectious virus particles may remain. Another factor that may influence the drying time is the nature and consistency of the soil load. That is why it is important to use a standardized soil load with predetermined drying times under ambient conditions.

Two randomly selected fingerpads are then eluted at the end of the drying period to determine the amount of infectious virus surviving the drying and the loss in virus infectivity due to drying; this virus titer serves as the "baseline" for any loss in virus infectivity as a result of any subsequent treatment. The dried inocula on at least two randomly selected fingerpads are exposed for the required contact time (normally no longer than 10 seconds) to 1 mL of the control or test formulation. The remaining fingerpads can be used to determine the degree of mechanical removal with standard hard water or serve for comparison with other handwash agents. Using the fingerpads allows replicates within the same experiment and can also test in parallel different products against the same virus or one product against more than one virus. We consider this statistical efficiency to be one of the most important points in favor of using the fingerpads. It is also a closed system, allowing for the determination of virus removal/inactivation at every major step of the handwash procedure.

Virus remaining on the treated area of the fingerpad after treatment can be recovered without any rinsing to determine virus inactivation alone or after rins-

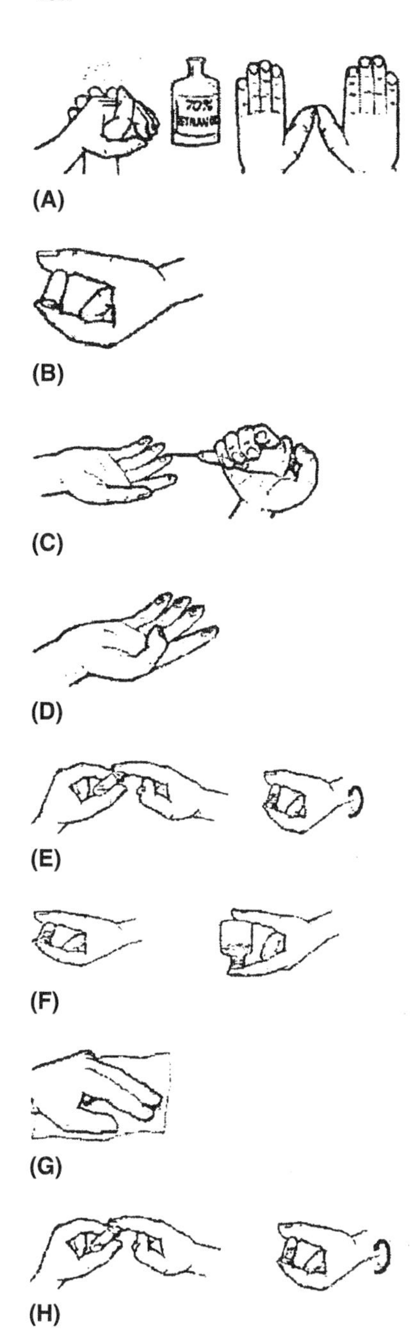
70%
(A)
(B)
(C)
(D)
(E)
(F)
(G)
(H)

ing it in water contained in a tube or bottle to determine the combined effect of inactivation and removal. Similarly, the virus on the rinsed area can be recovered without any drying or after drying it with paper, cloth, or warm-air drying in order to determine what effect the drying of washed hands has in the overall handwashing procedure. The controls and test eluates are titrated for virus infectivity using an appropriate permissive cell line.

The fingerpad procedure is, therefore, capable of assessing, separately, virus elimination after exposure to the handwash agent, posttreatment water rinse, and the drying of washed hands. At the end of the experiment, the volunteer's fingerpads are pressed over tissue soaked in a product known to inactivate the virus under test. The volunteer is then required to wash his or her hands thoroughly with an ordinary liquid soap and running tap water before leaving the laboratory.

The virus elution procedure in this protocol, which includes scraping of the fingerpad against the inside lip of the plastic elution vial after the treatment of the contaminated area with the product under test and after the posttreatment water rinse, can recover as much as 90% of the infectious virus remaining on the fingerpads at the end of the drying period. Through this action, friction is regularly applied during the test procedure. Although the purpose of this scraping

---

Figure 1 Procedure for fingerpad test. (A) The panelist washes hands with nongermicidal soap and water and dries them with paper towel. About 5 mL of 70–75% (v/v) ethanol is placed on the hands and they are rubbed together till dry. (B) Each digit is pressed against the mouth of a cryovial (8 mm inside diam.) to demarcate the target area. (C) 10 μL of virus with soil load is placed at center of each demarcated area. Inoculum from thumbpads is eluted immediately (H below) to act as ''input'' control for virus. (D) Inoculum on fingerpads allowed to become visibly dry (20–25 minutes). Two randomly selected fingerpads are eluted immediately (H below) at the end of drying (''baseline'' control). (E) Dried inoculum on at least two randomly selected fingerpads is exposed to 1 mL of test product or control fluid in a cryovial for desired contact time, with specified number of full inversions; skin scraped against inside lip of vial to collect as much fluid as possible. For waterless handwash agents or to determine virus elimination after exposure to the product alone, fingerpads can be eluted (H below) without further treatment. (F) To simulate posttreatment rinsing of hands, fingerpads are exposed to 1–15 mL of water for 5–10 seconds. Virus can be eluted (H below) at this stage or after drying of hands. (G) To determine virus removal after the drying of washed hands, they can be dried in air or with paper or cloth towel for specified time and virus recovered from them. (H) To elute virus, the digit is placed on the mouth of a cryovial with 1 mL of eluent and subjected to 20 full inversions; skin is scraped against inside lip of vial to collect as much fluid as possible. The eluates and controls are titrated for virus and $\log_{10}$ reductions calculated.

is to recover as much of the fluid as possible from the fingerpad, we consider this degree of friction to be sufficient, because any other kind of friction step is always difficult to standardize and may therefore compromise the reproducibility of a protocol. More rigorous friction is perhaps only important in dealing with reductions in resident flora by surgical scrubs because such bacteria may actually grow in the crevices in the skin. This is not the case with viruses and other transient microflora. Furthermore, during hygienic handwashing the friction that is applied is at best very gentle and may not in itself play an important role in the activity of topicals against viruses. The experiments of Schürmann and Eggers [59] have clearly demonstrated that the lubricating effect of ''soapy'' handwashing agents reduces the impact of friction applied normally during the lathering of hands and that the rubbing of hands with sand was required to properly dislodge poliovirus from the skin of experimentally contaminated hands.

The fingerpad protocol is a simple, reproducible, and quantitative method to study the survival of viruses and bacteria on human hands and their transfer between hands and inanimate objects and to assess the activity of handwash agents against viruses and bacteria. This protocol can be controlled better than the methods using the whole hand and presents a much lower risk to study participants. It has now been used in our laboratory to test a variety of handwash agents against a human rhinovirus, a human rotavirus, hepatitis A virus, a human adenovirus, a human parainfluenzavirus, a poliovirus, as well as *Escherichia coli* and *Staphylococcus aureus* [22,64,68,73]. In fact, in one study [68] we used a mixture of hepatitis A virus and poliovirus type 1 (Sabin) for simultaneous testing of a variety of handwash agents against the two viruses.

4. Ex Vivo Tests Using Animate Carriers

Because of safety reasons, the in vivo methods mentioned above cannot be used for working with formulations where the safety of the chemical ingredients is unknown and/or with high-risk infectious agents such as HIV. To overcome this, we have developed ex vivo methods using either human skin removed during cosmetic surgery or pieces of human umbilical cord as detailed below and applied them to study the virucidal activity of topicals (Table 4).

Testing Using Human Skin Fragments. Pieces of skin removed during cosmetic surgery can be kept metabolically active in the lab for several hours, and disks cut from them can be mounted on stainless steel holders for use in pathogen survival and inactivation studies [73]. This method has already been used to work with human herpesvirus 2 and human adenovirus type 4 [72,76]. However, the supply of human skin can be unpredictable, and, when available, it would come mostly from females of a certain age group. Therefore, the behavior of pathogens on such carriers may not be truly representative of the general population.

Whereas the skin of a variety of experimental animals can be or has been used in testing hygienic handwash agents against viruses of human origin, we are not aware of any systematic efforts to develop such a model for general application; the use of such substrates is subject to many limitations when compared to human skin and may be suitable only as a screening mechanism for product testing. The advent of in vitro culture of human keratinocytes offers interesting possibilities for screening and testing of topicals but suffers from high costs and difficulties of standardization. Such skin fragments are also difficult to manipulate and may not truly represent the topography of mature intact skin.

Testing Using the Umbilical Cord as a Model for Mucous Membranes. The topography, moisture level, microbial ecology, and surface temperature of mucous surfaces are quite different from those of skin. It would, therefore, be safe to assume that the behavior of microbial pathogens on mucous membranes may also be different from that on skin. Each of these differences may also influence the germicidal activity of an antiseptic applied to mucous membranes.

What little information that is available on the activity of topicals against viruses on mucous membranes comes from experiments using animals [13,14]. Ethical considerations, costs, and the general difficulties of working with animals models severely limit a wider application of such test protocols, especially in the initial development and screening of formulations. The recent introduction of the human umbilical cord as a surrogate for mucous membranes is an attempt at addressing this gap [3], and preliminary experiments with this model show its potential for application in studies on virus survival and inactivation on mucous membranes [72].

Human umbilical cord tissue is relatively readily available in most locations, and the ability to fragment it into multiple similar carriers allows for replicates and statistical efficiency in experimental design. Ethical clearance and informed consent for its use in experimentation are required.

Testing Using Animal Skin. Human skin is unique in the thickness of its stratum corneum, density of hair follicles, and the nature of its sweat glands [77]. Pieces of skin from animals such as pigs are frequently used in testing the activity of topicals against bacteria [78], but only limited published information is available on the application of this model to viruses [26].

5. Other Substrates as Carriers

Whereas skin from human cadavers, membranes made out of collagen, cultured corneal fibroblasts [79], and human skin grown in vitro [80] could also be used as substrates in testing the activity of topicals against viruses, they all suffer from a variety of limitations. For example, the viability and barrier integrity of cadaveric skin are compromised [77], layers of cultured cells are too fragile for the handling required in a germicidal test, and collagen membranes are devoid of any of the characteristics of viable skin.

VI. POLICY AND RESEARCH NEEDS

There are many issues here that need clear answers. Some of these may represent policy matters, whereas others will require further data through properly conducted research.

A. Is In Situ Inactivation of Viruses by Handwash Products Necessary?

The available data show that many products now on the market cannot inactivate nonenveloped viruses, but such viruses may be able to be removed by detergent and/or flushing action. Based on the limited data we have obtained so far and in view of other published studies, alcohols and alcohol-containing products stand out as superior to most other formulations. If requirement for virucidal activity is made mandatory in order to register an antiseptic claim for handwash agents, then many products now sold as handwashes may not be able to meet such a claim. Some products that can kill nonenveloped viruses on the skin are likely to be unsafe for regular and repeated use in hygienic handwashing and will most likely be disqualified on grounds of toxicity. On the other hand, plain alcohols have been very widely used in Europe, usually in conjunction with emollient(s), for many years now with relatively few problems. Moreover, nonenveloped viruses clearly pose a hazard to the young, the elderly, and the immunocompromised, and a clear and obvious means to permit antiseptic choice for caregivers and others would be beneficial. The discharge of infectious virus particles released from hands being washed should not be a major concern because it would represent an extremely small proportion of infectious virus input in the wastewater stream from the flushing of toilets, for example.

One possible compromise here would be to develop an index based on the performance of a neutral agent that only mechanically removes viruses from hands. In the past, nonmedicated soaps have often been used for this purpose. However, soaps on the market differ widely, especially in properties such as pH and detergent action. Tap water also differs at different geographical locations. Therefore, a simple, safe, and readily available solution such as standard hard water (e.g., with 200 ppm hardness) could be used to establish the reference point for this index. Product efficacy claims could then be allowed at a certain differential above the mechanical virus removal with hard water. Manufacturers could publish the performance index of their formulation on the label to aid in product selection. To prevent minor differences in indices being used as a sales feature, a simple product classification scheme could be developed.

B. What Should Be the Criterion of a Product's Potency as a Hygienic Handwash Agent for the Purposes of Its Registration?

In many in vitro carrier test protocols for chemical germicides [33,34], a product must reduce the infectivity titer of the test virus by at least 4 $\log_{10}$ to be considered effective. This is too high a requirement for most hygienic handwash agents to meet in vivo tests using the whole-hand method or the fingerpad protocol. Alcohols and alcohol-based products often achieve virus reduction levels between 2 and 3 $\log_{10}$. On the other hand, water or soap and water, as well as many other products, may achieve only up to 1 $\log_{10}$ reduction in contaminating virus. Perhaps a level of virus reduction of no less than 1 $\log_{10}$ above that achieved for mechanical removal may be considered as appropriate for allowing an effectiveness claim for handwash or other topical products against viruses. Any such criterion is arbitrary by nature, but the ultimate objective here is the reduction of the risk of disease spread through hands without discouraging compliance with handwashing.

C. Is Testing of Various Components in a Formulation Necessary?

Testing of individual components for their virus-killing/removal potency using suitable test protocols may be feasible with some formulations, whereas with others it may be virtually impossible. For example, the addition of certain chemicals is sometimes needed simply to dissolve or emulsify an antimicrobial component in a product. Therefore, a policy that treats products differently would be difficult to justify and implement. The current requirement for listing only the active ingredients on the product label needs to be reviewed because many categories of ''inert'' ingredients are capable of potentiating the actives in a formulation. This is a particularly important consideration when dealing with the antiviral activity of handwash agents, because a product with a good surfactant could reduce the virus titer on experimentally contaminated hands by its detergent action alone and the ''active ingredient'' in the formulation may indeed be quite inert in its activity against the test virus. We would therefore prefer to see the entire formulation tested in a manner in which it is recommended for use.

D. Should the Testing Determine Virus Elimination by the Product Alone or by the Process of Hand Decontamination as a Whole?

In carrier tests to determine the virucidal activity of other types of chemical germicides, there is no provision to include any virus loss due to precleaning of an object or its posttreatment rinsing. In the case of semi-critical medical devices, such as flexible fiber-optic endoscopes, precleaning, and posttreatment rinsing are integral parts of the disinfection process. On the other hand, precleaning and rinsing may or may not occur in the chemical disinfection of environmental surfaces. In the case of hygienic handwashing, one normally prewets hands with tap water, rinses them again in water to wash off the handwash agent, and then dries them using one of several possible means. Our studies have shown that there is an incremental reduction in the infectivity titer of the virus on hands at least after the posttreatment water rinse and the drying of washed hands [64,81]. The fingertip [69] and whole-hand [64] methods described earlier determine virus reduction only as a combined action of all the steps in the handwashing process. Is this appropriate? Does such an approach really determine the virus-eliminating potential of a product, or does it assess the efficiency of the handwashing/drying process as a whole? How does this approach then compare the potency of conventional handwashing agents with products meant for ''waterless'' washing of hands?

E. Would Multitiered Testing of Topicals Against Viruses Be Desirable?

Any suspension and/or in vitro carrier tests on hygienic handwash agents should be only for the preliminary screening of the formulations for their antiviral activity. However, we have observed that products such as ethanol may work more efficiently on human hands than on inanimate materials. We believe that this is related to the level of moisture and, possibly, the temperature of human skin. The next level could be ex vivo tests using cultured human skin or fragments of human foreskin: neither of these may properly simulate human skin in the assessment of handwash agents. In our view, the only required level of testing for product registration should be based on the fingerpad protocol using suitably rigorous surrogates and properly selected adult human volunteers. Discretionary testing with additional viruses increases unnecessarily the risk to the study participants, the confusion of the user community, and the costs to manufacturers. With the possible exception of papillomaviruses, the use of experimental animals in the routine testing of hygienic handwash agents is considered both unnecessary and potentially invalid. It should, therefore, be avoided as far as possible, but in some cases it may be necessary to validate in animal models the efficacy of topicals designed for use on mucous membranes.

VII. CONCLUDING REMARKS

Viruses continue to be important pathogens, but in spite of their considerable impact on human health, our understanding of the actual mechanisms of spread of many viral infections in hospitals and other such settings remains weak. This makes it difficult to design and apply proper strategies to prevent and control nosocomial outbreaks of viral infections. Hands are universally recognized as vehicles for the spread of a number of viruses, but compliance with handwashing, proper handwashing techniques, and perhaps the use of ineffective handwash agents continue to undermine the full potential of infection control measures in this regard. The ease with which washed hands can pick up infectious viruses upon contact with contaminated environmental surfaces and objects [6] suggests that the emphasis on handwashing should be combined with an awareness of the need for proper and regular decontamination of those surfaces and objects that come in frequent contact with washed hands.

Globally, viruses cause millions of cases of morbidity and mortality in humans every year, and thus far the development of chemotherapy against them has met with very limited success. Also, many enteric and respiratory infections due to viruses remain refractory to prevention by vaccination. Therefore, interruption of the spread of viral infections through practices such as regular and proper washing of hands continues to be essential for personal hygiene and general public health. But the use of handwash agents without proven activity against viruses may only create a false sense of security. There is, therefore, an urgent need to develop and introduce scientifically sound test methods and a suitable regulatory framework to allow manufacturers to make reasonable label claims against viruses to give the needed confidence in such claims.

The distinction between resident and transient microflora of the skin is particularly important when dealing with viruses and hygienic handwash agents because, apart from herpes-, pox-, and papillomaviruses in infected individuals, human skin is not known to carry viruses (with the possible exception of bacteriophages) as members of its resident flora. As far as we are aware, postexposure antisepsis of wounds is important in rabies only. In contrast to this, many types of viruses can be picked up by hands where they may survive for up to a few hours. There is limited direct, but plenty of strong circumstantial evidence for the role of hands as vehicles for human pathogenic viruses. Use of handwashing agents with broad-spectrum activity against viruses has proven helpful in prevention and control of outbreaks of respiratory and enteric infections caused by viruses.

Our findings have clearly demonstrated that transfer of infectious virus particles to and from hands can readily occur on contact with other animate and inanimate surfaces [68,82]. This suggests that touching or handling virus-contaminated objects with washed hands can lead to their immediate recontamination. Proper disinfection of environmental surfaces [83] and washing of hands [69] with certain types of agents can interrupt virus transfer to clean surfaces. It is therefore important to remember that handwashing and environmental surface decontamination reinforce each other, particularly in critical-care areas and food-handling establishments.

The safety and testing requirements for topical antiseptics should fall somewhere in between antiviral drugs and other types of chemical germicides. Even though the topic of chemical germicides in general has been the subject of several conferences and symposia in the past two decades, the specific issue of the activity of topicals against viruses still remains to be discussed. The U.S. Food and Drug Administration's (FDA) tentative final monograph on topical antimicrobials [84] does not mention viruses at all, while FDA's Center for Food Safety and Applied Nutrition [85] regards enteric viruses as important targets for preventing the spread of infections by the hands of foodhandlers.

Regulators, manufacturers, and users alike are seeking information and directions, respectively, for the registration, marketing and purchase of hygienic handwash agents. Therefore, this issue needs addressing through research and development, as well as a dialogue between the stakeholders, in order to bring safe and effective products to the market. We hope that this chapter will serve as a springboard for further discussions in this regard. Efforts currently underway to harmonize test requirements for chemical germicides at the regional and international levels could eventually compromise the quality of the products if less stringent test protocols are adopted. This points to the need for the development and introduction of test methodology based on solid scientific grounds and sound reasoning rather than political expediency.

ACKNOWLEDGMENTS

We wish to thank Ms. Susan Springthorpe and Mr. Jason Tetro for their valuable input and help in the preparation of this chapter.

REFERENCES

1. BN Fields, DM Knipe, PM Howley, eds. Fields Virology, 3rd ed. Philadelphia: Lippincott-Raven, 1996.
2. SA Sattar, JA Tetro, VS Springthorpe. Impact of changing societal trends on the spread of infectious diseases in American and Canadian homes. Am J Infect Control 27:S4–S21, 1999.
3. SA Sattar, VS Springthorpe. Methods for testing the virucidal activity of chemicals. In: SS Block, ed. Disinfection, Sterilization, and Preservation. New York: Lippincott Williams & Wilkins, 2001, pp. 1391–1412.
4. SA Sattar, VS Springthorpe. Transmission of viral infections through animate and inanimate surfaces and infection control through chemical disinfection. In: C Hurst, ed. Modeling Disease Transmission and Its Prevention by Disinfection. Cambridge, United Kingdom: Cambridge University Press, 1996, pp. 224–257.
5. JCN Westwood, SA Sattar. The minimal infective dose. In: G Berg et al., eds. Vi-

ruses in Water. Washington, DC: American Public Health Association. 1976, pp. 61–69.
6. FV Rheinbaden, S Schunemann, T Gross, MH Wolff. Transmission of viruses via contact in a household setting: experiments using bacteriophage straight phiX174 as a model virus. J Hosp Infect 46:61–66, 2000.
7. X Jiang, X Dai, S Goldblatt, C Buescher, TM Cusack, DO Matson, LK Pickering. Pathogen transmission in child care settings studied by using a cauliflower virus DNA as a surrogate marker. J Infect Dis 177:881–888, 1998.
8. SE Reed. An investigation of possible transmission of rhinovirus colds through indirect contact. J Hyg 75:249–258, 1975.
9. F Pancic, DC Carpentier, PE Came. Role of infectious secretions in the transmission of rhinovirus. J Clin Microbiol 12:567–571, 1980.
10. JO Hendley, JM Gwaltney, Jr. Mechanisms of transmission of rhinovirus infections. Epidemiol Rev 10:242–258, 1988.
11. JM Gwaltney, PB Moskalski, JO Hendley. Interruption of experimental rhinovirus transmission. J Infect Dis 142:811–815, 1980.
12. RL Ward, DI Bernstein, DR Knowlton, JR Sherwood, EC Young, TM Cusack, JR Rubino, GM Shiff. Prevention of surface-to-human transmission of rotaviruses by treatment with disinfectant spray. J Clin Microbiol 29:1991–1996, 1991.
13. JN Mbithi, VS Springthorpe, JR Boulet, SA Sattar. Survival of hepatitis A virus on human hands & its transfer on contact with animate and inanimate surfaces. J Clin Microbiol 30:757–763, 1992.
14. P Greenhead, P Hayes, PS Watts, KG Laing, GE Griffin, RJ Shattock. Parameters of human immunodeficiency virus infection of human cervical tissue and inhibition by vaginal virucides. J Virol 74:5577–5586, 2000.
15. C Tevi-Benissan, C Makuva, M Morelli, MC George-Courbot, M Matta, A Georges, L Belec. Protection of cynomolgus macaque against cervicovaginal transmission of SIVmac251 by the spermicide benzalkonium chloride. J AIDS 24:147–153, 2000.
16. ER Kern. Importance of animal models for evaluation of topical microbicides against genital herpesvirus infections. In: N Biswal et al., eds. Proc 1st Workshop on Antiviral Claims for Topical Antiseptics. May–June, 1994, Food & Drug Admin, Rockville, MD: U.S. Government Printing Press, 1994, pp. 67–84.
17. C Marwick. International plan focuses on eradication of polio and containment of the virus. J Am Med Assoc 283:1553–1554, 2000.
18. MV Jones, K Bellamy, R Alcock, R Hudson. The use of bacteriophage MS2 as a model system to evaluate virucidal hand disinfectants. J Hosp Infect 17:279–285, 1991.
19. J-Y Maillard. Bacteriophages: a model system for human viruses. Lett Appl Microbiol 23:273–274, 1996.
20. J-Y Maillard, AD Russell. Viricidal activity and mechanisms of action of biocides. Sci Prog 80:287–315, 1997.
21. SA Sattar, JA Tetro, VS Springthorpe, A Guilivi. Preventing the spread of hepatitis B and C viruses: Where are germicides relevant? Am J Infect Control 29:187–197, 2001.
22. SA Sattar, M Abebe, A Bueti, H Jampani, J Newman. Determination of the activity of an alcohol-based hand gel against human adeno-, rhino-, and rotaviruses using the fingerpad method. Infect Control Hosp Epidemiol 21:516–519, 2000.

23. JN Mbithi, VS Springthorpe, SA Sattar. Chemical disinfection of hepatitis A virus on environmental surfaces. Appl Environ Microbiol 56:3601–3604, 1990.
24. SA Sattar, VS Springthorpe. New methods for efficacy testing of disinfectants and antiseptics. In: WA Rutala, ed. Disinfection, Antisepsis and Sterilization: Principles and Practices in Healthcare Facilities. Washington, DC: Association of Practitioners in Infect Control. 2001, pp. 173–186.
25. SA Sattar, RA Raphael, H Lochnan, VS Springthorpe. Rotavirus inactivation by chemical disinfectants and antiseptics used in hospitals. Can J Microbiol 29:1464–1469, 1983.
26. JD Woolwine, JL Gerberding. Effect of testing method on the apparent activities of antiviral disinfectants and antiseptics. Antimicrob Agents Chemother 39:921–923, 1995.
27. JN Mbithi, VS Springthorpe, SA Sattar. Effect of relative humidity and air temperature on survival of hepatitis A virus on environmental surfaces. Appl Environ Microbiol 57:1394–1399, 1991.
28. EL Larson. APIC guideline for handwashing and hand antisepsis in health care settings. Am J Infect Control 23(4):251–269, 1995.
29. ZA Quraishi, M McGuckin, FX Blais. Duration of handwashing in intensive care units: a descriptive study. Am J Infect Control 12:83–87, 1984.
30. EA Jonczy, J Daly, GJ Kotwal. A novel approach using an attenuated recombinant vaccinia virus to test the antipoxviral effects of handsoaps. Antiviral Res 45:149–153, 2000.
31. B Rodu, F Lakeman. In vitro virucidal activity by components of a topical film-forming medication. J Oral Pathol 17:324–326, 1988.
32. N Lloyd-Evans, VS Springthorpe, SA Sattar. Chemical disinfection of human rotavirus-contaminated inanimate surfaces. J Hyg 97:163–173, 1986.
33. U.S. Environmental Protection Agency. Efficacy data requirements: Virucides. Document #DIS/TSS-7. Washington, DC: U.S. EPA, November 1981.
34. American Society for Testing and Materials (ASTM). A standard test for determining the virus-eliminating effectiveness of liquid handwash agents using the fingerpads of adult volunteers. Designation E 1838-96, West Conshohocken, PA.
35. JH Blackwell, JHS Chen. Effects of various germicidal chemicals on H.Ep.2 cell culture and herpes simplex virus. J Assoc Off Anal Chem 53:1229–1236, 1970.
36. JC Doultree, JD Druce, CJ Birch, DS Bowden, JA Marshall. Inactivation of feline calicivirus, a Norwalk virus surrogate. J Hosp Infect 41(1):51–57, 1999.
37. V Gordon, S Parry, K Bellamy, R Osborne. Assessment of chemical disinfectants against human immunodeficiency virus: overcoming the problem of cytotoxicity and the evaluation of selected actives. J Virol Methods 45:247–257, 1993.
38. B Damery, A Cremieux. Virucidal activity against herpes and vaccinia virus of 8 antiseptic formulations. Int J Pharmaceut 49:205–208, 1989.
39. GC Lavelle, SL Gubbe, JL Neveaux, BJ Bowden. Evaluation of an antimicrobial soap formula for virucidal efficacy in vitro against human immunodeficiency virus in a blood-virus mixture. Antimicrob Agents Chemother 33:2034–2036, 1989.
40. MA Kennedy, VS Melion, G Caldwell, ND Potgieter. Virucidal efficacy of the newer quaternary ammonium compounds. J Am Animal Hosp Assoc 31:254–258, 1995.

41. M Boudouma, L Enjalbert, J Didier. A simple method for the evaluation of antiseptic and disinfectant virucidal activity. J Virol Methods 9:271–276, 1984.
42. G Garrigue, L Enjalbert, C Hengy, M Boudouma, A Boucays. In vitro virucidal activity of antiseptics and disinfectants. III—Technique by dilution-ultrafiltration-reconcentration. Path Biol 32:647–650, 1984.
43. S Valot, D Edert, A Le Faou. A simple method for the in vitro study of the virucidal activity of disinfectants. J Virol Methods 86:21–24, 2000.
44. Canadian General Standards Board. Assessment of Efficacy of Antimicrobial Agents for Use on Environmental Surfaces and Medical Devices Document Number: CGSB 2.161-97, 1997.
45. A Wood, D Payne. The action of three antiseptics/disinfectants against enveloped and non-enveloped viruses. J Hosp Infect 38:283–293, 1998.
46. A Bailey, M Longson. Virucidal activity of chlorhexidine on strains of Herpesvirus hominis, poliovirus, and adenovirus. J Clin Pathol 25:76–78, 1972.
47. CH Carter, JO Hendley, LA Mika, M Gwaltney. Rhinovirus inactivation by aqueous iodine in vitro and on skin. Proc Soc Exp Biol Med 165:380–383, 1980.
48. JB Kurtz, TW Lee, AJ Parsons. The action of alcohols on rotavirus, astrovirus and enterovirus. J Hosp Infect 1:321–325, 1980.
49. JA Tan, RD Schnagl. Inactivation of a rotavirus by disinfectants. Med J Aust 10: 19–23, 1981.
50. WS Croughan, AM Behbehani. Comparative study of inactivation of herpes simplex virus types 1 and 2 by commonly used antiseptic agents. J Clin Microbiol 26:213–215, 1988.
51. B Rodu, F Lakeman. In vitro virucidal activity by components of a topical film-forming medication. J Oral Pathol 17:324–326, 1988.
52. GC Lavelle, SL Gubbe, JL Neveaux, BJ Bowden. Evaluation of an antimicrobial soap formula for virucidal efficacy in vitro against human immunodeficiency virus in a blood-virus mixture. Antimicrob Agents Chemother 33:2034–2036, 1989.
53. R Tyler, GA Ayliffe, C Bradley. Virucidal activity of disinfectants: studies with the poliovirus. J Hosp Infect 15:339–345, 1990.
54. LR Krilov, SH Harkness. Inactivation of respiratory syncytial virus by detergents and disinfectants. Pediatr Infect Dis J 12:582–584, 1993.
55. R Kawana, T Kitamura, O Nakagomi, I Matsumoto, M Arita, N Yoshihara, K Yanagi, A Yamada, O Morita, Y Yoshida, Y Furuya, S Chiba. Inactivation of human viruses by povidone-iodine in comparison with other antiseptics. Dermatology 195(suppl 2):29–35, 1997.
56. PA Contreras, IR Sami, ME Darnell, MG Ottolini, GA Prince. Inactivation of respiratory syncytial virus by generic hand dishwashing detergents and antibacterial hand soaps. Infect Control Hosp Epidemiol 20:57–58, 1999.
57. EA Jonczy, J Daly, GJ Kotwal. A novel approach using an attenuated recombinant vaccinia virus to test the antipoxviral effects of handsoaps. Antiviral Res 45:149–153, 2000.
58. SA Sattar, VS Springthorpe, Y Karim, P Loro. Chemical disinfection of non-porous inanimate surfaces experimentally contaminated with four human pathogenic viruses. Epidemiol Infect 102:493–505, 1989.
59. W Schürmann, HJ Eggers. Antiviral activity of an alcoholic hand disinfectant. Com-

parison of the in vitro suspension test with in vivo experiments on hands, and on individual fingertips. Antiviral Res 3:25–41, 1983.
60. DO Cliver, KD Kostenbader, Jr. Disinfection of virus on hands for prevention of food-borne disease. Intl J Food Microbiol 1:75–87, 1984.
61. GF Hayden, D DeForest, JO Hendley, JM Gwaltney. Inactivation of rhinovirus on human fingers by virucidal activity of glutaric acid. Antimicrob Agents Chemother 26:928–929, 1984.
62. O Bydzovska. Screening the viricidal efficiency of antisepsis, disinfection and chemical sterilization—a draft methodology for practice. J Hyg Epidemiol Microbiol Immunol 31:375–380, 1987.
63. RG Faix. Comparative efficacy of handwashing agents against cytomegalovirus. Infect Control 8:158–162, 1987.
64. SA Ansari, SA Sattar, VS Springthorpe, GA Wells, W Tostowaryk. In vivo protocol for testing efficacy of hand-washing agents against viruses and bacteria: experiments with rotavirus and *Escherichia coli.* Appl Environ Microbiol 55:3113–3118, 1989.
65. T Ueno, K Saijo. Prevention of adenovirus infection and antiviral activity of a hand disinfectant, Welpas. Nippon Ganka Gakkai Zasshi 94:44–48, 1990.
66. HJ Eggers. Experiments on antiviral activity of hand disinfectants. Some theoretical and practical considerations. Zentralbl Bakteriol 273:36–51, 1990.
67. JG Davies, JR Babb, CR Bradley, GA Ayliffe. Preliminary study of test methods to assess the virucidal activity of skin disinfectants using poliovirus and bacteriophages. J Hosp Infect 25:125–131, 1993.
68. JN Mbithi, VS Springthorpe, SA Sattar. Comparative in vivo efficiencies of hand-washing agents against hepatitis A virus (HM-175) and poliovirus type 1 (Sabin). Appl Environ Microbiol 59:3463–3469, 1993.
69. K Bellamy, R Alcock, JR Babb, JG Davies, GAJ Ayliffe. A test for the assessment of hygienic hand disinfection using rotavirus. J Hosp Infect 24:201–210, 1993.
70. DL Prince, HN Prince, O Thraenhart, E Muchmore, E Bonder, J Pugh. Methodological approaches to disinfection of human hepatitis B virus. J Clin Microbiol 31:3296–3304, 1993.
71. J Steinmann, R Nehrkorn, A Meyer, K Becker. Two in-vivo protocols for testing virucidal efficacy of handwashing and hand disinfection. Zentralbl Hyg Umweltmed 196:425–436, 1995.
72. ML Graham, VS Springthorpe, SA Sattar. Ex vivo protocol for testing virus survival on human skin: experiment with herpesvirus 2. Appl Environ Microbiol 62:4252–4255, 1996.
73. SA Ansari, VS Springthorpe, SA Sattar, S Rivard, M Rahman. Potential role of hands in the spread of respiratory viral infections: studies with human parainfluenza virus 3 and rhinovirus 14. J Clin Microbiol 29:2115–2119, 1991.
74. American Society for Testing & Materials (ASTM). Standard Test Method for Evaluation of Handwashing Formulations for Virus-Eliminating Activity Using the Entire Hand. Designation E-2011-99, West Conshohocken, PA.
75. ML Rotter, W Koller, G Wewalka, HP Werner, GA Ayliffe, JR Babb. Evaluation of procedures for hygienic hand-disinfection: controlled parallel experiments on the Vienna test model. J Hyg (Lond) 96:27–37, 1986.
76. ML Graham. Development of an ex vivo model to study the survival and inactivation

of pathogens on human skin. M.Sc. thesis, Dept of Microbiology and Immunology, University of Ottawa, Ottawa, ON, Canada, 1997.
77. RL Bronaugh. Determination of percutaneous absorption by in vitro techniques. In: HI Maibach, ed. Percutaneous Absorption: Mechanisms, Methodology, Drug Delivery. New York: Marcel Dekker, 1989, pp. 239–258.
78. LM Bush, LM Benson, JH White. Pig skin as test substrate for evaluating topical antimicrobial activity. J Clin Microbiol. 3:343–348, 1986.
79. S Valluri, TP Fleming, KA Laycock, IS Tarle, MA Goldberg, FJ Garcia-Ferrer, LR Essary, JS Pepose. In vitro and in vivo effects of polyhexamethylene biguanide against herpes simples virus infection. Cornea 16:556–559, 1997.
80. JT Schultz, RG Tomkins, JF Burke. Artificial skin. Annu Rev Med 51:231–211, 2000.
81. SA Ansari, VS Springthorpe, SA Sattar, W Tostowaryk, GA Wells. Comparison of cloth, paper, and warm air drying in eliminating viruses and bacteria from washed hands. Am J Infect Control 19:243–249, 1991.
82. SA Ansari, SA Sattar, VS Springthorpe, GA Wells, W Tostowaryk. Rotavirus survival on human hands and transfer of infectious virus to animate and non-porous inanimate surfaces. J Clin Microbiol 26:1513–1518, 1988.
83. SA Sattar, H Jacobson, VS Springthorpe, TM Cusack, JR Rubino. Chemical disinfection to interrupt transfer of rhinovirus type 14 from environmental surfaces to hands. Appl Environ Microbiol 59:1579–1585, 1993.
84. U.S. Food and Drug Administration. Tentative Final Monograph for Health-Care Antiseptic Products. Washington, DC: U.S. FDA, June 1994.
85. U.S. Food and Drug Administration. No Bare Hand Contact. Report of the Nat. Advisory Committee for the Microbiological Criteria for Food. Center for Food Safety and Applied Nutrition. Washington, DC: U.S. FDA, November 1999.

Index

9 780824 707880